AF581802

Nanoparticles in the Environment and Nanotoxicology

Nanoparticles in the Environment and Nanotoxicology

Editor

Vivian Hsiu-Chuan Liao

MDPI • Basel • Beijing • Wuhan • Barcelona • Belgrade • Manchester • Tokyo • Cluj • Tianjin

Editor
Vivian Hsiu-Chuan Liao
National Taiwan University,
Taipei, Taiwan

Editorial Office
MDPI
St. Alban-Anlage 66
4052 Basel, Switzerland

This is a reprint of articles from the Special Issue published online in the open access journal *Nanomaterials* (ISSN 2079-4991) (available at: https://www.mdpi.com/journal/nanomaterials/special_issues/Environment_Nanotoxicology).

For citation purposes, cite each article independently as indicated on the article page online and as indicated below:

LastName, A.A.; LastName, B.B.; LastName, C.C. Article Title. *Journal Name* **Year**, *Volume Number*, Page Range.

ISBN 978-3-0365-7230-7 (Hbk)
ISBN 978-3-0365-7231-4 (PDF)

Contents

Editorial

Nanoparticles in the Environment and Nanotoxicology

Vivian Hsiu-Chuan Liao

Department of Bioenvironmental Systems Engineering, National Taiwan University, Taipei 106, Taiwan; vivianliao@ntu.edu.tw; Tel.: +886-2-33665239

Nanomaterials, including engineered nanoparticles and microplastics/nanoplastics, have attracted increasing concern as they might potentially release into the environment, leading to potential risks to ecosystems. However, there are still huge gaps and unknown knowledge about the fate, behavior, and toxicity of nanoparticles in the environment. This Special Issue aims to provide recent novel research findings of nanoparticles on their determination, detection, and degradation in the environment as well as the toxicity of nanoparticles. This Special Issue provides ten outstanding papers comprising two review articles and eight research articles with a range of original contributions detailing various types of nanoparticles in terms of their toxicity [1–5], removal of antibiotics [2,6], and behavior in natural environments [3,7,8].

This Special Issue collected various research papers addressing the toxic effects of various types of nanoparticles using different tested organisms in various media. For example, How and Huang [1] utilized the soil nematode *Caenorhabditis elegans* to investigate neurotoxicity caused by zinc oxide nanoparticles (ZnO-NPs) exposure. They reported that ZnO-NPs particulates attribute to the neurotoxicity in *C. elegans* via dietary transfer [1]. Le et al. [4] synthesized nano-sized artificial black carbon (aBC), examining its toxicity on A549 human lung cells, and found that aBC with an increased content of the oxygen functional group displayed higher toxicity to A549 cells. Interestingly, exposure to Ce oxide nanoparticles (nCeO_2), even at high concentrations, did not result in negative effects on spontaneous plant species, the monocot *Holcus lanatus* and dicots *Lychnis-flos-cuculi* and *Diplotaxis tenuifolia* [5]. However, studies on the effects resulting from the exposure of terrestrial species to nanomaterials are limited, and this study highlights the importance of research in this field.

It is worth noting that the behavior of engineered nanoparticles in natural aquatic environments plays a vital role in determining their toxicity and associated risks [9]. Lee et al. [3] investigated the behavior of silver nanoparticles (Ag-NPs) and ZnO-NPs in natural waters to determine their toxicity in zebrafish embryos. They observed that the acute toxicity of AgNPs or ZnONPs in nature water was much lower than that of Milli-Q (MQ) water [3]. This suggests a possible interaction and transformation between AgNPs or ZnONPs and components in the natural water environment, leading to reduced toxicity [3]. The continuous flow dissolution method was established to measure dissolution rates of Ag-NPs in environmentally relevant water [7]. This work might support the Organization for Economic Co-operation and Development (OECD) testing guidelines focusing on natural aquatic environments [7]. Furthermore, a wide range of commercial products containing nano-enabled products (NEPs) were assessed for the release of engineered nanomaterials (ENMs) and their characteristics in environmental waters [8].

The contamination of pharmaceuticals and personal care products (PPCPs) has driven attention due to their intensive usage and widespread into the environment as well as their potential risk to humans and ecosystems [10]. In this Special Issue, the potential of using nanomaterials to remove PPCPs was reported [2,6]. Capsoni et al. [2] synthesized halloysite nanotubes (HNT) with magnetic Fe_3O_4 nanoparticles as adsorbents to quantitatively reduce the concentration of antibiotic ofloxacin in tap, river, and effluent waters and further evaluated the acute toxicity with the freshwater organism *Daphnia magna*. Sulfamethazine

Citation: Liao, V.H.-C. Nanoparticles in the Environment and Nanotoxicology. *Nanomaterials* **2023**, *13*, 1053. https://doi.org/10.3390/nano13061053

Received: 1 March 2023
Accepted: 3 March 2023
Published: 15 March 2023

(SMZ), one of the most commonly used antibiotics, was found to experience degradation that could be promoted by graphene oxide (GO) under UV light in a water environment [6].

Besides engineered nanoparticles, ubiquitous microplastics/nanoplastics have gained concern regarding their distribution, source, and fate in the environment, as well as the toxicity of these plastics to the environment and humans. Yang et al. [11] summarized the distribution, source, occurrence, and fate of microplastics in the atmosphere, which was less reviewed compared to that in oceans, freshwater, and soil. The possible health impacts of microplastics/nanoplastics on humans were viewed by Yee et al. [12]. Currently, the distribution and concentrations of microplastics/nanoplastics in the environments are largely unknown due to the restriction of the detection limit, the lack of validated methods, and no effective methods for the determination and quantification [12–14]. This makes an accurate assessment of human health and ecological risks of microplastics/nanoplastics a scientific challenge. Nevertheless, in-depth future research should continue to seek to understand the mechanisms of toxicity and potential risks of chronic exposure to various types of nanoparticles at relevant concentrations in the environment.

Conflicts of Interest: The author declares no conflict of interest.

References

1. How, C.M.; Huang, C.W. Dietary transfer of zinc oxide nanoparticles induces locomotive defects associated with GABAergic motor neuron damage in *Caenorhabditis elegans*. *Nanomaterials* **2023**, *13*, 289. [CrossRef] [PubMed]
2. Capsoni, D.; Lucini, P.; Conti, D.M.; Bianchi, M.; Maraschi, F.; De Felice, B.; Bruni, G.; Abdolrahimi, M.; Peddis, D.; Parolini, M.; et al. Fe_3O_4-halloysite nanotube composites as sustainable adsorbents: Efficiency in ofloxacin removal from polluted waters and ecotoxicity. *Nanomaterials* **2022**, *12*, 4330. [CrossRef] [PubMed]
3. Lee, Y.L.; Shih, Y.S.; Chen, Z.Y.; Cheng, F.Y.; Lu, J.Y.; Wu, Y.H.; Wang, Y.J. Toxic effects and mechanisms of silver and zinc oxide nanoparticles on zebrafish embryos in aquatic ecosystems. *Nanomaterials* **2022**, *12*, 717. [CrossRef] [PubMed]
4. Le, Y.T.; Youn, J.S.; Moon, H.G.; Chen, X.Y.; Kim, D.I.; Cho, H.W.; Lee, K.H.; Jeon, K.J. Relationship between cytotoxicity and surface oxidation of artificial black carbon. *Nanomaterials* **2021**, *11*, 1455. [CrossRef] [PubMed]
5. Lizzi, D.; Mattiello, A.; Piani, B.; Fellet, G.; Adamiano, A.; Marchiol, L. Germination and early development of three spontaneous plant species exposed to nanoceria (nCeO2) with different concentrations and particle sizes. *Nanomaterials* **2020**, *10*, 2534. [CrossRef] [PubMed]
6. Liu, F.F.; Li, M.R.; Wang, S.C.; Zhang, Y.X.; Liu, G.Z.; Fan, J.L. Phototransformation of graphene oxide on the removal of sulfamethazine in a water environment. *Nanomaterials* **2021**, *11*, 2134. [CrossRef] [PubMed]
7. Stetten, L.; Mackevica, A.; Tepe, N.; Hofmann, T.; von der Kammer, F. Towards standardization for determining dissolution kinetics of nanomaterials in natural aquatic environments: Continuous flow dissolution of Ag nanoparticles. *Nanomaterials* **2021**, *12*, 519. [CrossRef] [PubMed]
8. Lehutso, R.F.; Thwala, M. Assessment of nanopollution from commercial products in water environments. *Nanomaterials* **2021**, *11*, 2537. [CrossRef] [PubMed]
9. Espinasse, B.P.; Geitner, N.K.; Schierz, A.; Therezien, M.; Richardson, C.J.; Lowry, G.V.; Ferguson, L.; Wiesner, M.R. Comparative persistence of engineered nanoparticles in a complex aquatic ecosystem. *Env. Sci. Technol.* **2018**, *52*, 4072–4078. [CrossRef] [PubMed]
10. Ebele, A.J.; Abou-Elwafa Abdallah, M.; Harrad, S. Pharmaceuticals and personal care products (PPCPs) in the freshwater aquatic environment. *Emerg. Contam.* **2017**, *3*, 1–16. [CrossRef]
11. Yang, H.; He, Y.; Yan, Y.; Junaid, M.; Wang, J. Characteristics, toxic effects, and analytical methods of microplastics in the atmosphere. *Nanomaterials* **2021**, *11*, 2747. [CrossRef] [PubMed]
12. Yee, M.S.; Hii, L.W.; Looi, C.K.; Lim, W.M.; Wong, S.F.; Kok, Y.Y.; Tan, B.K.; Wong, C.Y.; Leong, C.O. Impact of microplastics and nanoplastics on human health. *Nanomaterials* **2021**, *11*, 496. [CrossRef] [PubMed]
13. Shen, M.; Zhang, Y.; Zhu, Y.; Song, B.; Zeng, G.; Hu, D.; Wen, X.; Ren, X. Recent advances in toxicological research of nanoplastics in the environment: A review. *Environ. Pollut.* **2019**, *252*, 511–521. [CrossRef] [PubMed]
14. Wang, L.; Wu, W.M.; Bolan, N.S.; Tsang, D.C.W.; Li, Y.; Qin, M.; Hou, D. Environmental fate, toxicity and risk management strategies of nanoplastics in the environment: Current status and future perspectives. *J. Hazard. Mater.* **2021**, *401*, 123415. [CrossRef]

 *nanomaterials*

Article

Dietary Transfer of Zinc Oxide Nanoparticles Induces Locomotive Defects Associated with GABAergic Motor Neuron Damage in *Caenorhabditis elegans*

Chun Ming How [1] and Chi-Wei Huang [2,*]

1 Department of Bioenvironmental Systems Engineering, National Taiwan University, Taipei 10617, Taiwan
2 Department of Marine Environmental Engineering, National Kaohsiung University of Science and Technology, Kaohsiung 81157, Taiwan
* Correspondence: chiweihuang@nkust.edu.tw; Tel.: +886-7-3617141 (ext. 23760)

Abstract: The widespread use of zinc oxide nanoparticles (ZnO-NPs) and their release into the environment have raised concerns about the potential toxicity caused by dietary transfer. However, the toxic effects and the mechanisms of dietary transfer of ZnO-NPs have rarely been investigated. We employed the bacteria-feeding nematode *Caenorhabditis elegans* as the model organism to investigate the neurotoxicity induced by exposure to ZnO-NPs via trophic transfer. Our results showed that ZnO-NPs accumulated in the intestine of *C. elegans* and also in *Escherichia coli* OP50 that they ingested. Additionally, impairment of locomotive behaviors, including decreased body bending and head thrashing frequencies, were observed in *C. elegans* that were fed *E. coli* pre-treated with ZnO-NPs, which might have occurred because of damage to the D-type GABAergic motor neurons. However, these toxic effects were not apparent in *C. elegans* that were fed *E. coli* pre-treated with zinc chloride ($ZnCl_2$). Therefore, ZnO-NPs particulates, rather than released Zn ions, damage the D-type GABAergic motor neurons and adversely affect the locomotive behaviors of *C. elegans* via dietary transfer.

Keywords: ZnO nanoparticles; trophic transfer; neurotoxicity; GABAergic motor neuron; neuron damage

Citation: How, C.M.; Huang, C.-W. Dietary Transfer of Zinc Oxide Nanoparticles Induces Locomotive Defects Associated with GABAergic Motor Neuron Damage in *Caenorhabditis elegans*. *Nanomaterials* **2023**, *13*, 289. https://doi.org/10.3390/nano13020289

Academic Editor: Alexey Pestryakov

Received: 27 December 2022
Revised: 6 January 2023
Accepted: 8 January 2023
Published: 10 January 2023

1. Introduction

Engineered nanoparticles (ENPs), which are extensively used in industrial and domestic products, are emitted into the environment [1]. This has raised concerns due to their potential toxicity to biota [1,2]. Zinc oxide nanoparticles (ZnO-NPs), with a global production of approximately 550–33,400 tons, are one of the most widely used ENPs in the electrical and manufacturing industries and personal care products [3,4]. In addition, due to their biocompatibility and cost-effectiveness, ZnO-NPs have been increasingly used in the biomedical field, such as in antibacterial, antifungal, antiviral, antidiabetic, and wound healing applications [5]. These advantages have also resulted in the growing usage of ZnO-NPs in dentistry [6]. Moreover, ZnO-NPs exhibit luminescent properties, making ZnO-NPs an excellent candidate for bioimaging [7]. ZnO-NPs are also commonly used as medicine in livestock, aquaculture, and pet animals despite the occasionally occurring side effects [8]. ZnO-NPs are also used in the synthesis of hybrid nanomaterials, which show excellent sorbent properties for the removal of pollutants [9–11]. As a result, the production, manufacturing, and consumption of ZnO-NPs result in their emission via industrial effluents [12]. The amount of ZnO-NPs entering sewage treatment plants by the European Union is estimated to be 1.05 million kg per year (1.7–45 μg/L in the effluent of sewage treatment plants) [12,13]. Considering that environmental ZnO-NPs levels are likely to rise due to the growing demand [13], this contaminant is a cause for concern because of its potential adverse effects on biota [14].

Dietary transfer is an important pathway for pollutant transfer and accumulation, which might cause harmful effects in organisms at higher trophic levels [15]. The inclusion of dietary transfer as a factor in ecotoxicological risk assessment for aquatic species is recommended considering the significant toxicity caused by dietary exposure [16]. Due to the long-lasting and non-biodegradable properties of metals, the effects of the accumulation of metal-based ENPs due to dietary transfer require urgent investigation [2]. ENPs, such as ZnO, silver (Ag), and titanium dioxide (TiO_2), are transferred through food chains and accumulate in the higher trophic levels [17–19]. The enhanced accumulation of metal-based ENPs, including Ag, silicon dioxide (SiO_2), tin oxide (SnO_2), cerium oxide (CeO_2), and magnetite (Fe_3O_4), has several sublethal adverse effects on reproduction, development, and locomotion [20–22]. Although ZnO-NPs are one of the most widely used ENPs, studies related to their toxic effects induced by dietary intake are limited. In addition, whether ZnO-NPs from dietary transfer affect other target tissues, such as the neurons, remains unclear.

Metal-based ENPs accumulate in various organs or tissues after ingestion and induce toxicity predominantly via reactive oxygen species (ROS) [23]. Recent research has suggested that neuromuscular defects are induced by ENPs in the nematode *Caenorhabditis elegans* [24,25]. Furthermore, locomotion in zebrafish and *C. elegans* is known to be adversely affected by ZnO-NPs [26,27]. Previous studies showed that D-type GABAergic motor neurons, which control locomotive behaviors, are potential targets of toxicants, including quantum dots and heavy metals [28,29]. The altered behaviors and neurotoxicity may reduce the fitness of organisms and result in ecotoxicity [30]. However, the involvement of the dietary transfer of ZnO-NPs in neurotoxicity and its underlying mechanisms remain unknown.

To investigate the toxic effects and mechanisms of the dietary transfer of ZnO-NPs, we used *C. elegans* as the model organism and *Escherichia coli* OP50 as its food source to establish a dietary transfer assay model. *C. elegans* is regarded as a useful model in environmental toxicology and neurotoxicology due to its short life cycle, ease of maintenance, and availability of mutants and transgenic strains [31]. A 72 h exposure period, which covers all developmental stages until puberty, may be considered long-term exposure in *C. elegans* [32]. Our study aims to investigate the toxic effects of the long-term dietary transfer of ZnO-NPs in the *E. coli*–*C. elegans* food chain model by examining their accumulation in these organisms and assessing the impairment of locomotive behaviors and D-type GABAergic motor neurons.

2. Materials and Methods

2.1. Chemicals, Characterization, and Strains Maintenance

All of the chemicals used in this study were purchased from Sigma-Aldrich Chemicals Co. (St. Louis, MO, USA). Zinc chloride ($ZnCl_2$) was used as a control to distinguish the effects of dissolved zinc ions (Zn^{2+}). ZnO-NPs (<50 nm, assay grade, >97% purity) were sonicated and freshly prepared in 1000 mg/L stock solution before the assays. Characteristics of ZnO-NPs, including morphology and hydrodynamic diameter, have been previously documented [26]. The TEM image of ZnO-NPs used in the present study was shown in Figure S1. A previous study also reported the TEM image and X-ray diffraction (XRD) results of ZnO-NPs [33]. The concentrations of Zn released from ZnO-NPs in a Luria–Bertani broth (LB) medium during the 8 h *E. coli* exposure period were determined at 0, 4, 6, and 8 h at 37 °C. The initial concentration of the ZnO-NPs suspension prepared in the LB medium was 5 mg/L. The samples were filtered through Amicon Ultra-15 Ultracel-3 centrifuge tubes (3 kDa cutoff $\approx$ 0.9 nm, Millipore, Billerica, MA, USA) to remove the remaining ZnO-NPs, and the concentrations of Zn^{2+} in the aqueous phase were measured using inductively coupled plasma atomic emission spectroscopy (ICP-AES, Spectro Ciros 120, Kleve, Germany).

E. coli OP50 and *C. elegans* strains, including wild-type N2 and transgenic EG1285 strain [*unc-47*p::GFP + *lin-15*(+)], were purchased from the *Caenorhabditis* Genetics Center (CGC, University of Minnesota, MN, USA). The freshly prepared overnight bacterial culture from a single colony was used for ZnO-NP exposure and dietary transfer assay. For maintaining the *C. elegans* strains, all of the worms were cultured at 20 °C on a nematode growth medium (NGM) agar plate covered with a bacterial lawn of *E. coli* OP50.

2.2. Measurement of Zn Concentration in E. coli OP50

Saturated *E. coli* OP50 were freshly prepared and subcultured in the LB medium with various concentrations of ZnO-NPs for 8 h at 37 °C according to the methods of a previous study [20]. Subsequently, the bacterial culture was washed three times using deionized water by centrifugation (15 min at 3000× *g*). The *E. coli* OP50 pellet that was collected was digested using concentrated nitric acid (HNO_3), and the Zn concentration was analyzed using ICP-AES.

2.3. Dietary Transfer Assay

The dietary transfer assay was adapted from a previous study with minor modifications [20]. *E. coli* OP50 treated with ZnO-NPs or $ZnCl_2$ for 8 h were washed and re-suspended in deionized water. Subsequently, they were spread on the NGM plates onto which the *C. elegans* L1-larvae were placed. After 72 h of exposure, the worms were washed using M9 buffer for future assays.

To characterize the distribution and localization of ZnO-NPs in *C. elegans*, rhodamine B (RhoB) was used to label the ZnO-NPs, and unbound RhoB was removed by performing dialysis for 24 h in deionized water (regenerated cellulose dialysis tubing, MWCO: 6000–8000; Orange Scientific, Braine-l'Alleud, Belgium) [34]. The freshly prepared RhoB-labeled ZnO-NPs (RhoB/ZnO-NPs) were administered to the *E. coli* OP50 for 8 h. The RhoB and deionized water without ZnO-NPs (RhoB/deionized water) were treated using dialysis for 24 h and used as the control. Wild-type N2 *C. elegans* L1-larvae were fed with RhoB/ZnO-NPs or RhoB pre-treated *E. coli* OP50 for 96 h. After following the exposure and washing steps, the worms were anesthetized in 50 mM sodium azide on 2% agarose gel mounted on a glass slide. Fluorescent images of the worms were acquired using an epifluorescence microscope (Leica, Wetzlar, Germany) with a suitable filter set (excitation, 550 ± 30 nm; emission, 615 ± 45 nm) and a digital camera. Fluorescence intensities were analyzed and quantified using ImageJ [35]. The tests were conducted at least three times independently, and 25 worms were scored per treatment in each replicate.

2.4. Locomotive Behaviors Tests

Wild-type N2 *C. elegans* were washed after exposure via dietary transfer and subjected to locomotive behavior tests. Locomotive behaviors, including body bending and head thrashing frequencies, were studied by modifying the methods used in a previous study [36]. The frequencies of body bending during 20 s and head thrashing during 1 min were manually measured and calculated. The tests were conducted at least three times independently, and 20 worms were scored per treatment in each replicate.

2.5. GABAergic Neuron Toxicity Assay

Transgenic EG1285 strain *C. elegans* were washed after exposure through dietary transfer. The morphology of green fluorescent protein (GFP)-labeled GABAergic neurons was characterized using an epifluorescence microscope. Fluorescent images of the worms were captured for the determination of neuron cell and gap numbers in GABAergic D-type motor neurons. The abnormality of GABAergic motor neurons (%) was defined as the total counts of GABAergic motor neuron cell loss and gap numbers divided by the total GABAergic motor neuron cells and the connections between the neuron cells in normal conditions. The tests were conducted at least three times independently, and 20 worms were scored per treatment in each replicate.

2.6. Statistical Analysis

SPSS Statistics for Windows, version 22.0 (IBM Corp., Armonk, NY, USA) was used to conduct the statistical analysis. The data were expressed as the mean ± standard error of the means (SEM). Statistical significance was determined using a one-way analysis of variance (ANOVA) with a least significant difference (LSD) post-hoc test and indicated using different lowercase letters ($p < 0.05$).

3. Results and Discussion

3.1. Concentration of Zn Released from ZnO-NPs in LB Medium

Metal nanoparticles exhibit distinct characteristics under different environmental conditions because of their interactions with abiotic and biotic factors [37]. Additionally, the ZnO-NPs are substantially affected by the background medium, resulting in the release of varying amounts of Zn^{2+} [26]. Therefore, before *E. coli* exposure, we determined the concentrations of Zn^{2+} released from the ZnO-NPs in the LB medium at different time points. The Zn^{2+} concentration in the LB medium remained constant at approximately 1.8 mg/L after 0.5, 2, 4, and 8 h (Figure 1). Furthermore, the Zn^{2+} concentration reached a steady state immediately after the ZnO-NPs were added to the LB medium (Figure 1).

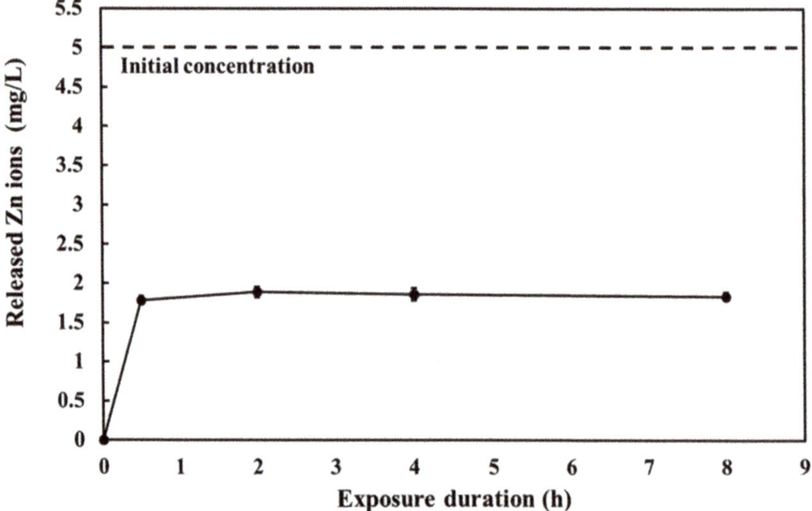

Figure 1. Released Zn ions from ZnO-NPs in LB medium. Initial concentration of ZnO-NPs suspension was 5 mg/L prepared in LB medium. Samples were incubated at 37 °C and analyzed at different time points at 0, 0.5, 2, 4, and 8 h. Samples were filtered to remove undissolved ZnO-NPs, and then the concentrations of Zn ions in the aqueous phase were measured using ICP-AES. The data are shown as the mean ± SEM.

Figure 1 shows that Zn^{2+} constituted approximately 35% of the total ZnO-NPs in the LB medium, implying that compared to the concentration of the particulate form, that of the ionic form of ZnO-NPs was relatively lower in the LB medium. Recent studies have also demonstrated the low solubility and stability of ZnO-NPs in the LB medium (5–51% Zn^{2+} of total ZnO-NPs) [38,39], which may have been due to the presence of organic matter in the LB medium. Some types of organic matter may inhibit the release of free Zn^{2+} from ZnO-NPs [40,41]. Moreover, the dissolution rate of ZnO-NPs is inversely correlated to aliphatic carbon content and hydrogen/carbon ratio [42]. Therefore, both ionic and particulate forms of ZnO-NPs can potentially transfer to *E. coli* OP50, and the accumulation was further investigated.

3.2. Zn Accumulation in E. coli OP50

To investigate the accumulation of ZnO-NPs in *E. coli* OP50 that might further transfer to *C. elegans*, we exposed *E. coli* OP50 to different concentrations of ZnO-NPs (5, 10, 50, and 100 mg/L) and analyzed the Zn concentrations in the bacteria. The concentrations were designed based on the minimal inhibitory concentration (MIC) of 400 mg/L for ZnO-NPs in *E. coli* strain [43]. We selected concentrations below the MIC, and serial dilution was applied to establish the dose-response relationships for the toxicological endpoints.

We found that Zn concentrations in bacterial cells increased in a dose-dependent manner (Figure 2). Exposure to 100 mg/L of ZnO-NPs caused a cellular burden of approximately 200 μg/10^8 cells, which was 100 times higher than that due to 50 mg/L of ZnO-NP exposure (Figure 2). Exposure to $ZnCl_2$ also resulted in a substantial accumulation of Zn in the bacterial cells; 50 mg/L $ZnCl_2$ led to a cellular burden of approximately 2.5 μg/10^8 cells, whereas 100 mg/L $ZnCl_2$ caused a cellular burden of approximately 500 μg/10^8 cells (Figure 2).

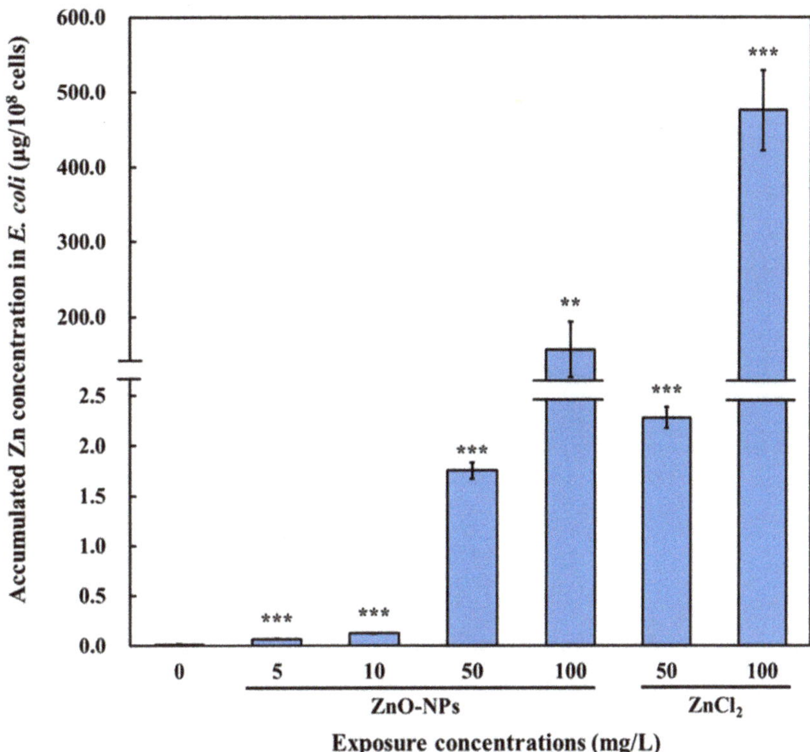

Figure 2. Accumulated Zn concentration in *E. coli* OP50 upon ZnO-NP or $ZnCl_2$ exposure. Saturated *E. coli* OP50 were diluted and incubated in LB medium with various concentrations of ZnO-NPs or $ZnCl_2$ for 8 h at 37 °C. Subsequently, *E. coli* OP50 pellet was washed and collected for Zn concentration analysis using ICP-AES. The data are shown as the mean ± SEM. Statistical significance was determined by ANOVA with LSD post-hoc test to compare to the control (0 mg/L). (**: $p < 0.01$, ***: $p < 0.001$).

Thus, ZnO-NPs accumulated in *E. coli* OP50 after 8 h of exposure at all of the examined concentrations (Figure 2). Metal nanoparticles, such as TiO_2 and Ag, accumulated in the bacteria, including *E. coli* and *Pseudomonas aeruginosa*, and then were transferred to higher trophic levels [20,44]. Several studies have shown that ZnO-NPs can damage the bacterial cell wall and enhance membrane permeability, thereby resulting in their accumulation in bacteria [45,46]. Additionally, the internalization of ZnO-NPs by *E. coli* and other bacterial cells has been observed previously [47,48]. The high concentration (100 mg/L) of ZnO-NPs and $ZnCl_2$ largely increased the Zn accumulation compared with 50 mg/L, which may be due to the membrane damage that facilitated higher Zn accumulation in the cytoplasm [49]. Therefore, our results suggest that ZnO-NPs can accumulate in prey (*E. coli*) and potentially be transferred to higher trophic levels through dietary intake.

3.3. Distribution and Accumulation of ZnO-NPs in C. elegans via Dietary Transfer

To further assess the dietary transfer of ZnO-NPs from *E. coli* to *C. elegans*, the worms were exposed to *E. coli* OP50 pre-treated with ZnO-NPs. The control worms were fed with *E. coli* pre-treated with RhoB/deionized water. The fluorescent dye RhoB was used to label the ZnO-NPs to visualize the distribution and accumulation of the ZnO-NPs due to dietary transfer. Compared with the controls (RhoB/deionized water), the RhoB-labeled ZnO-NPs accumulated mainly in the pharynx and intestine of *C. elegans* (Figure 3A). Quantification of the fluorescence intensity showed that the background levels in the controls were approximately 6 RFU/worm, which significantly increased to approximately 15 RFU/worm in the presence of RhoB-labeled ZnO-NPs (Figure 3B).

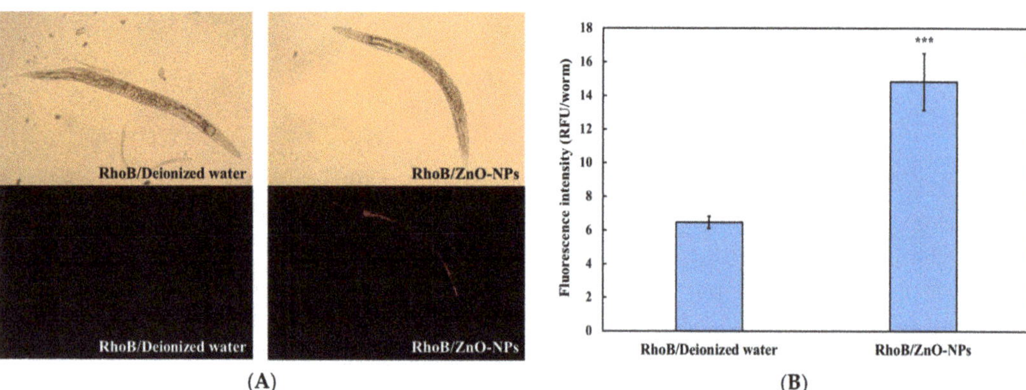

Figure 3. Accumulated ZnO-NPs in *C. elegans* through dietary transfer. Wild-type N2 *C. elegans* L1-larvae were fed with rhodamine B (RhOB)-labeling ZnO-NPs (RhoB/ZnO-NPs) pretreated *E. coli* OP50 for 96 h. RhoB/Deionized water was used as the control. After exposure and washing, (**A**) fluorescence images of worms were taken, and (**B**) fluorescence intensity was analyzed using ImageJ. The data are shown as the mean ± SEM. The tests were conducted at least three times independently, and 25 worms were scored per treatment in each replicate. Statistical significance was determined by ANOVA with LSD post-hoc test to compare to the control. (***: $p < 0.001$).

While the dietary transfer of ZnO-NPs in aquatic food chains has been demonstrated in several studies [19,50–52], little is known about their distribution. Trophic transfer of ZnO-NPs occurs in simple food chains involving algae (*Chlorella ellipsoidea*) and clams (*Corbicula fluminea*) [53]. Furthermore, goldfish fed with brine shrimp pre-exposed to ZnO-NPs showed significant accumulation of Zn in the intestine [54]. In *C. elegans*, the intestine is the primary target of nanomaterials, including SiO_2, carbon nanotubes, graphene oxide, and Ag [20,55–57]. This may be the reason for the significant ZnO-NP accumulation in the intestine and pharynx of *C. elegans* due to dietary transfer from *E. coli* OP50 (Figure 3A).

3.4. Effects of Dietary Transfer of ZnO-NPs on Locomotive Behaviors of C. elegans

The sublethal endpoints, including body bending and head thrashing frequencies, of *C. elegans* have been used to assess neurotoxicity [31,58,59]. Moreover, we previously found that aquatic exposure to ZnO-NPs in simulated surface water (EPA water) significantly impairs locomotive behaviors, indicating that neurotoxicity is a potential result of ZnO-NP exposure [26]. Therefore, the effects of the dietary transfer of ZnO-NPs on *C. elegans* were investigated using locomotive behavior tests. In addition, $ZnCl_2$ was used to differentiate between toxicity due to ionic Zn and that caused by ZnO-NPs.

The body bending frequency decreased in *C. elegans* fed with ZnO-NPs pre-treated *E. coli* OP50 in a dose-dependent manner (Figure 4A). In contrast, there was no significant difference in the body bending frequency of *C. elegans* fed with relatively high concentrations (50 and 100 mg/L) of $ZnCl_2$ pre-treated *E. coli* OP50 and that of the controls (fed with 0 mg/L $ZnCl_2$ pre-treated *E. coli* OP50) (Figure 4A). Tests pertaining to the head thrashing frequency exhibited similar results (Figure 4B). *E. coli* pre-treated with 50 and 100 mg/L of ZnO-NPs demonstrated a 9–10% reduction in head thrashing and a 17–18% reduction in body bending (Figure 4A,B), implying that body bending frequency may be a more sensitive endpoint than head thrashing.

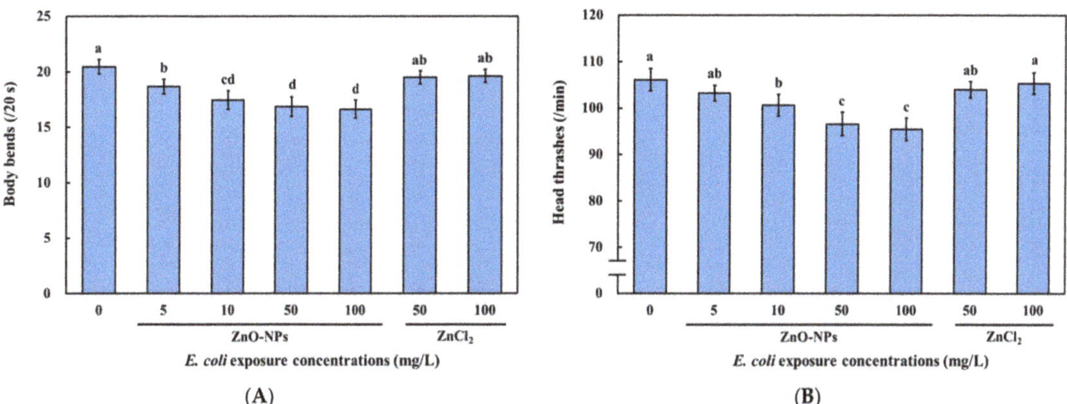

Figure 4. Locomotive behavior defects resulted from dietary transfer of ZnO-NPs in *C. elegans*. Wild-type N2 *C. elegans* L1-larvae were fed with ZnO-NPs or $ZnCl_2$ pretreated *E. coli* OP50 for 72 h. After exposure and washing, (**A**) body bends and (**B**) head thrashes of worms were determined. The data are shown as the mean ± SEM. The tests were conducted at least three times independently, and 20 worms were scored per treatment in each replicate. Statistical significance was determined by one-way ANOVA with LSD post-hoc test and indicated by different lowercase letters ($p < 0.05$).

Our results suggest that the predicted environmental concentration of ZnO-NPs (76 μg/L) [60] may be harmful to locomotive behavior and cause ecotoxicity. Altered behaviors caused by environmental toxicants can reduce the fitness and population of organisms, indicating the potential impact of neurotoxicity on the ecosystem [30]. A previous study showed that exposure to ZnO-NPs impaired motor functions in mice [11]. Liquid exposure to ZnO-NPs shows impairment of locomotive behaviors in *C. elegans* and zebrafish, which is more significant than that caused by $ZnCl_2$ [26,27]. Thus, the biological actions of ZnO-NPs and Zn^{2+} are different, and the impairment of motor functions induced by ZnO-NPs may be similar in different species. It has been shown that the sensitivity of body bends to the liquid exposure of ZnO-NPs was higher than head thrashes in *C. elegans*, which is in agreement with our results [26]. Similarly, the direct liquid exposure of 500 μg/L of ZnO-NPs caused higher toxic effects on body bends than 500 μg/L of $ZnCl_2$ [26]. Moreover, our results showed that impaired locomotive behaviors and neurotoxicity in *C. elegans* were due to the dietary transfer of ZnO-NPs (Figure 4A,B). The dietary transfer of Ag nanoparticles is known to disrupt the locomotion of springtails (Collembola) [21]. Additionally, copper (Cu) nanoparticles accumulate in *Daphnia magna* by dietary transfer and impair their feeding rate [61]. Therefore, various metal nanoparticles in the environment can produce toxic effects when they are transferred to higher trophic levels. Although the dietary transfer of ZnO-NPs through trophic levels has been reported, its neurotoxicity due to dietary transfer has rarely been reported. Our findings suggest that ZnO-NPs can accumulate in *C. elegans* via trophic transfer, thereby impairing locomotion behaviors. However, the mechanisms behind the locomotion defects due to the dietary transfer of metal-based nanoparticles are unclear.

3.5. Effects of Dietary Transfer of ZnO-NPs on D-Type GABAergic Motor Neurons of C. elegans

The D-type GABAergic motor neurons are inhibitory motor neurons that control locomotive behaviors in *C. elegans* [31,62]. Therefore, we investigated the effect of the dietary transfer of ZnO-NPs on these neurons (Figure 5A) and found that they decreased the cell bodies of D-type GABAergic motor neurons in a dose-dependent manner (Figure 5B). The representative images of GABAergic neuron in *C. elegans* with different treatments were shown in Figure S2. In addition, ZnO-NP treatment caused greater degeneration of D-type GABAergic motor neurons than that caused by $ZnCl_2$ (Figure 5B). Moreover, a gap formation on the cords of the D-type GABAergic motor neurons increased due to the dietary exposure of ZnO-NPs in a dose-dependent manner, which was more significant than that caused by the $ZnCl_2$ treatment (Figure 5B). Thus, dietary exposure to ZnO-NPs significantly damages D-type GABAergic neurons in *C. elegans*. In addition, we have checked the cholinergic motor neurons, which are the other motor neurons controlling the locomotive behavior of *C. elegans*. The results of the gap numbers showed that there is no significant difference between the control and treatment with ZnO-NPs or $ZnCl_2$ (Figure S3).

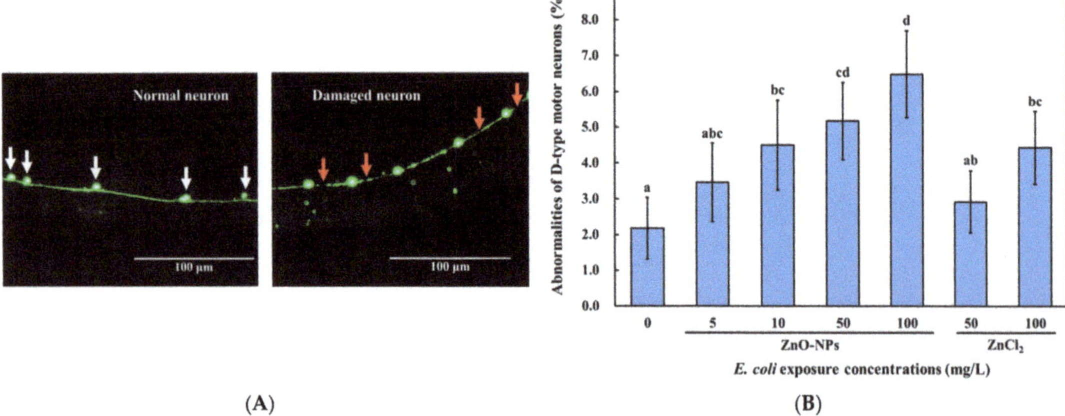

Figure 5. GABAergic neuron damage resulted from dietary transfer of ZnO-NPs in *C. elegans*. Transgenic strain EG1285 (*unc-47*p::GFP) L1-larvae were fed with ZnO-NPs or $ZnCl_2$ pretreated *E. coli* OP50 for 72 h. After exposure and washing, fluorescence images of worms were taken to determine neural abnormalities, including neuron loss and degenerating commissures of GABAergic D-type motor neurons. (**A**) Representative image for determination of GABAergic neuronal damage. White arrows indicate normal cell bodies; red arrows indicate gaps on the neuronal cord. (**B**) GABAergic neuron damage resulted from dietary transfer of ZnO-NPs. The data are shown as the mean ± SEM. The tests were conducted at least three times independently, and 20 worms were scored per treatment in each replicate. Statistical significance was determined by one-way ANOVA with LSD post-hoc test and indicated by different lowercase letters ($p < 0.05$).

Toxic metals and nanoparticles, such as quantum dots, have been shown to adversely affect locomotive behaviors and damage D-type GABAergic motor neurons [28,29]. Nevertheless, other nanomaterials, including graphene-based Ag ones, do not harm GABAergic neurons [25,55]. This may be because of different toxicity mechanisms. Our results confirmed that the D-type GABAergic motor neuron system was damaged in *C. elegans* that were fed with ZnO-NPs pre-treated *E. coli* OP50 (Figure 5B), which might have contributed to the impairment of the locomotive behaviors (Figure 4A,B). A previous study found that exposure to ZnO-NPs also induced dopaminergic neuronal damage in zebrafish brains [63]. Additionally, exposure to ZnO-NPs resulted in neuronal damage in the brains of Wistar rats [64]. These suggest that ZnO-NP-induced neuronal damage might not be exclusive to *C. elegans*. A previous study showed that cancer cell lines were more sensitive to the cytotoxic effects of ZnO-NPs than normal cell lines [65]. Despite the growing body of evidence demonstrating the promising potential of ZnO-NPs in biomedical applications, the non-selective cytotoxic effects against the various cell lines of ZnO-NPs remain controversial [66]. Our findings suggest that precautions should be taken as dietary ZnO-NPs might result in neuronal damage in vivo.

Notably, the adverse effects on the neuron system and locomotive behaviors were not observed in the *C. elegans* that were fed $ZnCl_2$ pre-treated *E. coli* OP50 (Figures 4 and 5), indicating that the effects of the dietary transfer of ZnO-NPs were primarily due to particulate forms rather than Zn^{2+}. The particle-specific effects of the dietary transfer of ZnO-NPs were also observed in several studies, especially those on bioaccumulation [19,51]. The present study further suggests that neurotoxicity caused by the dietary transfer of ZnO-NPs is more significant than that due to $ZnCl_2$ exposure. Previous studies have mainly reported the accumulation of dietary ZnO-NPs in the gut, as well as mortality, reproduction, and oxidative stress response [19,50–52]. We provided evidence that dietary ZnO-NPs can also adversely affect the locomotion of *C. elegans* by causing neuronal damage in vivo. Therefore, our findings reveal rarely reported neuronal damage induced by dietary ZnO-NP exposure.

4. Conclusions

In summary, the present study provides evidence regarding the dietary transfer of ZnO-NPs from *E. coli* OP50 to *C. elegans* and their toxic effects on locomotive behaviors and D-type GABAergic motor neurons. Additionally, these effects were found to be more significant in *C. elegans* that were fed ZnO-NPs pre-treated *E. coli* OP50 than in those exposed to $ZnCl_2$ pre-treated *E. coli* OP50, which indicates that the neurotoxicity caused by the dietary transfer of ZnO-NPs is mostly due to nanoparticles rather than Zn^{2+}. The results of the present study highlight the potential neurotoxicity and toxic mechanisms of ZnO-NPs transferred through food chains.

Supplementary Materials: The following supporting information can be downloaded at: https://www.mdpi.com/article/10.3390/nano13020289/s1, Figure S1: TEM image of ZnO nanoparticles (ZnO-NPs) used in this study; Figure S2 Representative images of GABAergic neurons in *C. elegans* (A–C); Figure S3. Effects of ZnO-NPs and ZnCl2 on cholinergic motor neurons.

Author Contributions: Conceptualization, C.M.H. and C.-W.H.; methodology, C.-W.H.; validation, C.M.H.; investigation, C.M.H. and C.-W.H.; data curation, C.-W.H.; writing—original draft preparation, C.M.H.; writing—review and editing, C.-W.H.; visualization, C.M.H.; supervision, C.-W.H. All authors have read and agreed to the published version of the manuscript.

Funding: This research was funded by the Ministry of Science and Technology (Taiwan), grant number MOST104-2221-E002-034-MY3. The APC was funded by the National Kaohsiung University of Science and Technology (grant number 111TSD00-2).

Data Availability Statement: The data would be available on request to the corresponding author.

Conflicts of Interest: The authors declare no conflict of interest.

References

1. Bundschuh, M.; Filser, J.; Luderwald, S.; Mckee, M.S.; Metreveli, G.; Schaumann, G.E.; Schulz, R.; Wagner, S. Nanoparticles in the environment: Where do we come from, where do we go to? *Environ. Sci. Eur.* **2018**, *30*, 6. [CrossRef] [PubMed]
2. Tangaa, S.R.; Selck, H.; Winther-Nielsen, M.; Khan, F.R. Trophic transfer of metal-based nanoparticles in aquatic environments: A review and recommendations for future research focus. *Environ. Sci. Nano* **2016**, *3*, 966–981. [CrossRef]

3. Krol, A.; Pomastowski, P.; Rafinska, K.; Railean-Plugaru, V.; Buszewski, B. Zinc oxide nanoparticles: Synthesis, antiseptic activity and toxicity mechanism. *Adv. Colloid Interface Sci.* **2017**, *249*, 37–52. [CrossRef]
4. Rajput, V.D.; Minkina, T.M.; Behal, A.; Sushkova, S.N.; Mandzhieva, S.; Singh, R.; Gorovtsov, A.; Tsitsuashvili, V.S.; Purvis, W.O.; Ghazaryan, K.A.; et al. Effects of zinc-oxide nanoparticles on soil, plants, animals and soil organisms: A review. *Environ. Nanotechnol. Monit. Manag.* **2018**, *9*, 76–84. [CrossRef]
5. Mandal, A.K.; Katuwal, S.; Tettey, F.; Gupta, A.; Bhattarai, S.; Jaisi, S.; Bhandari, D.P.; Shah, A.K.; Bhattarai, N.; Parajuli, N. Current research on zinc oxide nanoparticles: Synthesis, characterization and biomedical applications. *Nanomaterials* **2022**, *12*, 3066. [CrossRef]
6. Moradpoor, H.; Safaei, M.; Mozaffari, H.R.; Sharifi, R.; Imani, M.M.; Golshah, A.; Bashardoust, N. An overview of recent progress in dental applications of zinc oxide nanoparticles. *RSC Adv.* **2021**, *11*, 21189–21206. [CrossRef]
7. Jiang, J.; Pi, J.; Cai, J. The Advancing of zinc oxide nanoparticles for biomedical applications. *Bioinorg. Chem. Appl.* **2018**, *2018*, 1062562. [CrossRef]
8. Rahman, H.S.; Othman, H.H.; Abdullah, R.; Edin, H.; Al-Haj, N.A. Beneficial and toxicological aspects of zinc oxide nanoparticles in animals. *Vet. Med. Sci.* **2022**, *8*, 1769–1779. [CrossRef]
9. Goda, M.N.; Said, A.E.-A.A.; Abdelhamid, H.N. Highly selective dehydration of methanol over metal-organic frameworks (MOFs)-derived ZnO@Carbon. *J. Environ. Chem. Eng.* **2021**, *9*, 106336. [CrossRef]
10. Soliman, A.I.A.; Abdel-Wahab, A.A.; Abdelhamid, H.N. Hierarchical porous zeolitic imidazolate frameworks (ZIF-8) and ZnO@N-doped carbon for selective adsorption and photocatalytic degradation of organic pollutants. *RSC Adv.* **2022**, *12*, 7075–7084. [CrossRef] [PubMed]
11. Abdelhamid, H.N.; Al Kiey, S.A.; Sharmoukh, W. A high-performance hybrid supercapacitor electrode based on ZnO/nitrogen-doped carbon nanohybrid. *Appl. Organomet. Chem.* **2022**, *36*, e6486. [CrossRef]
12. Sun, T.Y.; Gottschalk, F.; Hungerbuhler, K.; Nowack, B. Comprehensive probabilistic modelling of environmental emissions of engineered nanomaterials. *Environ. Pollut.* **2014**, *185*, 69–76. [CrossRef]
13. Dumont, E.; Johnson, A.C.; Keller, V.D.J.; Williams, R.J. Nano silver and nano zinc-oxide in surface waters—Exposure estimation for Europe at high spatial and temporal resolution. *Environ. Pollut.* **2015**, *196*, 341–349. [CrossRef]
14. Chong, C.L.; Fang, C.M.; Pung, S.Y.; Ong, C.E.; Pung, Y.F.; Kong, C.; Pan, Y. Current updates on the *in vivo* assessment of zinc oxide nanoparticles toxicity using animal models. *BioNanoScience* **2021**, *11*, 590–620. [CrossRef]
15. Suedel, B.C.; Boraczek, J.A.; Peddicord, R.K.; Clifford, P.A.; Dillon, T.M. Trophic transfer and biomagnification potential of contaminants in aquatic ecosystems. *Rev. Environ. Contam. Toxicol.* **1994**, *136*, 21–89.
16. Fisher, N.S.; Hook, S.E. Toxicology tests with aquatic animals need to consider the trophic transfer of metals. *Toxicology* **2002**, *181–182*, 531–536. [CrossRef]
17. Kalman, J.; Paul, K.B.; Khan, F.R.; Stone, V.; Fernandes, T.F. Characterisation of bioaccumulation dynamics of three differently coated silver nanoparticles and aqueous silver in a simple freshwater food chain. *Environ. Chem.* **2015**, *12*, 662–672. [CrossRef]
18. Chen, J.Y.; Li, H.R.; Han, X.Q.; Wei, X.Z. Transmission and accumulation of nano-TiO_2 in a 2-step food chain (*Scenedesmus obliquus* to *Daphnia magna*). *Bull. Environ. Contam. Toxicol.* **2015**, *95*, 145–149. [CrossRef]
19. Skjolding, L.M.; Winther-Nielsen, M.; Baun, A. Trophic transfer of differently functionalized zinc oxide nanoparticles from crustaceans (*Daphnia magna*) to zebrafish (*Danio rerio*). *Aquat. Toxicol.* **2014**, *157*, 101–108. [CrossRef]
20. Luo, X.; Xu, S.M.; Yang, Y.N.; Li, L.Z.; Chen, S.P.; Xu, A.; Wu, L.J. Insights into the ecotoxicity of silver nanoparticles transferred from *Escherichia coli* to *Caenorhabditis elegans*. *Sci. Rep.* **2016**, *6*, 36465. [CrossRef]
21. Kwak, J.I.; An, Y.J. Trophic transfer of silver nanoparticles from earthworms disrupts the locomotion of springtails (Collembola). *J. Hazard. Mater.* **2016**, *315*, 110–116. [CrossRef]
22. Gambardella, C.; Gallus, L.; Gatti, A.M.; Faimali, M.; Carbone, S.; Antisari, L.V.; Falugi, C.; Ferrando, S. Toxicity and transfer of metal oxide nanoparticles from microalgae to sea urchin larvae. *Chem. Ecol.* **2014**, *30*, 308–316. [CrossRef]
23. Sengul, A.B.; Asmatulu, E. Toxicity of metal and metal oxide nanoparticles: A review. *Environ. Chem. Lett.* **2020**, *18*, 1659–1683. [CrossRef]
24. Von Mikecz, A. Lifetime eco-nanotoxicology in an adult organism: Where and when is the invertebrate *C. elegans* vulnerable? *Environ. Sci. Nano* **2018**, *5*, 616–622. [CrossRef]
25. Piechulek, A.; von Mikecz, A. Life span-resolved nanotoxicology enables identification of age-associated neuromuscular vulnerabilities in the nematode *Caenorhabditis elegans*. *Environ. Pollut.* **2018**, *233*, 1095–1103. [CrossRef]
26. Huang, C.W.; Li, S.W.; Liao, V.H.C. Chronic ZnO-NPs exposure at environmentally relevant concentrations results in metabolic and locomotive toxicities in *Caenorhabditis elegans*. *Environ. Pollut.* **2017**, *220*, 1456–1464. [CrossRef]
27. Chen, T.H.; Lin, C.C.; Meng, P.J. Zinc oxide nanoparticles alter hatching and larval locomotor activity in zebrafish (*Danio rerio*). *J. Hazard. Mater.* **2014**, *277*, 134–140. [CrossRef]
28. Du, M.; Wang, D. The neurotoxic effects of heavy metal exposure on GABAergic nervous system in nematode *Caenorhabditis elegans*. *Environ. Toxicol. Pharmacol.* **2009**, *27*, 314–320. [CrossRef]
29. Zhao, Y.L.; Wang, X.; Wu, Q.L.; Li, Y.P.; Tang, M.; Wang, D.Y. Quantum dots exposure alters both development and function of D-type GABAergic motor neurons in nematode *Caenorhabditis elegans*. *Toxicol. Res.* **2015**, *4*, 399–408. [CrossRef]
30. Legradi, J.B.; Di Paolo, C.; Kraak, M.H.S.; van der Geest, H.G.; Schymanski, E.L.; Williams, A.J.; Dingemans, M.M.L.; Massei, R.; Brack, W.; Cousin, X.; et al. An ecotoxicological view on neurotoxicity assessment. *Environ. Sci. Eur.* **2018**, *30*, 46. [CrossRef]

31. Leung, M.C.; Williams, P.L.; Benedetto, A.; Au, C.; Helmcke, K.J.; Aschner, M.; Meyer, J.N. *Caenorhabditis elegans*: An emerging model in biomedical and environmental toxicology. *Toxicol. Sci.* **2008**, *106*, 5–28. [CrossRef] [PubMed]
32. Comber, S.D.; Conrad, A.U.; Hoss, S.; Webb, S.; Marshall, S. Chronic toxicity of sediment-associated linear alkylbenzene sulphonates (LAS) to freshwater benthic organisms. *Environ. Pollut.* **2006**, *144*, 661–668. [CrossRef]
33. Amin, N.; Zulkifli, S.Z.; Azmai, M.N.A.; Ismail, A. Toxicity of zinc oxide nanoparticles on the embryo of Javanese medaka (*Oryzias javanicus* Bleeker, 1854): A comparative study. *Animals* **2021**, *11*, 2170. [CrossRef]
34. Zhang, L.M.; Xia, J.G.; Zhao, Q.H.; Liu, L.W.; Zhang, Z.J. Functional graphene oxide as a nanocarrier for controlled loading and targeted delivery of mixed anticancer drugs. *Small* **2010**, *6*, 537–544. [CrossRef]
35. Schneider, C.A.; Rasband, W.S.; Eliceiri, K.W. NIH Image to ImageJ: 25 years of image analysis. *Nat. Methods* **2012**, *9*, 671–675. [CrossRef]
36. Tsalik, E.L.; Hobert, O. Functional mapping of neurons that control locomotory behavior in *Caenorhabditis elegans*. *J. Neurobiol.* **2003**, *56*, 178–197. [CrossRef]
37. Gupta, G.S.; Dhawan, A. Chapter 17—Fate and potential hazards of nanoparticles in the environment. In *Nanoparticle Therapeutics*; Kesharwani, P., Singh, K.K., Eds.; Academic Press: Cambridge, MA, USA, 2022; pp. 581–602.
38. Matula, K.; Richter, L.; Adamkiewicz, W.; Akerstrom, B.; Paczesny, J.; Holyst, R. Influence of nanomechanical stress induced by ZnO nanoparticles of different shapes on the viability of cells. *Soft Matter* **2016**, *12*, 4162–4169. [CrossRef]
39. Gelabert, A.; Sivry, Y.; Gobbi, P.; Mansouri-Guilani, N.; Menguy, N.; Brayner, R.; Siron, V.; Benedetti, M.F.; Ferrari, R. Testing nanoeffect onto model bacteria: Impact of speciation and genotypes. *Nanotoxicology* **2016**, *10*, 216–225. [CrossRef]
40. Li, M.; Lin, D.H.; Zhu, L.Z. Effects of water chemistry on the dissolution of ZnO nanoparticles and their toxicity to *Escherichia coli*. *Environ. Pollut.* **2013**, *173*, 97–102. [CrossRef]
41. Miao, A.J.; Zhang, X.Y.; Luo, Z.; Chen, C.S.; Chin, W.C.; Santschi, P.H.; Quigg, A. Zinc oxide-engineered nanoparticles: Dissolution and toxicity to marine phytoplankton. *Environ. Toxicol. Chem.* **2010**, *29*, 2814–2822. [CrossRef]
42. Jiang, C.; Aiken, G.R.; Hsu-Kim, H. Effects of natural organic matter properties on the dissolution kinetics of zinc oxide nanoparticles. *Environ. Sci. Technol.* **2015**, *49*, 11476–11484. [CrossRef] [PubMed]
43. Xie, Y.; He, Y.; Irwin, P.L.; Jin, T.; Shi, X. Antibacterial activity and mechanism of action of zinc oxide nanoparticles against *Campylobacter jejuni*. *Appl. Environ. Microbiol.* **2011**, *77*, 2325–2331. [CrossRef]
44. Mielke, R.E.; Priester, J.H.; Werlin, R.A.; Gelb, J.; Horst, A.M.; Orias, E.; Holden, P.A. Differential growth of and nanoscale TiO_2 accumulation in *Tetrahymena thermophila* by direct feeding versus trophic transfer from *Pseudomonas aeruginosa*. *Appl. Environ. Microbiol.* **2013**, *79*, 5616–5624. [CrossRef]
45. Siddiqi, K.S.; Husen, A. Properties of zinc oxide nanoparticles and their activity against microbes. *Nanoscale Res. Lett.* **2018**, *13*, 141. [CrossRef] [PubMed]
46. Sirelkhatim, A.; Mahmud, S.; Seeni, A.; Kaus, N.H.M.; Ann, L.C.; Bakhori, S.K.M.; Hasan, H.; Mohamad, D. Review on zinc oxide nanoparticles: Antibacterial activity and toxicity mechanism. *Nano-Micro Lett.* **2015**, *7*, 219–242. [CrossRef] [PubMed]
47. Brayner, R.; Ferrari-Iliou, R.; Brivois, N.; Djediat, S.; Benedetti, M.F.; Fievet, F. Toxicological impact studies based on *Escherichia coli* bacteria in ultrafine ZnO nanoparticles colloidal medium. *Nano Lett.* **2006**, *6*, 866–870. [CrossRef]
48. Bandyopadhyay, S.; Peralta-Videa, J.R.; Plascencia-Villa, G.; Jose-Yacaman, M.; Gardea-Torresdey, J.L. Comparative toxicity assessment of CeO_2 and ZnO nanoparticles towards *Sinorhizobium meliloti*, a symbiotic alfalfa associated bacterium: Use of advanced microscopic and spectroscopic techniques. *J. Hazard. Mater.* **2012**, *241*, 379–386. [CrossRef] [PubMed]
49. Dimkpa, C.O.; Calder, A.; Britt, D.W.; McLean, J.E.; Anderson, A.J. Responses of a soil bacterium, *Pseudomonas chlororaphis* O6 to commercial metal oxide nanoparticles compared with responses to metal ions. *Environ. Pollut.* **2011**, *159*, 1749–1756. [CrossRef]
50. Jarvis, T.A.; Miller, R.J.; Lenihan, H.S.; Bielmyer, G.K. Toxicity of ZnO nanoparticles to the copepod *Acartia tonsa*, exposed through a phytoplankton diet. *Environ. Toxicol. Chem.* **2013**, *32*, 1264–1269. [CrossRef]
51. Bhuvaneshwari, M.; Iswarya, V.; Vishnu, S.; Chandrasekaran, N.; Mukherjee, A. Dietary transfer of zinc oxide particles from algae (*Scenedesmus obliquus*) to daphnia (*Ceriodaphnia dubia*). *Environ. Res.* **2018**, *164*, 395–404. [CrossRef]
52. Chen, Y.; Wu, F.; Li, W.; Luan, T.; Lin, L. Comparison on the effects of water-borne and dietary-borne accumulated ZnO nanoparticles on *Daphnia magna*. *Chemosphere* **2017**, *189*, 94–103. [CrossRef]
53. Huang, C.W.; Chang, C.H.; Li, S.W.; Yen, P.L.; Liao, V.H.C. Co-exposure to foodborne and waterborne ZnO nanoparticles in aquatic sediment environments enhances DNA damage and stress gene expression in freshwater Asian clam *Corbicula fluminea*. *Environ. Sci. Nano* **2020**, *7*, 1252–1265. [CrossRef]
54. Ates, M.; Arslan, Z.; Demir, V.; Daniels, J.; Farah, I.O. Accumulation and toxicity of CuO and ZnO nanoparticles through waterborne and dietary exposure of goldfish (*Carassius auratus*). *Environ. Toxicol.* **2015**, *30*, 119–128. [CrossRef] [PubMed]
55. Li, P.; Xu, T.; Wu, S.; Lei, L.; He, D. Chronic exposure to graphene-based nanomaterials induces behavioral deficits and neural damage in *Caenorhabditis elegans*. *J. Appl. Toxicol.* **2017**, *37*, 1140–1150. [CrossRef] [PubMed]
56. Pluskota, A.; Horzowski, E.; Bossinger, O.; von Mikecz, A. In *Caenorhabditis elegans* nanoparticle-bio-interactions become transparent: Silica-nanoparticles induce reproductive senescence. *PLoS ONE* **2009**, *4*, e6622. [CrossRef] [PubMed]
57. Wu, Q.L.; Li, Y.X.; Li, Y.P.; Zhao, Y.L.; Ge, L.; Wang, H.F.; Wang, D.Y. Crucial role of the biological barrier at the primary targeted organs in controlling the translocation and toxicity of multi-walled carbon nanotubes in the nematode *Caenorhabditis elegans*. *Nanoscale* **2013**, *5*, 11166–11178. [CrossRef] [PubMed]

58. Tseng, I.L.; Yang, Y.F.; Yu, C.W.; Li, W.H.; Liao, V.H.C. Phthalates induce neurotoxicity affecting locomotor and thermotactic behaviors and AFD neurons through oxidative stress in *Caenorhabditis elegans*. *PLoS ONE* **2014**, *9*, e99945. [CrossRef]
59. Li, Y.; Yu, S.; Wu, Q.; Tang, M.; Pu, Y.; Wang, D. Chronic Al_2O_3-nanoparticle exposure causes neurotoxic effects on locomotion behaviors by inducing severe ROS production and disruption of ROS defense mechanisms in nematode *Caenorhabditis elegans*. *J. Hazard. Mater.* **2012**, *219–220*, 221–230. [CrossRef]
60. Boxall, A.; Chaudhry, Q.; Sinclair, C.; Jones, A.; Aitken, R.; Jefferson, B.; Watts, C. *Current and Future Predicted Environmental Exposure to Engineered Nanoparticles*; Central Science Laboratory: York, UK, 2007.
61. Yu, Q.; Zhang, Z.; Monikh, F.A.; Wu, J.; Wang, Z.; Vijver, M.G.; Bosker, T.; Peijnenburg, W.J.G.M. Trophic transfer of Cu nanoparticles in a simulated aquatic food chain. *Ecotoxicol. Environ. Saf.* **2022**, *242*, 113920. [CrossRef]
62. Chen, P.; Martinez-Finley, E.J.; Bornhorst, J.; Chakraborty, S.; Aschner, M. Metal-induced neurodegeneration in *C. elegans*. *Front. Aging Neurosci.* **2013**, *5*, 18. [CrossRef]
63. Jin, M.; Li, N.; Sheng, W.; Ji, X.; Liang, X.; Kong, B.; Yin, P.; Li, Y.; Zhang, X.; Liu, K. Toxicity of different zinc oxide nanomaterials and dose-dependent onset and development of Parkinson's disease-like symptoms induced by zinc oxide nanorods. *Environ. Int.* **2021**, *146*, 106179. [CrossRef] [PubMed]
64. Liu, H.; Yang, H.; Fang, Y.; Li, K.; Tian, L.; Liu, X.; Zhang, W.; Tan, Y.; Lai, W.; Bian, L.; et al. Neurotoxicity and biomarkers of zinc oxide nanoparticles in main functional brain regions and dopaminergic neurons. *Sci. Total Environ.* **2020**, *705*, 135809. [CrossRef] [PubMed]
65. Babayevska, N.; Przysiecka, Ł.; Iatsunskyi, I.; Nowaczyk, G.; Jarek, M.; Janiszewska, E.; Jurga, S. ZnO size and shape effect on antibacterial activity and cytotoxicity profile. *Sci. Rep.* **2022**, *12*, 8148. [CrossRef] [PubMed]
66. Mishra, P.K.; Mishra, H.; Ekielski, A.; Talegaonkar, S.; Vaidya, B. Zinc oxide nanoparticles: A promising nanomaterial for biomedical applications. *Drug Discov.* **2017**, *22*, 1825–1834. [CrossRef]

 *nanomaterials*

Article

Fe_3O_4-Halloysite Nanotube Composites as Sustainable Adsorbents: Efficiency in Ofloxacin Removal from Polluted Waters and Ecotoxicity

Doretta Capsoni [1,2], Paola Lucini [1,2], Debora Maria Conti [1,2], Michela Bianchi [1], Federica Maraschi [1], Beatrice De Felice [3], Giovanna Bruni [1,2], Maryam Abdolrahimi [4,5], Davide Peddis [4,6], Marco Parolini [3], Silvia Pisani [7] and Michela Sturini [1,2,*]

1 Department of Chemistry, University of Pavia, 27100 Pavia, Italy
2 C.S.G.I. (Consorzio Interuniversitario per lo Sviluppo dei Sistemi a Grande Interfase) & Department of Chemistry, Physical Chemistry Section, University of Pavia, 27100 Pavia, Italy
3 Department of Environmental Science and Policy, University of Milan, 20133 Milan, Italy
4 Institute of Structure of Matter, National Research Council (CNR), Monterotondo Scalo, 00015 Rome, Italy
5 Dipartimento di Scienze, Università degli Studi Roma Tre, Via della Vasca Navale 84, 00146 Rome, Italy
6 Department of Chemistry and Industrial Chemistry, University of Genova, 16146 Genova, Italy
7 Department of Otorhinolaryngology, Fondazione IRCCS Policlinico San Matteo, 27100 Pavia, Italy
* Correspondence: michela.sturini@unipv.it; Tel.: +39-0382-987347

Abstract: The present work aimed at decorating halloysite nanotubes (HNT) with magnetic Fe_3O_4 nanoparticles through different synthetic routes (co-precipitation, hydrothermal, and sol-gel) to test the efficiency of three magnetic composites (HNT/Fe_3O_4) to remove the antibiotic ofloxacin (OFL) from waters. The chemical–physical features of the obtained materials were characterized through the application of diverse techniques (XRPD, FT-IR spectroscopy, SEM, EDS, and TEM microscopy, thermogravimetric analysis, and magnetization measurements), while ecotoxicity was assessed through a standard test on the freshwater organism *Daphnia magna*. Independently of the synthesis procedure, the magnetic composites were successfully obtained. The Fe_3O_4 is nanometric (about 10 nm) and the weight percentage is sample-dependent. It decorates the HNT's surface and also forms aggregates linking the nanotubes in Fe_3O_4-rich samples. Thermodynamic and kinetic experiments showed different adsorption capacities of OFL, ranging from 23 to 45 mg g^{-1}. The kinetic process occurred within a few minutes, independently of the composite. The capability of the three HNT/Fe_3O_4 in removing the OFL was confirmed under realistic conditions, when OFL was added to tap, river, and effluent waters at μg L^{-1} concentration. No acute toxicity of the composites was observed on freshwater organisms. Despite the good results obtained for all the composites, the sample by co-precipitation is the most performant as it: (i) is easily magnetically separated from the media after the use; (ii) does not undergo any degradation after three adsorption cycles; (iii) is synthetized through a low-cost procedure. These features make this material an excellent candidate for removal of OFL from water.

Keywords: magnetite-halloysite composites; magnetic sorbent materials; fluoroquinolone antibiotic; adsorption; wastewater treatment; magnetic remediation; emerging contaminants; ecotoxicity

Citation: Capsoni, D.; Lucini, P.; Conti, D.M.; Bianchi, M.; Maraschi, F.; De Felice, B.; Bruni, G.; Abdolrahimi, M.; Peddis, D.; Parolini, M.; et al. Fe_3O_4-Halloysite Nanotube Composites as Sustainable Adsorbents: Efficiency in Ofloxacin Removal from Polluted Waters and Ecotoxicity. *Nanomaterials* **2022**, *12*, 4330. https://doi.org/10.3390/nano12234330

Academic Editor: Vivian Hsiu-Chuan Liao

Received: 4 November 2022
Accepted: 2 December 2022
Published: 6 December 2022

1. Introduction

In the current scenario of water shortage, there is an urgent need to favor water loops. For this purpose, preserving and guaranteeing water quality is mandatory, as reclaimed water can be directly reused and re-enter natural water bodies [1]. A critical aspect of water quality is represented by xenobiotics, such as heavy metals, dyes, pesticides, etc., detected in natural water bodies, also at trace levels, because of their recalcitrance in conventional wastewater treatment plants (WWTPs) [2]. In particular, pharmaceuticals and personal care products (PPCPs) have attracted the attention of the scientific community and civil society because of their widespread diffusion in the environment and their potential toxicity towards humans and ecosystems [3,4]. Although the current levels of PPCPs in aquatic ecosystems can be considered as low, they pose a severe threat for aquatic organisms because of their high biological activity and peculiar mechanism(s) of toxic action [4]. Among PPCPs, a remarkable concern is due to antibiotics whose presence in water ecosystems has been identified to affect natural microbial communities and to stimulate multi-resistant bacteria and antibiotic resistance genes, which pose serious risks to human and veterinary health [4,5]. To tackle the rising threats induced by the release of antibiotics, a recent action plan has proposed developing innovative strategies to reduce the diffusion of these emerging contaminants [6]. Over the last years, many research efforts have been made to develop sustainable and low-cost processes, easily implementable to conventional WWTPs and efficient in antibiotic removal from wastewater [7]. In this context, adsorption is a convenient method in terms of low energy consumption, reuse of the adsorbent material, no production of toxic by-products, and reduced waste production after treatment [8,9].

Many materials, both bare and functionalized, have been tested for water and wastewater decontamination, including activated carbon, nanomaterials, biopolymers, clays, agriculture and industrial wastes, and other natural sorbents [10–13].

The use of natural sorbents in the adsorption process [14–16] offers even more advantages, as they are abundant, low-cost, non-toxic, easy to modify, and competitive in water remediation compared to most conventional adsorbents [8,12,14].

Nanoclay materials surely fit the advantages mentioned above as sorbents to remove various pollutants, such as heavy metals, pesticides and antibiotics [17,18]. Among nanoclays, those displaying a tubular structure are even more intriguing, due to their additional properties related to the nanoscale dimension, cylindrical hollow form, and porosity. The halloysite nanotubes (HNTs) pertain to these nanoclays. Halloysites are aluminosilicates belonging to the kaolin group, with the chemical formula $Al_2Si_2O_5(OH)_4 \cdot n\, H_2O$. Two halloysite forms are reported in the literature, depending on the moles of hydrating molecules and the d_{001} basal spacing: halloysite-(10 Å) is the di-hydrated form [19], and halloysite-(7 Å) is the anhydrous one. The latter form is the most common, due to the easy release of the halloysite water molecules at ambient conditions [20]. The halloysite structure is based on corner-sharing SiO_4 tetrahedra sheets connected via oxygens to edge-sharing AlO_6 octahedra ones [21,22]. The mismatch of the larger SiO_4 tetrahedra and the smaller AlO_6 octahedra accounts for the local stress on the atomic scale of the aluminosilicate layer, inducing its wrapping and the nanotubes' morphology [22]. The nanotube typically displays lengths of 0.4–1 µm, an outer diameter of 20–200 nm, and an inner lumen diameter of 10–70 nm [23]. The siloxane (Si-O-Si) groups form the negatively charged outer surface, and the aluminol groups (-OH and Al-OH) form the positively charged inner one [24,25]. The peculiar physical and chemical features reported above make the HNTs suitable candidates for applications in various fields, including controlled drug release, nanotemplating, and adsorption. They are also employed as catalyst support and nanocomposites [26].

It is well known that the separation of the nanosorbent phase after pollutants removal is not a trivial challenge. A feasible and low-cost approach is to decorate the adsorbent material with magnetic nanoparticles to make it easily magnetically recovered. Some examples on the synthesis of halloysite–magnetite composites by co-precipitation, thermal decomposition, and solvothermal approaches are reported in the literature [27–30], and these materials are not yet investigated for water depollution.

Another key point to optimize before the application of nanomaterials in water remediation processes concerns the investigation of potential environmental and human risks associated with their use. The characterization of nanomaterials should have to include not only the assessment of any transformation occurring in environmental media, from its inclusion into a polluted site to the removal (or degradation) after the remediation of the target pollutant [31], but also the potential toxicity towards aquatic organisms. Ecotoxicology can provide useful tools to assess the risk related to nanomaterials and to select eco-friendly and sustainable ones for water remediation [32,33]. The application of standard and/or novel ecotoxicological tests completes the characterization of nanomaterials through the identification of possible toxicological targets and sheds light on the mechanism(s) of toxic action in aquatic species at different levels of the ecological hierarchy [34].

In the present study, we synthesized HNT/Fe_3O_4 nanocomposites by using three different approaches: co-precipitation, sol-gel, and hydrothermal. Each material was characterized by FT-IR spectroscopy, X-ray powder diffraction (XRPD), scanning electron and transmission electron microscopy (SEM and TEM), energy dispersive spectroscopy (EDS), thermogravimetric analysis (TGA), and magnetization measurements. Moreover, the magnetite and halloysite amount in each sample was evaluated by EDS, TGA, and magnetization data. Lastly, potential ecotoxicity of these materials towards aquatic organisms was tested on the freshwater Cladoceran *Daphnia magna* according to the *Daphnia* sp. Acute Immobilization Test, OECD 202 guideline (OECD, 2004). Adsorption properties and mechanism of each nanocomposite were investigated, and compared with the commercial halloysite. The antibiotic ofloxacin (OFL) was chosen as the target molecule to assess the adsorption efficiency of HNT/Fe_3O_4 nanocomposites for different reasons: (i) it is a very useful antibacterial agent belonging to the last class of antibiotics; (ii) it is largely detected in wastewaters and surface waters [3]; (iii) it is a recalcitrant to biological degradation [35]; (iv) it maintains a certain antibacterial activity after the first steps of its degradation [35]; (v) it has been used in our previous studies regarding both fluoroquinolones' environmental fate and their removal by adsorption processes [36–41]. The suitability of three materials for OFL removal under environmental conditions, i.e., tap and river waters, and wastewater treatment plant (WWTP) effluent, was also verified.

2. Materials and Methods

2.1. Materials

All the chemicals employed were reagent grade or higher in quality. HNT, $FeCl_3{\cdot}6H_2O$, $FeSO_4{\cdot}7H_2O$, $Fe(NO_3)_3{\cdot}9H_2O$, ammonia solution (NH_3 H_2O), sodium acetate (CH_3COONa), ethylene glycol ($C_2H_6O_2$), ethanol (EtOH), glucose ($C_6H_{12}O_6$), and OFL were purchased from Merck (Milano, Italy).

High-performance liquid chromatography (HPLC) gradient-grade acetonitrile (ACN) was purchased by VWR International (Milano, Italy), H_3PO_4 (85% *w/w*), and water for liquid chromatography/mass spectrometry (LC/MS) by Carlo Erba Reagents (Cornaredo, Milano, Italy).

2.2. Synthesis

Halloysite nanotubes–magnetite composites (HNT/Fe_3O_4) and magnetite alone (Fe_3O_4) were synthesized by co-precipitation, sol-gel, and hydrothermal routes, as follows.

Table 1 summarizes the synthesis approaches and the sample names.

Table 1. Scheme of the synthesis procedures and samples names.

Synthesis Procedure	Sample	Sample Name
Coprecipitation	magnetite halloysite g–magnetite	Fe_3O_4-C HNT/Fe_3O_4-C
Sol-gel	magnetite halloysite–magnetite	Fe_3O_4-SG HNT/Fe_3O_4-SG
Hydrothermal	magnetite halloysite–magnetite	Fe_3O_4-H HNT/Fe_3O_4-H

2.2.1. Co-Precipitation Procedure

The HNT/Fe_3O_4-C sample was synthesized following the procedure of Xie et al. [27]. An amount of 0.5 g of HNT was added to an aqueous solution of 4.32 mmol of $FeCl_3 \cdot 6H_2O$ and 2.16 mmol of $FeSO_4 \cdot 7H_2O$. The suspension was heated at 60 °C under N_2 flux, and an 8 M ammonia solution was added dropwise to reach pH 9–10. The suspension was further heated for 4 h at 70 °C, then the solid was magnetically recovered, washed three times, and dried for 3 h at 100 °C. The same procedure was applied to synthesize the Fe_3O_4 alone (sample Fe_3O_4-C), by omitting the addition of HNTs.

2.2.2. Sol-Gel Procedure

The HNT/Fe_3O_4-SG sample was synthesized as reported by He et al. [29]. An amount of 1 g of HNT was dispersed in an ethanol solution containing 1.98 mmol of $Fe(NO_3)_3 \cdot 9H_2O$. The dispersion was sonicated, stirred 24 h at room temperature and dried for 24 h at 35 °C. An amount of 2 mL of ethylene glycol was added, and the sample was heated for 2 h at 400 °C under N_2 flux (N_2 99.999%; flow rate: 3 L h^{-1}; heating and cooling rate: 5 °C min^{-1}). The same procedure was applied to synthesize the Fe_3O_4 alone (sample Fe_3O_4-SG), by omitting the addition of HNTs.

2.2.3. Hydrothermal Procedure

HNT/Fe_3O_4-H sample was synthesized following the procedure of Tian et al. [30], with some modifications. The procedure consists of two hydrothermal steps: the former to prepare HNT enriched with a carbonaceous component, and the latter to decorate it with magnetite. An amount of 0.5 g of HNT was added to a glucose solution (10 g L^{-1}) and magnetically stirred. The dispersion was poured into a Teflon-lined stainless-steel autoclave and heated for 48 h at 160 °C. The obtained product was washed 5 times in ethanol, centrifuged, and dried for 18 h at 60 °C under vacuum. An amount of 0.5 g of the final product was added to a solution containing 3 mmol of $FeCl_3 \cdot 6H_2O$ in ethylene glycol. After stirring for 24 h, 1.8 g of sodium acetate and 0.5 g of ethylene glycol were added, and the dispersion was poured into a Teflon-lined stainless-steel autoclave and heated for 8 h at 200 °C. The obtained magnetic composite was washed with distilled water and dried for 12 h at 80 °C. The procedure of the second step was also applied to synthesize the Fe_3O_4 alone (sample Fe_3O_4-H), by omitting the addition of HNTs.

2.2.4. Characterization Techniques

X-ray powder diffraction measurements were performed using a Bruker D5005 diffractometer (Bruker, Karlsruhe, Germany) with the CuKα radiation, graphite monochromator, and scintillation detector. The patterns were collected in the 7–80° two-theta angular range, step size of 0.03°, and a counting time of 20 s/step. A silicon low-background sample holder was used.

FT-IR spectra were obtained with a Nicolet FT-IR iS10 Spectrometer (Nicolet, Madison, WI, USA) equipped with ATR (attenuated total reflectance) sampling accessory (Smart iTR with ZnSe plate) by co-adding 32 scans in the 4000–650 cm^{-1} range at 4 cm^{-1} resolution.

Thermogravimetric measurements were performed by a TGA Q5000 IR apparatus interfaced with a TA 5000 data station (TA Instruments, Newcastle, DE, USA). The samples were scanned at 10 °C min^{-1} under nitrogen flow (45 mL min^{-1}) in the 20–850 °C temperature range. Each measurement was repeated at least three times.

The specific surface area and porosity were investigated by N_2 adsorption using the BET method in a Sorptomatic 1990 Porosimeter (Thermo Electron, Waltham, MA, USA).

SEM measurements were performed using a Zeiss EVO MA10 (Carl Zeiss, Oberkochen, Germany) Microscope, equipped with an Energy Dispersive Detector for the EDS analysis. The SEM images were collected on gold-sputtered samples. HR-SEM images were taken from an FEG-SEM Tescan Mira3 XMU. Samples were mounted onto aluminum stubs using double sided carbon adhesive tape and were then made electrically conductive by coating in vacuum with a thin layer of Pt. Observations were made at 25 kV with an In-Beam SE detector at a working distance of 3 mm.

TEM micrographs were carried out on a JEOL JEM-1200 EX II (JEOL Ltd., Tokio, Japan) microscope operating at 100 kV high voltage (tungsten filament gun) and equipped with a TEM CCD camera Olympus Mega View III (Olympus soft imaging solutions (OSIS) GmbH, from 2015 EMSIS GmbH, Munster, Germany) with 1376 × 1032 pixel format. The samples were prepared by drop-casting the solution on nickel grids formvar/carbon coated.

Dynamic light scattering (DLS)—Nicomp 380 ZLS (Particle Sizing Systems, Lakeview Blvd. Fremont, CA, USA) was used. For analyses, samples were diluted 1:10 in MilliQ water. The main parameters set up were: channel 10, intensity 100 kHz, temperature 23 °C, viscosity 0.933 cPoise, and a liquid index of refraction 1.333. The values considered at the end of the analyses were: mean diameter (nm), standard deviation, and Zeta potential (mV).

To investigate the magnetic behavior of the materials, field dependence of magnetization was investigated using a vibrating sample magnetometer (VSM Model 10–Microsense) equipped with an electromagnetic producing magnetic field in the range ±2 T.

2.3. Adsorption Experiments and Analytical Measurements

2.3.1. Adsorption and Kinetic Experiments

OFL adsorption on HNT/Fe_3O_4-C, HNT/Fe_3O_4-SG, HNT/Fe_3O_4-H, and commercial HNT was studied by a batch method. For adsorption equilibrium experiments, 20 mg of each material was suspended in 10 mL of tap water spiked with OFL in the range of 25–200 mg L^{-1}. Flasks were wrapped with aluminum foil to prevent light-induced drug decomposition and shaken for 24 h at room temperature with an orbital shaker. Subsequently, the suspensions were magnetically separated, and the supernatants were filtered (0.22 μm) and analyzed by UV-vis spectrophotometer at 287 nm to determine the antibiotic concentration in solution at equilibrium (C_e). The adsorbed OFL amount at equilibrium (q_e, mg g^{-1}) was calculated by Equation (1):

$$q_e = \frac{(C_0 - C_e) \cdot V}{m} \quad (1)$$

where C_0 is the initial OFL concentration (mg L^{-1}), C_e is the drug concentration in solution at equilibrium (mg L^{-1}), V is the volume of the solution (L), and m is the amount of the sorbent material (g).

For the kinetic experiments, 20 mg of each material were suspended in 10 mL of 20 mg L^{-1} OFL tap water solution. Falcon tubes, wrapped with aluminum foil, were shaken by a roller shaker and, at selected times, the adsorbent was magnetically treated. Then, a few mL of the supernatant were collected, filtered (0.22 μm) in a quartz cuvette, and analyzed by a UV spectrophotometer at 287 nm. The analyzed solution was recovered to keep the suspension volume constant for all experiments. The adsorbed OFL amount at time t (q_t, mg g^{-1}) was calculated as (Equation (2)):

$$q_t = \frac{(C_0 - C_t) \cdot V}{m} \quad (2)$$

where C_0 is the initial OFL concentration (mg L^{-1}), C_t is the drug concentration in solution at time t (mg L^{-1}), V is the volume of the solution (L), and m is the amount of the sorbent material (g).

All experiments were performed in duplicate. The thermodynamic and kinetic parameters were estimated by dedicated software (OriginPro, Version 2019b. OriginLab Corporation, Northampton, MA, USA).

The well-known Langmuir's and Freundlich's isotherm models were applied to fit the experimental data. The Langmuir model (Equation (3)) describes the adsorption process that takes place on specific homogeneous sites and in a monolayer on the material surface:

$$q_e = \frac{q_m K_L C_e}{1 + K_L C_e} \quad (3)$$

where K_L is the Langmuir constant and q_m is the monolayer saturation capacity.

The Freundlich model defines non-ideal adsorption on the heterogeneous surface, and Equation (4) expresses it:

$$q_e = K_F C_e^{1/n} \tag{4}$$

where K_F is the empirical constant indicative of adsorption capacity, and n is the empirical parameter representing the adsorption intensity.

The time-dependent data were fitted by pseudo-first-order (Equation (5)) and pseudo-second-order kinetic (Equation (6)) models:

$$q_t = q_e(1 - e^{k_1 t}) \tag{5}$$

$$q_t = \frac{q_e^2 k_2 t}{1 + q_e k_2 t} \tag{6}$$

where q_t and q_e are the drug adsorbed amount at time t and equilibrium, respectively, and k_1 and k_2 are the pseudo-first-order and the pseudo-second-order rate constants.

2.3.2. Analytical Measurements

For OFL analysis at mg L^{-1}, a UV-vis UVmini-1240 spectrophotometer (Shimadzu Corporation) was used. The instrument was set at 287 nm, corresponding to the maximum OFL absorption. Calibration in the range of 1–10 mg L^{-1} yielded optimal linearity ($R^2 > 0.9988$). The quantification limit was 0.8 mg L^{-1}.

HPLC system consisting of a pump Series 200 (Perkin Elmer, Milano, Italy) equipped with a vacuum degasser and a programmable fluorescence detector (FD) was used for OFL analysis at µg L^{-1}. The fluorescence excitation/emission wavelengths selected were 280/450 nm. Fifty µL of each sample were filtered (0.22 µm nylon syringe filter) and injected into a 250 × 4.6 mm, 5 µm Ascentis RPAmide (Supelco-Merck Life Science, Milano, Italy) coupled with a similar guard-column. The mobile phase was 25 mM H_3PO_4—ACN (85:15), and the flow rate 1 mL min^{-1}. Calibration in the range 1–20 µg L^{-1} yielded optimal linearity ($R^2 > 0.9988$). The quantification limit was 0.9 µg L^{-1}.

2.4. Acute Toxicity Tests with Daphnia magna

The potential acute toxicity of the different materials, i.e., HNT, Fe_3O_4-C, and HNT/Fe_3O_4-C, was tested on the freshwater Cladoceran *Daphnia magna* according to the *Daphnia* sp. Acute Immobilization Test, OECD 202 guideline (OECD, 2004). Adult *Daphnia magna* individuals were cultured (30 individuals/L) in a commercial mineral water (San Benedetto®) under controlled laboratory conditions reported elsewhere [42]. Five replicates containing ten daphnids (i.e., <24 h old individuals) each were performed per each experimental condition, including control. In detail, daphnids were exposed for 48 h at 20 ± 0.5 °C and 16 h light: 8 h dark photoperiod under static, non-renewal conditions to 0.2 g L^{-1} of the materials. A single concentration mimicking the amount of residues in waters after depollution treatment was tested. This concentration reflected the amount of each material used in the experiments aimed at investigating their capability in the removal of OFL. The viability of individuals was tested after 24 and 48 h of exposure. Individuals were considered dead when they did not swim for over 15 s after a slight stirring of the solutions. After checking for viability, all the individuals were observed under a Leica Microsystem EZ4 Stereoscopic microscope to check for the ingestion of materials by daphnids.

3. Results and Discussion

First, structure, morphology, composition, magnetic behavior, adsorption capacity, and adsorption kinetics of the magnetic HNT composites and the commercial HNT were investigated. Then the materials were tested under environmental conditions to remove the antibiotic OFL chosen as being representative of emerging contaminants. In addition, their potential ecotoxic effects, along with reusability, were evaluated.

3.1. Morphological, Structural, and Magnetic Characterization

Figure 1a shows the XRPD pattern of the commercial halloysite. It compares to those reported in the literature [27,30,43,44] and deposited in JCPDS database (PDF# 028-1487). The peak detected at about 12° corresponds to the d_{001} basal spacing of 7.35 Å, peculiar of the anhydrous form (halloysite-(7 Å)). The (002) reflection is observed at about 24°. The peaks at 20° and 62.8° are typical of halloysites with nanotubular morphology [44,45]. No peaks are detected at about 8.8°, assigned to the d_{001} basal spacing of the di-hydrated halloysite (halloysite-(10 Å)). This is consistent with the easy loss of the interlayer water molecules near room temperature [46]. The very sharp reflections observed at 10.1, 26.6, and 27.3° are attributed respectively to the small amount of kaolinite 1A (PDF# 074-1786), quartz (PDF# 046-1045), and rutile (PDF# 021-1276); these impurity phases are often detected in halloysite clay minerals.

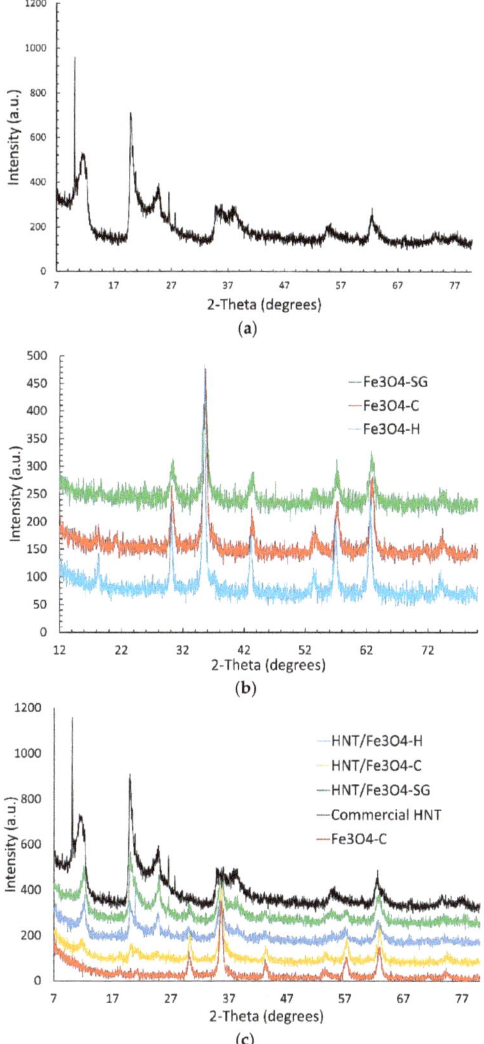

Figure 1. XRPD patterns of (**a**) commercial HNT, (**b**) Fe_3O_4, and (**c**) HNT/Fe_3O_4 composites, Fe_3O_4 and HNT.

Figure 1b displays the XRPD pattern of the Fe_3O_4 samples obtained by the three synthetic routes. The 2-theta reflection positions fairly agree with those expected for the magnetite structure (PDF# 088-0315). The iron oxide phase has been successfully synthesized, and no impurity phases are detected within the detection limit of the technique. The iron oxide samples are nanocrystalline: a crystallite size of 10, 13, and 8 nm was calculated for Fe_3O_4-C, Fe_3O_4-SG, and Fe_3O_4-H samples by applying the Scherrer equation to the 311 reflection.

Figure 1c displays the diffraction pattern of the magnetite–halloysite composites. The diffraction patterns of the commercial halloysite and the Fe_3O_4-C sample (chosen as reference for the magnetic phase), are also shown for comparison. The three composite samples display the peaks of both the magnetite and the halloysite phases, thus confirming the successful formation of the magnetite–halloysite adduct. An investigation of the magnetite crystallite size in the composites by applying the Scherrer equation could not be carried out, due to the strong overlap of the 311 reflection of the magnetite phase to the broad peaks of halloysite in the 33–40° 2 theta range. Nonetheless, the comparable peaks broadening of the magnetite phase in the composites and in the Fe_3O_4 samples suggests nanocrystalline magnetite is obtained also in the HNT/Fe_3O_4 samples. The peaks' intensity of halloysite and magnetite in the composite samples returns an idea on the phases amount in each sample. The peaks' intensity of halloysite decreases and Fe_3O_4 increases progressively from HNT/Fe_3O_4-SG to HNT/Fe_3O_4-H and HNT/Fe_3O_4-C, suggesting that the magnetite and halloysite amounts in the composite samples depend on the synthesis route.

The FT-IR spectra of the commercial halloysite and the HNT/Fe_3O_4 composites are shown in Figure 2. The spectrum of the commercial HNT well compares to the literature ones [27,29,30,43]. The bands centered at about 3622 and 3707 cm^{-1} are attributed to the stretching vibrations of the Al-OH of the HNT inner surface, while the small peaks at about 3545 and 1641 cm^{-1} to the stretching and banding of the H_2O molecules in the interlayer. This result puts into evidence the possible presence of small amount of the hydrated form (halloysite-(10 Å)) in the commercial halloysite, below the detection limit of XRPD. The bands at about 1031, 794, and 689 cm^{-1} are attributed to the Si-O stretching modes, the one at about 918 cm^{-1} to the Al-OH ones. In the FT-IR spectra of the HNT/Fe_3O_4 composites (Figure 2b), all the halloysite bands are detected. As for the Fe_3O_4 phase, only one broad band centered at about 3435 cm^{-1} attributed to OH-bending of hydroxyl groups was observed [43]. This broad band was not detected in the HNT/Fe_3O_4-SG sample, displaying a high amount of halloysite and a few magnetites (see XRPD results).

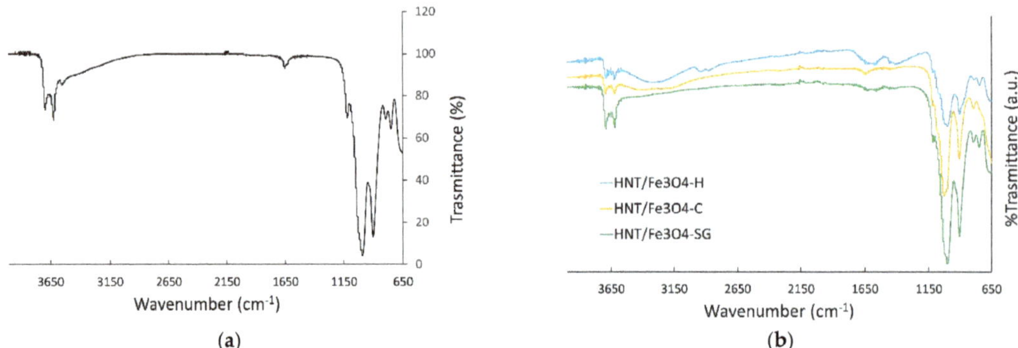

Figure 2. FT-IR spectra of (**a**) the commercial halloysite, and (**b**) the HNT/Fe_3O_4 composites.

The SEM images of the commercial HNT are shown in Figure S1a,b. The sample displayed 2–10 μm agglomerates of nanotubular particles, better highlighted in TEM micrographs (Figure 3a,b). The nanotubes exhibited an external diameter of 60–70 nm, a lumen of 20–30 nm, and variable length, from a few hundred nanometers to 1–2 μm. The

DLS results showed a bimodal particle size distribution. The mean particle size is reported in Table S1.

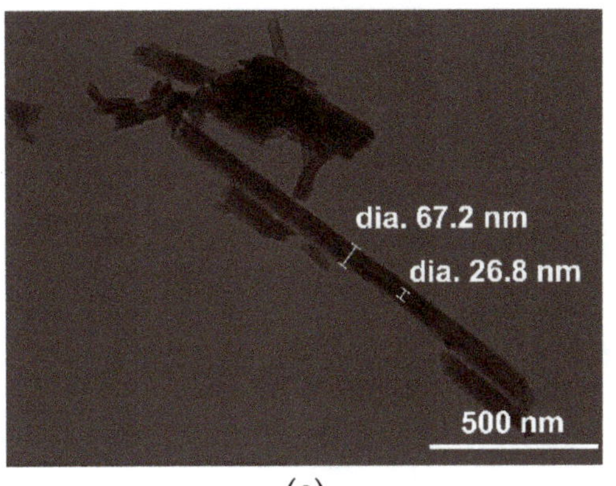

(a)

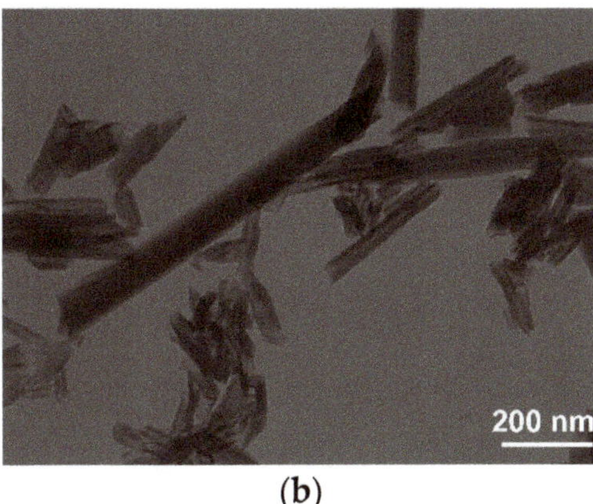

(b)

Figure 3. TEM images of the commercial halloysite sample at magnifications of (**a**) 75 kX and (**b**) 100 kX.

Figure S2 shows the SEM micrographs of the HNT/Fe_3O_4 composites synthesized by co-precipitation (Figure S2a,b), hydrothermal (Figure S2c,d), and sol-gel (Figure S2e,f) routes. All the composites displayed micrometric nanotubular particles, whose morphology well compares to the HNT sample one (Figure S1a,b). In addition, nanometric rounded aggregates, possibly due to the magnetite phase, were observed on the nanotubes surface and between the nanotubes, interconnecting them; they were mainly detected in the HNT/Fe_3O_4-C sample (Figure S2a,b) which was richer in magnetite, as suggested by XRPD and FT-IR results.

Figure 4 displays the TEM images of the HNT/Fe_3O_4 composites and Fe_3O_4 samples synthesized by co-precipitation (Figure 4a–c), hydrothermal (Figure 4d–f), and sol-gel (Figure 4g–i) routes. Independent of the applied synthesis, both rounded Fe_3O_4 nanometric particles and halloysite nanotubes were observed in the HNT/Fe_3O_4 composites. Noteworthy, the Fe_3O_4 amount was high in the HNT/Fe_3O_4-C sample (Figure 4a,b); it covered the nanotubes' surface, but also formed aggregates linking the nanotubes. This was also slightly observed in the HNT/Fe_3O_4-H sample. The Fe_3O_4 agglomerates were mainly observed on the tips of nanotubes. As reported by Tian et al. [30], the synthetic strategy based on the use of glucose in the first step favored the formation of carbon/organic groups on the HNT surface and on the tip of nanotubes, acting as nucleation centers for the Fe_3O_4 nanoparticles. As for the HNT/Fe_3O_4-SG sample, it displayed a lower amount of magnetite (see XRPD and FT-IR results), and the Fe_3O_4 nanoparticles only decorated the nanotubes' surface. The size and shape of the magnetite nanoparticles in the composites (about 10 nm) well compared to the Fe_3O_4 samples (Figure 4c,f,i) for the Fe_3O_4-C, Fe_3O_4-H, and Fe_3O_4-SG respectively), and fairly agreed with the crystallite size evaluated by XRPD data. In both the magnetite and composite samples, the Fe_3O_4 nanoparticles aggregate; particle size distribution was evaluated by DLS analysis and reported in Table S1. The Fe_3O_4-C sample displayed wide particle size distribution. The HNT/Fe_3O_4-C and NHT/Fe_3O_4-SG samples displayed particle size >900 nm, slightly similar to the larger ones of the commercial halloysite. Instead, the HNT/Fe_3O_4-H composite displayed lower particle size. To better characterize the tendency of particles to aggregate and to investigate particles' surface charge changes, zeta-potential was evaluated. Commercial HNT exhibits a negative zeta-potential of −31.77 mV; this value confirms that the outer nanotube surface is negatively charged and is in good agreement with the literature data [47].

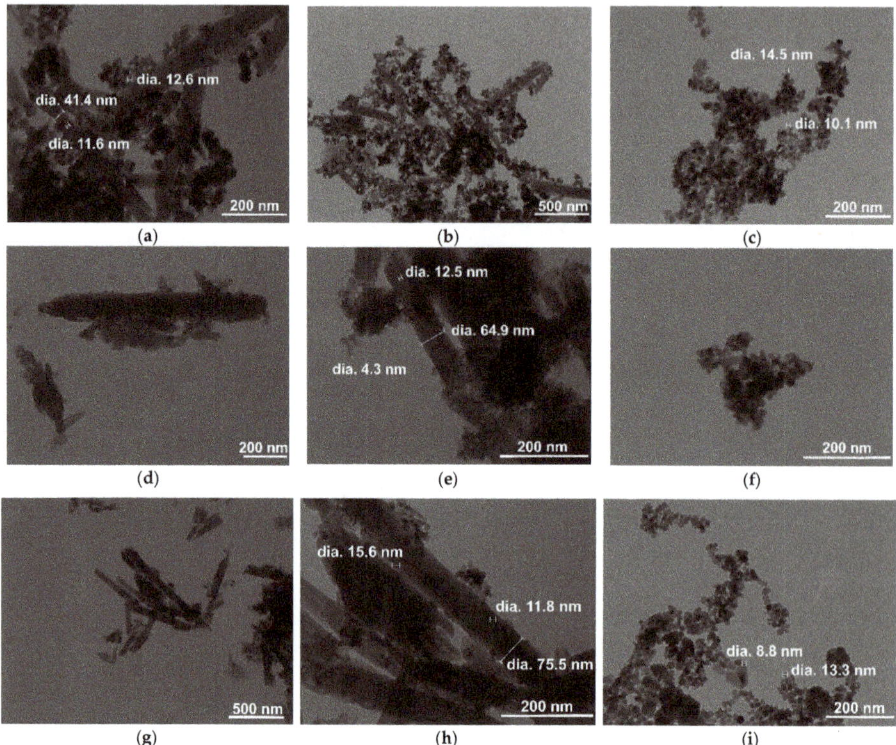

Figure 4. TEM images of the HNT/Fe_3O_4 and magnetite samples. HNT/Fe_3O_4-C at magnifications of (**a**) 150 kX and (**b**) 50 kX: Fe_3O_4-C (**c**) at 150 kX; HNT/Fe_3O_4-H at (**d**) 100 kX and (**e**) 200 kX; Fe_3O_4-H (**f**) at 200 kX; HNT/Fe_3O_4-SG at (**g**) 50 kX and (**h**) 200 kX; Fe_3O_4-SG (**i**) at 150 kX.

The Fe_3O_4-C sample (chosen as reference of the magnetite samples) exhibits a zeta-potential of −7.16 mV, comparable to the literature values [48]; this value is not sufficient to achieve a stable suspension, and justifies particle aggregation (see TEM and DLS results).

Zeta-potential values of −36.36, −12.89 and −112.02 mV are obtained for HNT/Fe_3O_4-C, HNT/Fe_3O_4-SG and HNT/Fe_3O_4-H composites. The sample prepared by the hydrothermal process displays the most negative zeta-potential value; this may be due to the carbonaceous component (see TEM results and Section 3.2.) and explains the improved stability of the suspension and the lower mean particle size, as shown by DLS results.

The EDS analysis was applied to display the distribution map of halloysite and magnetite in each composite sample and to evaluate the weight percentage. Figures S3–S5 show the distribution maps of Al, Fe, and Si for the HNT/Fe_3O_4-C, HNT/Fe_3O_4-H, and HNT/Fe_3O_4-SG samples. Independently of the synthetic route, Al and Si were detected in the same areas. The Fe distribution was rather homogeneous in the sol-gel and hydrothermal samples (Figures S4 and S5, respectively), but also in some regions in which Fe prevails were detected. In the co-precipitation composite, Fe prevailed in areas poor in Al and Si, thus confirming the presence of magnetite aggregates connecting the halloysite particles.

From the EDS analysis, the Al, Si, and Fe atomic percentages were evaluated. Al:Si:Fe molar ratios of 5.25:5.15:20.33, 5.11:5.03:4.62, and 12.36:13.35:3.46 were obtained for the HNT/Fe3O4-C, HNT/Fe_3O_4-H, and HNT/Fe_3O_4-SG samples, respectively. According to the halloysite chemical formula, equimolar values of Al and Si were detected in each sample. The molar ratios obtained by EDS were used to calculate halloysite and magnetite weight percentage in each composite: the results are shown in Table 2.

Table 2. Halloysite and magnetite weight percentages evaluated by EDS, TGA, and magnetization data.

Sample	Halloysite (wt%)			Magnetite (wt%)		
	EDS	TGA	Magnetization	EDS	TGA	Magnetization
HNT/Fe_3O_4-C	30	29	12	70	71	88
HNT/Fe_3O_4-H	65	65	68	35	35	32
HNT/Fe_3O_4-SG	85.5	83	93	14.5	17	7

The halloysite amount in the HNT/Fe_3O_4 composites was also calculated by thermogravimetric analyses. The thermograms of commercial HNT and composites are shown in Figure 5. The halloysite TG curve (Figure 5a) well compared to the literature data [27]. The mass loss detected at low temperature (below 250 °C) was ascribed to the release of physisorbed water molecules. The steep mass loss observed at about 450 °C gave more insight, as it is due to the dehydroxylation process of the structural Al-OH groups of the aluminosilicate layers. A weight loss of 13.95% was calculated from halloysite stoichiometry. The mass loss detected in the commercial HNT was about 14.60%, in fair agreement with the calculated value. Figure 5b–d show the thermograms of the HNT/Fe_3O_4-C, HNT/Fe_3O_4-H, and HNT/Fe_3O_4-SG samples, respectively. Different mass losses were detected at low temperature (below 250 °C), depending on the amount of the physisorbed water, then a sample-dependent steep mass loss occurs at about 450 °C. As reported by Xie et al. [27], this mass loss can be compared to the HNT sample one (14.60%) to evaluate the halloysite weight percentage in each composite. The results are reported in Table 2; the halloysite weight percentages well compared to the values obtained by EDS analysis.

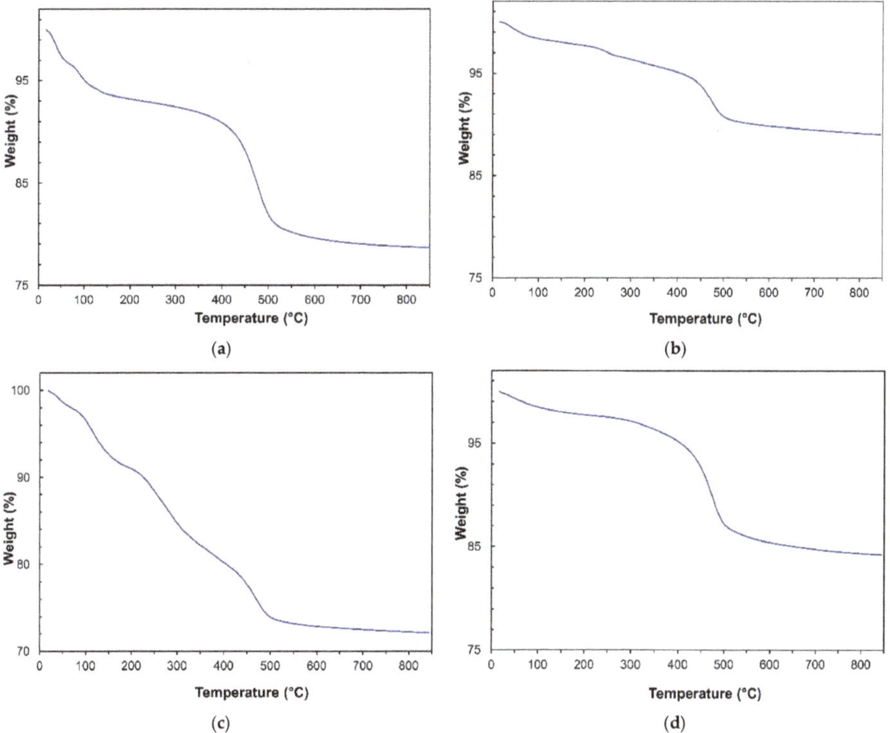

Figure 5. TGA curves of (**a**) commercial HNT (**b**) HNT/Fe_3O_4-C, (**c**) HNT/Fe_3O_4-H, and (**d**) HNT/Fe_3O_4-SG samples.

Field dependence of magnetization was investigated for all the samples at 300 K (Figure 6a,b).

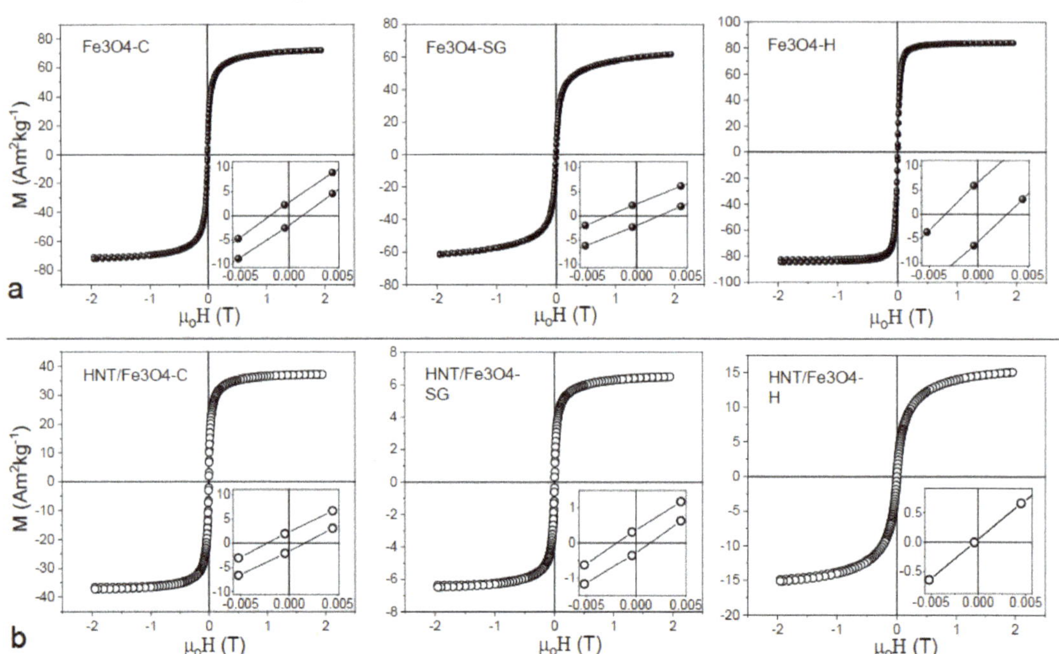

Figure 6. Field dependence of magnetization at 300 K for the (**a**) bare Fe_3O_4 nanoparticles synthetized by coprecipitation (Fe_3O_4-C) sol-gel (Fe_3O_4-SG) and hydrothermal methods (Fe_3O_4-H) and (**b**) HNTs/Fe_3O_4 nanocomposites. The insets are zoom of the coercive field region.

For bare nanoparticles prepared with co-precipitation and sol-gel synthesis methods (Figure 6a), negligible value of reduced remanence magnetization (Mr/Ms) and small value of coercivity were obtained (Table 3), suggesting that at 300 K most of the nanoparticles were in a superparamagnetic state and just a small fraction of nanoparticles showed a quasi-static behavior. While the zero coercivity in the nanoparticles synthesized with the hydrothermal procedure indicated that all nanoparticles were in a supermagnetic state. Fe_3O_4-C and Fe_3O_4-SG samples showed a weak non-saturating character at high field, with respect to the Fe_3O_4-H sample. Due to the small difference in size between the samples, a non-saturating character showed by samples prepared by sol-gel and co-precipitation techniques can be ascribed to an increase in surface anisotropy, probably due to the presence of magnetic disorder (i.e., canted spin) [49,50] at the particles' surface. This hypothesis was also confirmed by the decrease in M_S in SG and C samples. All the HNT nanocomposites showed a decrease in M_S with respect to bare nanoparticles in qualitative agreement with TGA and EDS measurements. This behavior confirmed that the amount of magnetic phase decreases along the order Fe_3O_4-C, Fe_3O_4-SG, and Fe_3O_4-H. From a quantitative point of view, if the agreement among magnetization measurements, TGA and EDS, was pretty good for Fe_3O_4-SG and Fe_3O_4-H, a difference was observed for Fe_3O_4-C nanocomposite. In particular, the particles prepared by co-precipitation looked to decrease their M_S when prepared as nanocomposites. This can be ascribed to a decrease in nanoparticles' crystallinity that can be observed in the co-precipitation synthesis with respect to hydrothermal and sol-gel syntheses [51,52].

Table 3. Saturation magnetization M_S, reduce remanence magnetization (Mr/M_S) and coercive field (μ_0H_C) of Fe_3O_4-C, Fe_3O_4-SG, Fe_3O_4-H, HNT/Fe_3O_4-C, HNT/Fe_3O_4-SG, and HNT/Fe_3O_4-H samples.

Samples	Ms ($Am^2\ kg^{-1}$)	Mr/Ms	μ_0H_C (Oe)
Fe_3O_4-C	70 (5)	0.03 (2)	16 (2)
HNT/Fe_3O_4-C	37 (4)	0.06 (2)	19 (4)
Fe_3O_4-SG	56 (3)	0.04 (2)	25 (4)
HNT/Fe_3O_4-SG	6 (2)	0.05 (2)	16 (3)
Fe_3O_4-H	83 (3)	0.06 (3)	32 (5)
HNT/Fe_3O_4-H	13 (5)	0	0

It is well known that the adsorption capacity of the materials is strictly related to their specific surface area [53]. The BET method was applied to investigate the specific surface area of the commercial halloysite and the three HNT/Fe_3O_4 composites. The values of 58.20, 57.66, 52.15, and 54.56 $m^2\ g^{-1}$ were obtained for the commercial HNT, HNT/Fe_3O_4-C, HNT/Fe_3O_4-H, and HNT/Fe_3O_4-SG samples, respectively. The pore specific volume was also evaluated, and values of 0.19, 0.26, 0.16, and 0.27 $cm^3\ g^{-1}$ were obtained. These results suggest that the deposition of the magnetite nanoparticles on the nanotubular halloysite surface did not affect the halloysite surface area and pore volumes. The obtained values fairly agreed with the literature data for halloysite nanotubes (surface areas: 22.1–81.6 $m^2\ g^{-1}$; pore volumes: 0.09–0.25 $cm^3\ g^{-1}$) [22].

3.2. Preliminary Adsorption Experiments

Before starting the adsorption experiments, control samples (20 mg HNT/Fe_3O_4 or HNT, 10 mL tap water), not containing OFL, were shaken for 24 h at room temperature. Then, the supernatants were magnetically separated for the pH measurement and analyzed by UV-vis spectrophotometer and HPLC-FD to check the instrumental baseline.

A pH value of 7.7–7.8, similar to that of natural waters, was measured in all samples, thus no additional pH adjustment was performed.

The background noise level was satisfactory for the commercial HNT, HNT/Fe_3O_4-C, and HNT/Fe_3O_4-H. On the contrary, HNT/Fe_3O_4-SG was rinsed with EtOH in an ultrasonic bath for 10 min, centrifuged for 5 min at 4000 rpm, separated, and dried at 50 °C for 1.5 h. The washing step was repeated twice to obtain a good signal-to-noise ratio.

3.3. Isotherm and Kinetic Studies

The behavior of the three magnetic HNT composites was evaluated through thermodynamic and kinetic experiments carried out under controlled conditions (see Section 2.3.1) and compared with the commercial HNT.

Adsorption isotherms are commonly used to describe the adsorption process in terms of maximum uptake and the relationship between the amount of adsorbed analyte (q_e) and its concentration in solution at equilibrium (C_e).

To fit the experimental data, the Langmuir and Freundlich models were considered.

As shown in Figure 7, the Langmuir model gave the best fitting of the experimental data.

Figure 7 shows that all materials were able to adsorb the antibiotic, although the maximum adsorption capacities were quite different. In detail, the highest value, 45 mg g^{-1}, was obtained for HNT/Fe_3O_4-H, while the lowest value was obtained for HNT/Fe_3O_4-C, which was equal to 23 mg g^{-1}. The HNT/Fe_3O_4-SG sample had an intermediate value of 31 mg g^{-1}, close to the commercial HNT (30 mg g^{-1}). This trend can be due to both the different amount of HNT present in the samples, ranging from about 30% in HNT/Fe_3O_4-C to more than 80% in HNT/Fe_3O_4-SG (see Table 2), and to the possible presence of some carbonaceous component related to the glucose added during HNT/Fe_3O_4-H synthesis. In fact, as reported by Tian et al. [30], the carbon/organic groups formed on the HNTs not only favor the Fe_3O_4 nanoparticle nucleation, but also may improve the analyte adsorption. On the contrary, no difference in the adsorption mechanism was observed among all materials. The Langmuir model, which describes a monolayer coverage, gives the best fitting of the experimental data, as confirmed by the good correlation coefficient R^2 and χ^2 values.

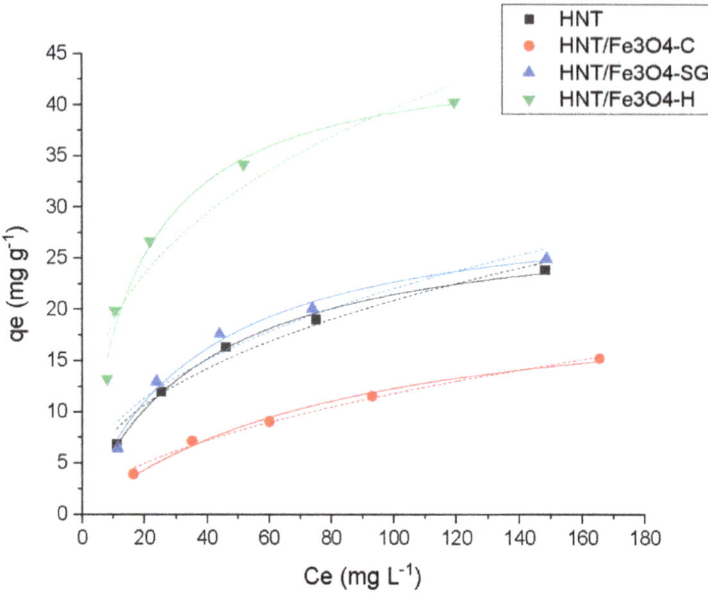

Figure 7. Adsorption profiles Langmuir (—) and Freundlich (. . .) for Ofloxacin (OFL) on HNT (■ black), HNT/Fe_3O_4-C (• red), HNT/Fe_3O_4-H (▲ blue) and HNT/Fe_3O_4-SG (▼ green) (Experimental conditions: Sorbent 20 mg, 10 mL OFL tap water solution from 25 to 200 mg L^{-1}, RSD < 10%).

The experimental q_{max} values of HNT/Fe_3O_4-C, HNT/Fe_3O_4-H, and HNT/Fe_3O_4-SG were in agreement with the calculated ones, and fell within the OFL adsorption range reported in the literature for other clays, i.e., 3.2 mg g^{-1} on kaolinite [54], 160.8 mg g^{-1} on calcined Verde-lodo bentonite clay [55]).

The isothermal parameters calculated by dedicated software are listed in Table 4.

Table 4. Isotherm parameters for OFL adsorption onto HNT, HNT/Fe_3O_4-C, HNT/Fe_3O_4-H, and HNT/Fe_3O_4-SG.

Adsorption Model	Isotherm Parameters	HNT	HNT/Fe_3O_4-C	HNT/Fe_3O_4-SG	HNT/Fe_3O_4-H
Langmuir	q_m (mg g^{-1})	29.6 (8)	23 (2)	31 (2)	45 (2)
	K_L (L mg^{-1})	0.026 (2)	0.012 (2)	0.028 (4)	0.063 (9)
	R^2	0.9970	0.9910	0.9881	0.9840
	χ^2	0.1739	0.2218	0.7893	2.5004
Freundlich	K_F (mg g^{-1}) (L $mg^{-1})^{1/n}$	3.1 (6)	1.1 (1)	3 (1)	9 (2)
	$1/n$	0.42 (4)	0.53 (3)	0.41 (7)	0.33 (5)
	R^2	0.9734	0.9931	0.9381	0.9304
	χ^2	1.5239	0.1712	4.1164	10.844

Concerning the kinetic aspect, quantitative adsorption occurred in less than five minutes in the presence of all the magnetic composites. As shown in Figure 8, a satisfactory fitting is obtained by applying the pseudo-second-order model, thus, considering a chemisorption process. For commercial HNT, the adsorption was instantaneous, thus, it was not possible to discriminate between the two models.

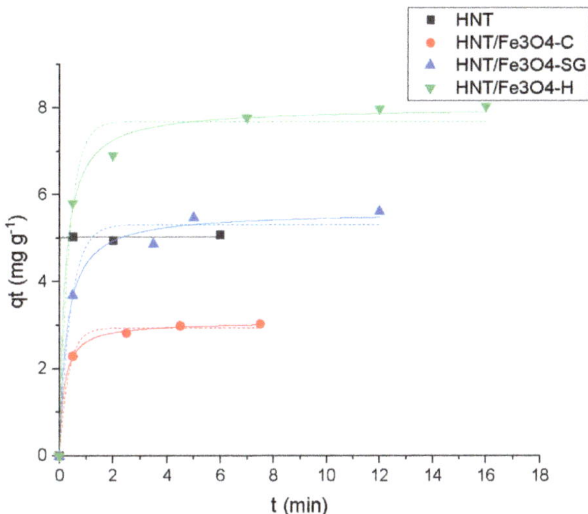

Figure 8. Kinetic profiles (pseudo-first-order (—), pseudo-second-order (. . .)) for OFL onto HNT (■ black), HNT/Fe_3O_4-C (● red), HNT/Fe_3O_4-H (▲ blue) and HNT/Fe_3O_4-SG (▼ green) (Experimental conditions: sorbent 20 mg, 10 mL tap water, OFL initial concentration 20 mg L^{-1}, RSD < 10%).

The calculated kinetic parameters are shown in Table 5.

Table 5. Kinetic parameters for OFL adsorption onto HNT, HNT/Fe_3O_4-C, HNT/Fe_3O_4-H, and HNT/Fe_3O_4-SG.

Kinetic Model	Kinetic Parameter	HNT	HNT/Fe_3O_4-C	HNT/Fe_3O_4-SG	HNT/Fe_3O_4-H
Pseudo-first order	q_e (mg g^{-1})	5.02 (4)	2.95 (5)	5.3 (2)	7.7 (2)
	k_1 (min^{-1})	124	3.0 (3)	2.4 (4)	2.8 (5)
	R^2	0.9996	0.9961	0.9854	0.9839
	χ^2	0.0041	0.0086	0.1052	0.1931
Pseudo-second order	q_e (mg g^{-1})	5.02 (6)	3.08 (3)	5.6 (2)	8.0 (1)
	k_2 (g mg^{-1} min^{-1})	3888	1.8 (2)	0.7 (2)	0.61 (9)
	R^2	0.9996	0.9992	0.9926	0.9965
	χ^2	0.0041	0.0017	0.0531	0.0418

3.3.1. Ofloxacin Removal from Real Waters Samples

Magnetic HNTs were also tested under environmental conditions, i.e., μg L^{-1} OFL concentration, tap and river waters, WWTP effluent (see Table S2 for the physicochemical parameters).

An amount of 20 mg of each material was suspended in 10 mL of each water sample, river water and WWTP effluent samples spiked with 10 μg L^{-1} OFL (C_0) and shaken for 24 h. Then, the suspensions were magnetically separated and the supernatants were filtered on a 0.22 μm nylon syringe filter before HPLC-FD analysis to quantify the drug content (C_e).

The removal efficiency ($R\%$) was calculated according to Equation (3):

$$R\% = \frac{C_0 - C_e}{C_0} \times 100 \quad (7)$$

where C_0 is the initial OFL concentration and C_e is the OFL concentration in solution at the equilibrium.

The obtained results were reported in Figure 9.

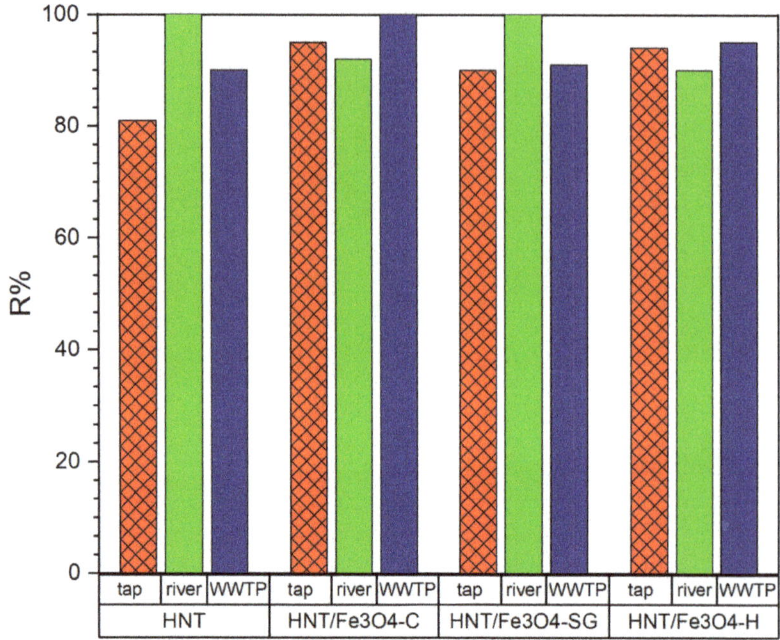

Figure 9. OFL removal (%) from tap and river water samples and effluent from WWTPS with HNT, HNT/Fe_3O_4-C, HNT/Fe_3O_4-H, and HNT/Fe_3O_4-SG (Experimental conditions: sorbent 20 mg, 10 mL tap water, OFL initial concentration 10 μg L^{-1}, $n = 3$, RSD < 10%).

The investigated HNT/Fe_3O_4 composites gained an antibiotic removal ≥90% despite different aqueous matrix constituents and other potential contaminants. The different amount of Fe_3O_4 in each composite did not affect the adsorption process; on the contrary, the Fe_3O_4 percent in HNT/Fe_3O_4-C, higher than in HNT/Fe_3O_4-H and HNT/Fe_3O_4-SG, favored its complete magnetic recovery from the media after the use with no additional centrifugation step.

3.3.2. Reusability and Post-Use Characterization of HNT/Fe_3O_4-C

Among the investigated magnetic HNTs, the HNT/Fe_3O_4-C sample ensured a quantitative OFL removal in different real water samples and excelled for its magnetic properties. For these reasons, its reusability was explored.

The HNT/Fe_3O_4-C sample was suspended in 10 mL tap water containing OFL 10 μg L^{-1}. After 1 h, HNT/Fe_3O_4-C was magnetically separated, and the supernatant was analyzed by HPLC-FD. Then the recovered sorbent material was suspended for a second time in 10 mL tap water samples containing OFL 10 μg L^{-1}. After 1 h contact, the suspended material was magnetically separated, and the OFL concentration in the solution was measured. A third cycle was carried out following the same procedure.

Figure 10 shows the adsorbed OFL percentage after each adsorption cycle. The adsorbed antibiotic amount slightly decreased from 95% after the first use to 75% after the third one.

This trend may be ascribed to a small loss of material during its magnetic separation from the sample solution and not to matrix interference, as XRPD analysis demonstrates.

The recovered sorbent material after three adsorption cycles was analyzed by XRPD and compared to the synthesized HNT/Fe_3O_4-C sample. The two diffraction patterns (Figure S6) are really comparable, confirming the sorbent material does not undergo degradation processes with use.

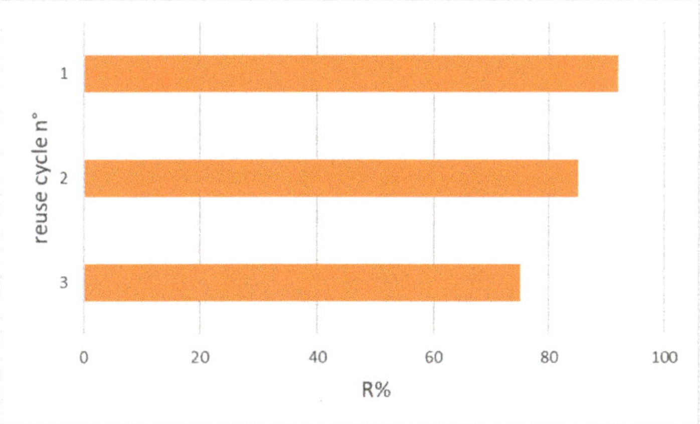

Figure 10. OFL removal after three reuse cycles with HNT/Fe_3O_4-C (Experimental conditions: sorbent 20 mg, 10 mL tap water, OFL initial concentration 10 μg L^{-1}).

3.3.3. Acute Toxicity Test with *Daphnia magna*

For the toxicity test, a single concentration, equal to 0.2 g L^{-1} of HNT, Fe_3O_4, and HNT/Fe_3O_4-C was tested. This concentration reflected a potential residual amount of each material in waters after depollution treatment.

All the individuals efficiently ingested the administered materials over 48 h of exposure (Figure 11), as shown by their presence in the digestive tract of exposed individuals.

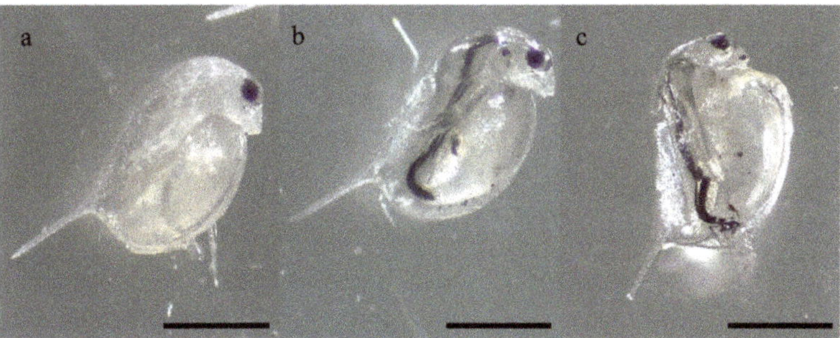

Figure 11. Individuals of *D. magna* showing their digestive tract full of HNT (**a**), Fe_3O_4 (**b**), and HNT/Fe_3O_4-C (**c**) after 48 h of exposure to 0.2 g L^{-1} (10 mg/50 mL) for each material. Scale bar = 500 μm.

No mortality occurred in the control group. Despite the ingestion of all the materials, the 48 h exposure to 0.2 g L^{-1} of HNT and HNT/Fe_3O_4-C did not induce the mortality of any daphnid, while the viability of the individuals included in the Fe_3O_4 experimental group was slightly decreased compared to the corresponding control, accounting for the $96 \pm 9\%$.

4. Conclusions

In the present work, magnetic halloysite nanotubes were successfully synthesized by three different approaches: co-precipitation, hydrothermal, and sol-gel method. The applied characterization techniques demonstrate that the nanometric-sized Fe_3O_4 (diameter of about 10 nm) were formed and connected to the HNT particles. Magnetic phase abundance depended on the synthetic route and was evaluated by EDS and TGA analyses, as well as by magnetization data. Thermodynamic and kinetic experiments suggested that HNT/Fe_3O_4 composites can be considered as performing materials for ofloxacin adsorption. All the investigated samples were able to quantitatively reduce the antibiotic concentration under realistic conditions and, more interestingly, the sample obtained by the co-precipitation synthetic approach—the most cost-effective—was also easily magnetically removed from the media after treatment and reused for three cycles with no degradation. The ecotoxicity test performed on the freshwater organism *D. magna* completed the characterization of this adsorbent material and confirmed that it might be safely applied in water depuration processes.

Supplementary Materials: The following supporting information can be downloaded at: https://www.mdpi.com/article/10.3390/nano12234330/s1, Figure S1: SEM images of the commercial halloysite at (a) 9 kX and (b) 200 kX.; Figure S2: SEM images of the HNT/Fe_3O_4 composites. (a,b): HNT/Fe_3O_4-C sample; (c) and (d): HNT/Fe_3O_4-H sample; (e,f): HNT/Fe_3O_4-SG sample. Magnification: 9 kX (left) and 200 kX (right); Figure S3: (a) investigated area and distribution maps of (b) Al, (c) Fe and (d) Si elements of the HNT/Fe_3O_4-C sample; Figure S4: (a) investigated area and distribution maps of (b) Al, (c) Fe and (d) Si elements of the HNT/Fe_3O_4-H sample; Figure S5: (a) investigated area and distribution maps of (b) Al, (c) Fe and (d) Si elements of the HNT/Fe_3O_4-SG sample; Figure S6: X-ray diffraction pattern of the HNT/Fe_3O_4-C sample as-prepared (black line) and after three cycles of OFL recover (red line); Table S1: Mean particle size and intensity determined by DLS analysis; Table S2: Physico-chemical characterization of tap and river water samples, and WWTP effluent.

Author Contributions: Conceptualization, D.C., M.S. and F.M.; formal analysis, D.M.C., P.L. and F.M.; investigation, P.L., D.M.C., M.B., B.D.F., M.A., G.B., S.P. and F.M.; resources, D.C.; writing—original draft preparation, M.S., D.C., D.P. and M.P.; writing—review and editing, D.C., M.S., D.P. and M.P.; visualization, D.M.C., P.L., M.A. and B.D.F., Project administration, D.C. and M.S. All authors have read and agreed to the published version of the manuscript.

Funding: This research received no external funding.

Data Availability Statement: The data presented in this study are available on request from the corresponding author.

Acknowledgments: The authors are grateful to Massimo Boiocchi for the support in the TEM analysis performed at the Centro Grandi Strumenti, University of Pavia, to Vittorio Berbenni for TGA, surface area and porosity measurements, and Alessandro Girella for the support in the HR-SEM analysis performed at the Arvedi Laboratory, CISRiC (Centro Interdipartimentale di Studi e Ricerche per la Conservazione del Patrimonio Culturale), University of Pavia.

Conflicts of Interest: The authors declare no conflict of interest.

References

1. Castiglioni, S.; Zuccato, E.; Fattore, E.; Riva, F.; Terzaghi, E.; Koenig, R.; Principi, P.; Di Guardo, A. Micropollutants in Lake Como water in the context of circular economy: A snapshot of water cycle contamination in a changing pollution scenario. *J. Hazard. Mater.* **2020**, *384*, 121441. [CrossRef] [PubMed]
2. Sousa, J.C.G.; Ribeiro, A.R.; Barbosa, M.O.; Pereira, M.F.R.; Silva, A.M.T. A review on environmental monitoring of water organic pollutants identified by EU guidelines. *J. Hazard. Mater.* **2018**, *344*, 146–162. [CrossRef] [PubMed]
3. Castiglioni, S.; Davoli, E.; Riva, F.; Palmiotto, M.; Camporini, P.; Manenti, A.; Zuccato, E. Data on occurrence and fate of emerging contaminants in a urbanised area. *Data Brief* **2018**, *17*, 533–543. [CrossRef] [PubMed]
4. Riva, F.; Castiglioni, S.; Fattore, E.; Manenti, A.; Davoli, E.; Zuccato, E. Monitoring emerging contaminants in the drinking water of Milan and assessment of the human risk. *Int. J. Hyg. Environ. Health* **2018**, *221*, 451–457. [CrossRef]
5. Grenni, P.; Ancona, V.; Barra Caracciolo, A. Ecological effects of antibiotics on natural ecosystems: A review. *Microchem. J.* **2018**, *136*, 25–39. [CrossRef]
6. EU Commision. *Communication from the Commission to the European Parliament, the Council, and the European Economic and Social Committee: European Union Strategic Approach to Pharmaceuticals in the Environment*; EU Commision: Brussels, Belgium, 2019.
7. Shahid, M.K.; Kashif, A.; Fuwad, A.; Choi, Y. Current advances in treatment technologies for removal of emerging contaminants from water—A critical review. *Coord. Chem. Rev.* **2021**, *442*, 213993. [CrossRef]
8. Mangla, D.; Annu; Sharma, A.; Ikram, S. Critical review on adsorptive removal of antibiotics: Present situation, challenges and future perspective. *J. Hazard. Mater.* **2022**, *425*, 127946. [CrossRef]

9. Sophia, A.C.; Lima, E.C. Removal of emerging contaminants from the environment by adsorption. *Ecotoxicol. Environ. Saf.* **2018**, *150*, 1–17. [CrossRef]
10. Awual, M.R. A novel facial composite adsorbent for enhanced copper(II) detection and removal from wastewater. *Chem. Eng. J.* **2015**, *266*, 368–375. [CrossRef]
11. Hasan, M.N.; Salman, M.S.; Islam, A.; Znad, H.; Hasan, M.M. Sustainable composite sensor material for optical cadmium(II) monitoring and capturing from wastewater. *Microchem. J.* **2021**, *161*, 105800. [CrossRef]
12. Abegunde, S.M.; Idowu, K.S.; Adejuwon, O.M.; Adeyemi-Adejolu, T. A review on the influence of chemical modification on the performance of adsorbents. *Resour. Environ. Sustain.* **2020**, *1*, 100001. [CrossRef]
13. Jjagwe, J.; Olupot, P.W.; Menya, E.; Kalibbala, H.M. Synthesis and Application of Granular Activated Carbon from Biomass Waste Materials for Water Treatment: A Review. *J. Bioresour. Bioprod.* **2021**, *6*, 292–322. [CrossRef]
14. Godage, N.H.; Gionfriddo, E. Use of natural sorbents as alternative and green extractive materials: A critical review. *Anal. Chim. Acta* **2020**, *1125*, 187–200. [CrossRef]
15. Kubra, K.T.; Salman, M.S.; Znad, H.; Hasan, M.N. Efficient encapsulation of toxic dye from wastewater using biodegradable polymeric adsorbent. *J. Mol. Liq.* **2021**, *329*, 115541. [CrossRef]
16. Saad, E.M.; Elshaarawy, R.F.; Mahmoud, S.A.; El-Moselhy, K.M. New Ulva Lactuca Algae Based Chitosan Bio-Composites for Bioremediation of Cd(II) Ions. *J. Bioresour. Bioprod.* **2021**, *6*, 223–242. [CrossRef]
17. Iravani, R.; An, C.; Adamian, Y.; Mohammadi, M. A Review on the Use of Nanoclay Adsorbents in Environmental Pollution Control. *Water Air Soil Pollut.* **2022**, *233*, 109. [CrossRef]
18. Soleimani, M.; Amini, N. *Remediation of Environmental Pollutants Using Nanoclays*; Springer: Cham, Switzerland, 2017; pp. 279–289.
19. Abdullayev, E.; Lvov, Y. Halloysite clay nanotubes as a ceramic "skeleton" for functional biopolymer composites with sustained drug release. *J. Mater. Chem. B* **2013**, *1*, 2894. [CrossRef]
20. Yu, L.; Wang, H.; Zhang, Y.; Zhang, B.; Liu, J. Recent advances in halloysite nanotube derived composites for water treatment. *Environ. Sci. Nano* **2016**, *3*, 28–44. [CrossRef]
21. Fizir, M.; Dramou, P.; Dahiru, N.S.; Ruya, W.; Huang, T.; He, H. Halloysite nanotubes in analytical sciences and in drug delivery: A review. *Microchim. Acta* **2018**, *185*, 389. [CrossRef]
22. Yuan, P.; Tan, D.; Annabi-Bergaya, F. Properties and applications of halloysite nanotubes: Recent research advances and future prospects. *Appl. Clay Sci.* **2015**, *112–113*, 75–93. [CrossRef]
23. Lvov, Y.; Aerov, A.; Fakhrullin, R. Clay nanotube encapsulation for functional biocomposites. *Adv. Colloid Interface Sci.* **2014**, *207*, 189–198. [CrossRef]
24. Peixoto, A.F.; Fernandes, A.C.; Pereira, C.; Pires, J.; Freire, C. Physicochemical characterization of organosilylated halloysite clay nanotubes. *Microporous Mesoporous Mater.* **2016**, *219*, 145–154. [CrossRef]
25. Yang, J.; Wu, Y.; Shen, Y.; Zhou, C.; Li, Y.-F.; He, R.-R.; Liu, M. Enhanced Therapeutic Efficacy of Doxorubicin for Breast Cancer Using Chitosan Oligosaccharide-Modified Halloysite Nanotubes. *ACS Appl. Mater. Interfaces* **2016**, *8*, 26578–26590. [CrossRef]
26. Rawtani, D.; Agrawal, Y.K. Halloysite as support matrices: A review. *Emerg. Mater. Res.* **2012**, *1*, 212–220. [CrossRef]
27. Xie, Y.; Qian, D.; Wu, D.; Ma, X. Magnetic halloysite nanotubes/iron oxide composites for the adsorption of dyes. *Chem. Eng. J.* **2011**, *168*, 959–963. [CrossRef]
28. Pan, J.; Yao, H.; Xu, L.; Ou, H.; Huo, P.; Li, X.; Yan, Y. Selective Recognition of 2,4,6-Trichlorophenol by Molecularly Imprinted Polymers Based on Magnetic Halloysite Nanotubes Composites. *J. Phys. Chem. C* **2011**, *115*, 5440–5449. [CrossRef]
29. He, J.; Zou, T.; Chen, X.; Dai, J.; Xie, A.; Zhou, Z.; Yan, Y. Magnetic organic–inorganic nanocomposite with ultrathin imprinted polymers via an in situ surface-initiated approach for specific separation of chloramphenicol. *RSC Adv.* **2016**, *6*, 70383–70393. [CrossRef]
30. Tian, X.; Wang, W.; Tian, N.; Zhou, C.; Yang, C.; Komarneni, S. Cr(VI) reduction and immobilization by novel carbonaceous modified magnetic Fe_3O_4/halloysite nanohybrid. *J. Hazard. Mater.* **2016**, *309*, 151–156. [CrossRef] [PubMed]
31. Nowack, B.; Ranville, J.F.; Diamond, S.; Gallego-Urrea, J.A.; Metcalfe, C.; Rose, J.; Horne, N.; Koelmans, A.A.; Klaine, S.J. Potential scenarios for nanomaterial release and subsequent alteration in the environment. *Environ. Toxicol. Chem.* **2012**, *31*, 50–59. [CrossRef]
32. Corsi, I.; Cherr, G.N.; Lenihan, H.S.; Labille, J.; Hassellov, M.; Canesi, L.; Dondero, F.; Frenzilli, G.; Hristozov, D.; Puntes, V.; et al. Common Strategies and Technologies for the Ecosafety Assessment and Design of Nanomaterials Entering the Marine Environment. *ACS Nano* **2014**, *8*, 9694–9709. [CrossRef]
33. Corsi, I.; Winther-Nielsen, M.; Sethi, R.; Punta, C.; Della Torre, C.; Libralato, G.; Lofrano, G.; Sabatini, L.; Aiello, M.; Fiordi, L.; et al. Ecofriendly nanotechnologies and nanomaterials for environmental applications: Key issue and consensus recommendations for sustainable and ecosafe nanoremediation. *Ecotoxicol. Environ. Saf.* **2018**, *154*, 237–244. [CrossRef] [PubMed]
34. Corsi, I.; Grassi, G. The Role of Ecotoxicology in the Eco-Design of Nanomaterials for Water Remediation. In *Ecotoxicology of Nanoparticles in Aquatic Systems*; CRC Press: Boca Raton, FL, USA, 2019; pp. 219–229.
35. Pretali, L.; Maraschi, F.; Cantalupi, A.; Albini, A.; Sturini, M. Water Depollution and Photo-Detoxification by Means of TiO_2: Fluoroquinolone Antibiotics as a Case Study. *Catalysts* **2020**, *10*, 628. [CrossRef]
36. Sturini, M.; Speltini, A.; Maraschi, F.; Profumo, A.; Tarantino, S.; Gualtieri, A.F.; Zema, M. Removal of fluoroquinolone contaminants from environmental waters on sepiolite and its photo-induced regeneration. *Chemosphere* **2016**, *150*, 686–693. [CrossRef] [PubMed]

37. Maraschi, F.; Sturini, M.; Speltini, A.; Pretali, L.; Profumo, A.; Pastorello, A.; Kumar, V.; Ferretti, M.; Caratto, V. TiO_2-modified zeolites for fluoroquinolones removal from wastewaters and reuse after solar light regeneration. *J. Environ. Chem. Eng.* **2014**, *2*, 2170–2176. [CrossRef]
38. Rivagli, E.; Pastorello, A.; Sturini, M.; Maraschi, F.; Speltini, A.; Zampori, L.; Setti, M.; Malavasi, L.; Profumo, A. Clay minerals for adsorption of veterinary FQs: Behavior and modeling. *J. Environ. Chem. Eng.* **2014**, *2*, 738–744. [CrossRef]
39. Sturini, M.; Puscalau, C.; Guerra, G.; Maraschi, F.; Bruni, G.; Monteforte, F.; Profumo, A.; Capsoni, D. Combined Layer-by-Layer/Hydrothermal Synthesis of Fe_3O_4@MIL-100(Fe) for Ofloxacin Adsorption from Environmental Waters. *Nanomaterials* **2021**, *11*, 3275. [CrossRef] [PubMed]
40. Capsoni, D.; Guerra, G.; Puscalau, C.; Maraschi, F.; Bruni, G.; Monteforte, F.; Profumo, A.; Sturini, M. Zinc Based Metal-Organic Frameworks as Ofloxacin Adsorbents in Polluted Waters: ZIF-8 vs. $Zn_3(BTC)_2$. *Int. J. Environ. Res. Public Health* **2021**, *18*, 1433. [CrossRef]
41. Belviso, C.; Guerra, G.; Abdolrahimi, M.; Peddis, D.; Maraschi, F.; Cavalcante, F.; Ferretti, M.; Martucci, A.; Sturini, M. Efficiency in Ofloxacin Antibiotic Water Remediation by Magnetic Zeolites Formed Combining Pure Sources and Wastes. *Processes* **2021**, *9*, 2137. [CrossRef]
42. De Felice, B.; Sabatini, V.; Antenucci, S.; Gattoni, G.; Santo, N.; Bacchetta, R.; Ortenzi, M.A.; Parolini, M. Polystyrene microplastics ingestion induced behavioral effects to the cladoceran *Daphnia magna*. *Chemosphere* **2019**, *231*, 423–431. [CrossRef]
43. Amjadi, M.; Samadi, A.; Manzoori, J.L. A composite prepared from halloysite nanotubes and magnetite (Fe_3O_4) as a new magnetic sorbent for the preconcentration of cadmium(II) prior to its determination by flame atomic absorption spectrometry. *Microchim. Acta* **2015**, *182*, 1627–1633. [CrossRef]
44. Aytekin, M.T.; Hoşgün, H.L. Characterization studies of heat-treated halloysite nanotubes. *Chem. Pap.* **2020**, *74*, 4547–4557. [CrossRef]
45. Lvov, Y.; Abdullayev, E. Functional polymer–clay nanotube composites with sustained release of chemical agents. *Prog. Polym. Sci.* **2013**, *38*, 1690–1719. [CrossRef]
46. Abdullayev, E.; Lvov, Y. *Halloysite for Controllable Loading and Release*; Elsevier: Amsterdam, The Netherlands, 2016; pp. 554–605.
47. Suner, S.S.; Sahiner, M.; Akcali, A.; Sahiner, N. Functionalization of halloysite nanotubes with polyethyleneimine and various ionic liquid forms with antimicrobial activity. *J. Appl. Polym. Sci.* **2020**, *137*, 48352. [CrossRef]
48. Zhu, A.; Yuan, L.; Liao, T. Suspension of Fe_3O_4 nanoparticles stabilized by chitosan and o-carboxymethylchitosan. *Int. J. Pharm.* **2008**, *350*, 361–368. [CrossRef] [PubMed]
49. Peddis, D.; Mansilla, M.V.; Mørup, S.; Cannas, C.; Musinu, A.; Piccaluga, G.; D'Orazio, F.; Lucari, F.; Fiorani, D. Spin-Canting and Magnetic Anisotropy in Ultrasmall $CoFe_2O_4$ Nanoparticles. *J. Phys. Chem. B* **2008**, *112*, 8507–8513. [CrossRef]
50. Peddis, D.; Cannas, C.; Piccaluga, G.; Agostinelli, E.; Fiorani, D. Spin-glass-like freezing and enhanced magnetization in ultra-small $CoFe_2O_4$ nanoparticles. *Nanotechnology* **2010**, *21*, 125705. [CrossRef] [PubMed]
51. Suber, L.; Peddis, D. Approaches to Synthesis and Characterization of Spherical and Anisometric Metal Oxide Magnetic Nanomaterials. In *Nanotechnologies for the Life Sciences*; Wiley-VCH Verlag GmbH & Co. KGaA: Weinheim, Germany, 2011.
52. Slimani, S.; Meneghini, C.; Abdolrahimi, M.; Talone, A.; Murillo, J.P.M.; Barucca, G.; Yaacoub, N.; Imperatori, P.; Illés, E.; Smari, M.; et al. Spinel Iron Oxide by the Co-Precipitation Method: Effect of the Reaction Atmosphere. *Appl. Sci.* **2021**, *11*, 5433. [CrossRef]
53. Peng, H.; Pan, B.; Wu, M.; Liu, Y.; Zhang, D.; Xing, B. Adsorption of ofloxacin and norfloxacin on carbon nanotubes: Hydrophobicity- and structure-controlled process. *J. Hazard. Mater.* **2012**, *233–234*, 89–96. [CrossRef] [PubMed]
54. Li, Y.; Bi, E.; Chen, H. Sorption Behavior of Ofloxacin to Kaolinite: Effects of pH, Ionic Strength, and Cu(II). *Water Air Soil Pollut.* **2017**, *228*, 46. [CrossRef]
55. Antonelli, R.; Martins, F.R.; Malpass, G.R.P.; da Silva, M.G.C.; Vieira, M.G.A. Ofloxacin adsorption by calcined Verde-lodo bentonite clay: Batch and fixed bed system evaluation. *J. Mol. Liq.* **2020**, *315*, 113718. [CrossRef]

 MDPI

Article

Toxic Effects and Mechanisms of Silver and Zinc Oxide Nanoparticles on Zebrafish Embryos in Aquatic Ecosystems

Yen-Ling Lee [1,2,†], Yung-Sheng Shih [1], Zi-Yu Chen [1,†], Fong-Yu Cheng [3], Jing-Yu Lu [1], Yuan-Hua Wu [4,*] and Ying-Jan Wang [1,5,*]

1 Department of Environmental and Occupational Health, College of Medicine, National Cheng Kung University, Tainan 70428, Taiwan; yenpig8291@gmail.com (Y.-L.L.); a08916@yahoo.com.tw (Y.-S.S.); q781001@gmail.com (Z.-Y.C.); annie880308@gmail.com (J.-Y.L.)
2 Department of Oncology, Tainan Hospital, Ministry of Health and Welfare, Tainan 70101, Taiwan
3 Department of Chemistry, Chinese Culture University, Taipei 11114, Taiwan; zfy3@ulive.pccu.edu.tw
4 Department of Oncology, National Cheng Kung University Hospital, College of Medicine, National Cheng Kung University, Tainan 70428, Taiwan
5 Department of Medical Research, China Medical University Hospital, China Medical University, Taichung 40402, Taiwan
* Correspondence: wuyh@mail.ncku.edu.tw (Y.-H.W.); yjwang@mail.ncku.edu.tw (Y.-J.W.); Tel.: +886-6-235-3535 (ext. 5804) (Y.-J.W.)
† These authors contributed equally to this work.

Abstract: The global application of engineered nanomaterials and nanoparticles (ENPs) in commercial products, industry, and medical fields has raised some concerns about their safety. These nanoparticles may gain access into rivers and marine environments through industrial or household wastewater discharge and thereby affect the ecosystem. In this study, we investigated the effects of silver nanoparticles (AgNPs) and zinc oxide nanoparticles (ZnONPs) on zebrafish embryos in aquatic environments. We aimed to characterize the AgNP and ZnONP aggregates in natural waters, such as lakes, reservoirs, and rivers, and to determine whether they are toxic to developing zebrafish embryos. Different toxic effects and mechanisms were investigated by measuring the survival rate, hatching rate, body length, reactive oxidative stress (ROS) level, apoptosis, and autophagy. Spiking AgNPs or ZnONPs into natural water samples led to significant acute toxicity to zebrafish embryos, whereas the level of acute toxicity was relatively low when compared to Milli-Q (MQ) water, indicating the interaction and transformation of AgNPs or ZnONPs with complex components in a water environment that led to reduced toxicity. ZnONPs, but not AgNPs, triggered a significant delay of embryo hatching. Zebrafish embryos exposed to filtered natural water spiked with AgNPs or ZnONPs exhibited increased ROS levels, apoptosis, and lysosomal activity, an indicator of autophagy. Since autophagy is considered as an early indicator of ENP interactions with cells and has been recognized as an important mechanism of ENP-induced toxicity, developing a transgenic zebrafish system to detect ENP-induced autophagy may be an ideal strategy for predicting possible ecotoxicity that can be applied in the future for the risk assessment of ENPs.

Keywords: silver nanoparticles; zinc oxide nanoparticles; developmental toxicity; reactive oxidative stress; apoptosis; autophagy

Citation: Lee, Y.-L.; Shih, Y.-S.; Chen, Z.-Y.; Cheng, F.-Y.; Lu, J.-Y.; Wu, Y.-H.; Wang, Y.-J. Toxic Effects and Mechanisms of Silver and Zinc Oxide Nanoparticles on Zebrafish Embryos in Aquatic Ecosystems. *Nanomaterials* **2022**, *12*, 717. https://doi.org/10.3390/nano12040717

Academic Editor: Vivian Hsiu-Chuan Liao

Received: 28 January 2022
Accepted: 19 February 2022
Published: 21 February 2022

1. Introduction

Currently, engineered nanomaterials and nanoparticles (ENPs) are used in a wide variety of industrial and commercial applications, including catalysts, environmental remediation, personal care products, and cosmetics. In addition, ENPs also show great promise in medicine, such as imaging and drug delivery [1]. Examples of two widely utilized ENPs are silver nanoparticles (AgNPs) and zinc oxide nanoparticles (ZnONPs) [2]. AgNPs are used in cosmetics, textiles, antibacterial agents, the food industry, paints, and medical devices, whereas ZnONPs are widely used in sunscreen, cosmetics, paints, and antibacterial ointments [3,4]. The global usage of both AgNPs and ZnONPs has increased exponentially, thereby increasing the amount of AgNPs and ZnONPs that enter the aquatic environment through various routes [3]. The continuous release of ENPs into the environment from consumer products, industrial waste, and sewage sludge raises concerns about their distribution, behavior, and characteristics in natural water and adverse effects on ecosystems [5]. Nevertheless, the comprehensive toxicological effects of AgNPs and ZnONPs present in the environment have still not been studied in detail. Thus, an analysis of the effects of AgNPs and ZnONPs on ecological systems and human health is important.

In general, a detailed understanding of the behavior of ENPs in aquatic environments is crucial to determine their final distribution and the associated risks [5–7]. Due to their small size, some ENPs, such as AgNPs and ZnONPs, release ions in aquatic systems that are responsible for the major causes of toxicity. However, they also have a confined solid phase similar to other poorly soluble compounds in the aquatic system [7]. In a real water system, natural organic matter (NOM), certain biological macromolecules, and the environmental conditions, including pH and ionic strength, will modify the ENP behavior [5]. Predicting the fate of ENPs in natural surface water systems continues to be a challenge because ENPs can heteroaggregate with natural organic and mineral suspended matter. For example, NOM has been proposed to dominate ENP surface chemistry, and inorganic compounds such as sulfate also play important roles in the modifications of ENP structures and in reducing their toxicity [5]. Transformations of ENPs, such as agglomeration, aggregation, dissolution, and surface change, by interacting with natural water components might substantially alter their environmental release and toxicity [2]. Thus, defining the behavior of ENPs in aquatic systems and their toxic effects on living organisms can facilitate the establishment of scientific foundations for the risk assessment of ENPs in aquatic ecosystems.

The zebrafish (*Danio rerio*) embryo (ZFE) is an ideal model to assess the hazards of both conventional chemicals and ENPs in (eco)toxicology [8]. This model possesses advantages, including rapid development and optical transparency, allowing easy observations of phenotypic responses at lethal, acute, and sublethal toxicological endpoints. In addition, large amounts of embryos can be generated rapidly at low cost, permitting them to serve as a high-throughput assay for the study of developmental processes upon exposure to ENPs [8]. We have recently reported the possible targets and mechanisms of the toxic effects of AgNPs on a ZFE model. Our major finding is that exposure to AgNPs alters lysosomal activity (an indicator of autophagy) and leads to a greater number of apoptotic cells distributed among the developmental organs of the embryo [8,9]. From the (eco)toxicological perspective, the toxic effects induced by AgNPs and ZnONPs, including cytotoxicity, hematotoxicity, immunotoxicity, hepatotoxicity, and embryotoxicity, either in vitro or in vivo, are quite similar [10]. The well-known toxicity induced by ZnONPs is predominantly mediated by the formation of ROS. Excessive ROS generation may damage mitochondria, which subsequently leads to inflammasome activation and cell death through apoptosis and autophagy, which might be a novel mechanism modulating ZnONP-induced inflammatory and cytotoxic effects [11]. ENPs-induced oxidative stress is considered the initiator of the disruption of mitochondrial membrane potential and apoptosis and/or autophagy. Although extensive research has been conducted on ENP applications and toxic mechanisms, research on the environmental transport behavior and ecotoxicity of emerging materials such as AgNPs or ZnONPs is still limited and is needed for sustainable environmental implementation.

We conducted experiments in both complex natural original waters (lake, reservoir, and river) and filtrates from natural waters produced with different pore sizes of filters (1 µm, 0.45 µm, 0.22 µm, and 0.1 µm) to understand the behavior, characteristics, and embryotoxicity of AgNPs and ZnONPs in aquatic systems. Although results obtained from natural waters may more realistically approximate actual ecosystems, they are often unable to provide sufficient information due to the complexity of water chemistry. The results obtained from filtrates of natural waters and MQ water (control group), which have a reduced complexity, can improve our understanding of the processes involved but are not necessarily representative of natural systems. The objective of this study was to characterize aggregates of AgNPs and ZnONPs in natural waters and determine whether they are toxic to developing zebrafish embryos. Different toxic effects and mechanisms were investigated here to evaluate ENP toxicity by measuring the survival rate, hatching rate, body length, oxidative stress, apoptosis, and autophagy. The results from those experiments are essential for a better understanding of AgNPs and ZnONPs in aquatic ecosystems and provide important underlying mechanisms for ecological risk assessments of them and other nanoparticles.

2. Materials and Methods

2.1. Natural Water Preparation and Sampling

We collected natural water samples from different water bodies, including the Erren River, Zengwun Reservoir, and Cheng Kung Lake located at the campus of National Cheng Kung University (NCKU), in spring (May) and autumn (September) in Taiwan. Among them, the Erren River was considered a polluted water body, whereas Zengwun reservoir water was the source of drinking water for the public. Natural water was collected in sterilized sampling bags. Before storage, all the original water samples were filtered with a 53 μm filter to remove large impurities and microorganisms, and then the water samples were stored at 4 °C.

2.2. Preparation and Characterization of AgNPs and ZnONPs

The principle of AgNP synthesis was based on the $NaBH_4$ reduction of the Ag ion. First, the silver nitrate solution (50 mL, 20 mM), sodium citrate solution (40 mL, 80 mM). and 890 mL deionized water were fully mixed. Then, a $NaBH_4$ solution (20 mL, 100 mM) was added slowly to promote the Ag ion reduction. The solution was stirred vigorously for 2 h, allowing a full reductive reaction. After synthesizing a sufficient amount of AgNP solution, it was centrifuged at 7500× *g* for 30 min, and the supernatant was collected. Furthermore, the solution was centrifuged again at 12,500× *g* for 2 h, and the precipitate was removed. The synthesized AgNP solution was stored at 4 °C in the dark.

We dissolved zinc acetate dehydrate ($Zn(CH_3COO)_2(H_2O)$) in 3.35 mM ethanol and stirred the solution vigorously at 60 °C to synthesize ZnONPs. Furthermore, the solution was slowly titrated with a 6.59 mM potassium hydroxide (KOH) solution into a zinc acetate dehydrate solution along the edge of the beaker and reacted for 1.5 h. Next, the solution was incubated at room temperature for 2 h. The synthetic ZnONP solution was centrifuged at 10,000 rpm for 10 min, and then the supernatant was removed. The precipitate was collected and subsequently washed twice with 50 mL of ethanol. Then, 0.25 mL of (3-aminopropyl)triethoxysilane (APTES), 0.5 mL of deionized water, and 0.05 mL of ammonia (25 wt%) were mixed immediately with the ZnONP solution and reacted at room temperature for 20 min. After the reaction, the solution was centrifuged at 10,000 rpm for 15 min, and then we collected the precipitates. The precipitates were washed 2 times with ethanol and resuspended in water to obtain amine-coated ZnONPs. The aim of the surface modification was to improve the dispersion of nanoparticles in various solutions. Precisely, the formal name of ZnONPs we applied in this study was aminopropyl silica-coated ZnONPs, which is abbreviated as ZnONPs. The primary size and morphology of NPs were observed by transmission electron microscopy (TEM, JEOL Co., Akishima, Tokyo, Japan). The elemental analysis of NPs and NPs spiked in natural water were detected through electron dispersive X-ray (EDX) (JEOL Co., Akishima, Tokyo, Japan). The hydrodynamic diameter and polydispersity index were analyzed via dynamic light scattering (DLS, Delsa™ Nano C, Beckman Coulter, Inc., Brea, CA, USA). The zeta potential of the NPs was measured by phase analysis light scattering (PALS, Delsa™ Nano C, Beckman Coulter, Inc., Brea, CA, USA). The NP stability was investigated by ultraviolet visible spectroscopy (UV-Vis, Thermo Fisher Scientific, Waltham, MA, USA).

2.3. Fish Husbandry and Egg Spawning

Zebrafish embryos (*Danio rerio*) were obtained from the Taiwan Zebrafish Core Facility. Zebrafish were raised and maintained in a thermostatic culture system at 28–30 °C with a photoperiod of 14 h of light/10 h of darkness. Fish were fed twice a day with brine shrimp. Male and female fish were placed in the mating box. Spawning was triggered when the light was turned on in the morning. At 4 hpf, zebrafish embryos were collected and rinsed several times to remove residues on the surface. The dead, unfertilized, and abnormal embryos were removed. Healthy embryos were randomly selected for experiments.

2.4. The Acute Toxicity Test on Zebrafish Embryos

The fish embryo acute toxicity test conducted in this study was based on OECD Test Guideline TG No. 236, released in 2013. All experiments were performed using a semi-static system. First, the fertilized embryos were gently collected with a dropper at 4 hpf, and healthy embryos were randomly divided and placed in 12-well plates (10 embryos/well). Embryos (30 embryos/group) were treated with AgNPs and ZnONPs. The positive controls (4 mg/L 3,4-dichloroaniline) were observed until 96 hpf, and mortality and deformities were recorded daily. The AgNPs and ZnONPs solutions were refreshed every 24 h until the end of the experiment to avoid ENP aggregation during exposure. All acute toxicity assays were performed in triplicate.

2.5. ROS Analyses

Dichlorodihydrofluorescein diacetate (DCFH-DA) (Sigma-Aldrich, St. Louis, MO, USA) was used to measure reactive oxygen species (ROS) levels in the zebrafish. The zebrafish embryos were exposed to AgNPs and ZnONPs solutions beginning at 4 hpf, and the DCFH-DA assay was conducted at 96 hpf. In the experimental procedure, we initially performed three washes with deionized water to remove solutions of AgNPs and ZnONPs and impurities from the surface of the zebrafish. Then, live samples were incubated with 5 μg/mL DCFH-DA at 28 °C for 1 h and coated with aluminum foil to protect them from light. After the incubation, the living fish were rinsed with deionized water three times. The embryos were fixed with 3% methyl cellulose to prevent movement, and images were captured immediately.

2.6. LysoTracker RED DND-99 Analyses

LysoTracker RED DND-99 (Sigma-Aldrich, St. Louis, MO, USA) reagent was used to evaluate lysosomes and autophagy in the zebrafish. The procedure was performed according to the manufacturer's protocol. Briefly, the samples were collected at 72 hpf and rinsed with deionized water three times. Then, the embryos were incubated with 10 μM lysosomal probe for 1 h at 28 °C. After the incubation, we used deionized water to wash the embryos three times and then fixed the embryos with 3% methyl cellulose. Embryos were observed and imaged using a fluorescence microscope (BX51, Olympus Co., Tokyo, Japan).

2.7. Whole-Mount TUNEL Assay

Whole-mount TUNEL staining was performed to evaluate apoptosis in zebrafish at 72 hpf. We washed the zebrafish with deionized water three times after exposure to Ag/ZnONPs. Zebrafish larvae were fixed with 4% paraformaldehyde for 1 h and then washed with PBS three times. Next, the embryos were incubated with blocking buffer on a shaker for 30 min and subsequently rinsed with PBS three times for 5 min. Then, the embryos were incubated with a permeabilization solution (0.1% Triton X-100 in 0.1% sodium citrate) for 30 min on ice and washed with PBS three times on a shaker for 5 min each. Finally, the TUNEL reaction mixture (labeling solution:enzyme solution = 9:1) was incubated with embryos in a water bath at 37 °C for 1 h, and we observed the embryos using a confocal fluorescence microscope (Nikon TE2000EPFS-C1-Si, Tokyo, Japan).

2.8. Statistics

Each experiment was independently repeated at least three times. Statistical analyses were performed with one-way ANOVA and a two-tailed Student's *t*-test. $p < 0.05$ was considered statistically significant between the experimental and control groups. All data were presented as the mean ± SEM.

3. Results

3.1. The Physical and Chemical Properties of AgNPs and ZnONPs

Two types of nanoparticles were synthesized in this study to evaluate the environmental toxicity of ENP-spiked natural water. We synthesized citrate-coated AgNPs and amine-modified ZnONPs and fully characterized their physicochemical properties. As shown in Figure 1, The TEM images present the mean primary sizes of AgNPs and ZnONPs, which were 13 and 27 nm, respectively, and the morphology was spherical (Figure 1a,c). Element analysis indicated that AgNPs contained silver and ZnONPs were composed of zinc and oxygen (Figure 1b). The Si element came from a synthesized process which used APTES, and the Cu came from carbon support films on copper grids, which were used in TEM analysis. Dynamic light scattering (DLS) results revealed that the hydrodynamic diameters of the AgNPs and ZnONPs were 16 nm and 43 nm, respectively, and the NPs exhibited a homogeneous dispersion (Figure 1c). In addition, AgNPs and ZnONPs represented negative (−18.6) and positive surface charges (+25.4), respectively (Figure 1c).

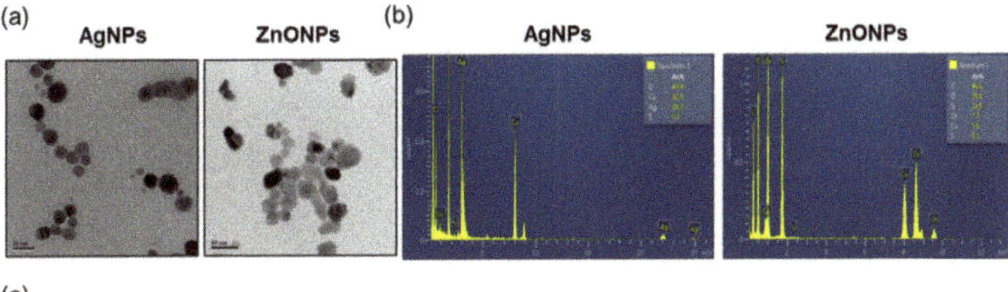

(c)

	Measurement	AgNPs	ZnONPs
Surface coating	-	Citrate	Amine
Primary size (nm)	TEM	13.3 ± 4.2	27.8 ± 7.9
Hydrodynamic diameter (nm)	DLS	16.3 ± 3.4	43.2 ± 11.5
Polydispersity index	DLS	0.101	0.233
Surface charges (mV)	PALS	-18.6	25.4
Maximum absorbance (nm)	UV-Vis	391 nm	370 nm

Figure 1. Physical and chemical properties of AgNPs and ZnONPs. (**a**) TEM images of AgNPs and ZnONPs. (**b**) EDX analysis of AgNPs and ZnONPs. (**c**) Analysis of the physicochemical properties of AgNPs and ZnONPs. AgNPs, silver nanoparticles; ZnONPs, zinc oxide nanoparticles; TEM, transmission electron microscopy; EDX, electron dispersive X-ray; DLS, dynamic light scattering; PALS, phase analysis light scattering; UV–Vis, ultraviolet visible spectroscopy.

3.2. The Deposition of AgNPs and ZnONPs Spiked in Natural Water

Several studies have shown that the physicochemical characteristics of ENPs change when they are spiked in natural water bodies [12,13]. We obtained water from NCKU Lake, Zengwun Reservoir, and Erren River to determine the physicochemical properties of Ag/ZnONPs in natural water. NCKU Lake is located at National Cheng Kung University and possesses a high species richness. Zengwun Reservoir is a relatively clean water source that provides drinking water. The Erren River is one of the contaminated rivers in Taiwan. To date, the Erren River is still considered moderately polluted.

AgNPs and ZnONPs were spiked into the water samples from NCKU Lake, Zengwun Reservoir, and Erren River, and deposition was observed after 30 min. For the NCKU Lake water sample, the AgNPs were still suspended in water, but the color quickly changed. ZnONPs spiked in NCKU Lake water were deposited. In addition, water from the Zengwun Reservoir and Erren River caused a substantial aggregation of AgNPs and ZnONPs upon spiking into the natural water (Figure 1a,b). Furthermore, the physicochemical characteristics of these nanoparticles spiked in natural water were analyzed by determining their particle size, hydrodynamic diameter, and dispersity, as summarized in Figure 2c–e. Both AgNPs and ZnONPs spiked in NCKU Lake water exhibited a larger particle size and hydrodynamic diameter and decreased dispersity (Figure 2c,e). After spiking in the Zengwun Reservoir and Erren River water, AgNPs and ZnONPs aggregated and were not suspended in solution, and thus the data were undetectable (Figure 2c,e). These results were attributed to the chemical reaction of these nanoparticles with complex water components after spiking in natural water, which led to aggregation and precipitation.

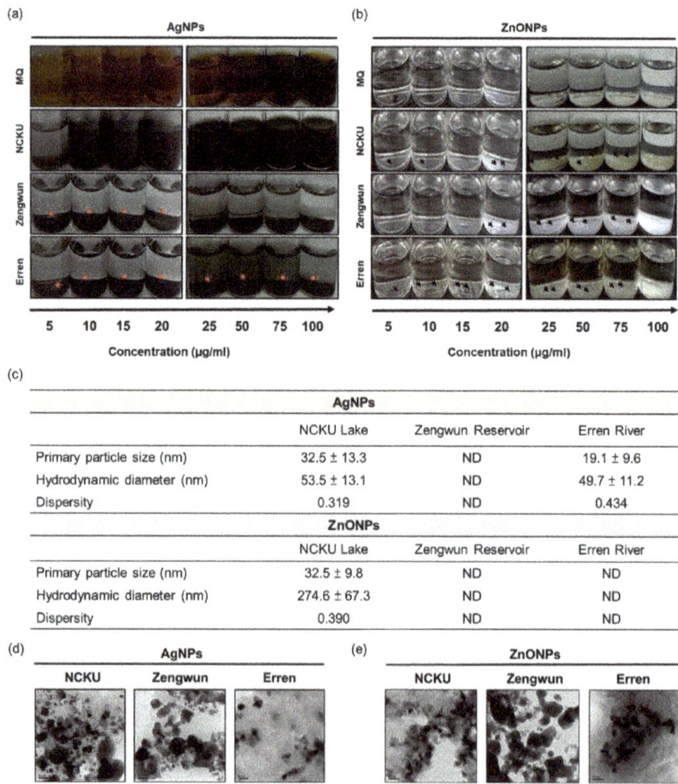

AgNPs	NCKU Lake	Zengwun Reservoir	Erren River
Primary particle size (nm)	32.5 ± 13.3	ND	19.1 ± 9.6
Hydrodynamic diameter (nm)	53.5 ± 13.1	ND	49.7 ± 11.2
Dispersity	0.319	ND	0.434
ZnONPs	**NCKU Lake**	**Zengwun Reservoir**	**Erren River**
Primary particle size (nm)	32.5 ± 9.8	ND	ND
Hydrodynamic diameter (nm)	274.6 ± 67.3	ND	ND
Dispersity	0.390	ND	ND

Figure 2. Deposition of AgNPs and ZnONPs spiked in natural water. The deposition of 5, 10, 15, 20, 25, 50, 75, and 100 μg/mL (**a**) AgNPs and (**b**) ZnONPs spiked in natural water from NCKU Lake, Zengwun Reservoir, and Erren River. Ag/ZnONPs aggregated and deposited at the bottom of the vessel. The arrow indicates Ag/ZnONPs deposition. (**c**) The physical and chemical properties of AgNPs and ZnONPs spiked in natural water. (**d**,**e**) TEM images of AgNPs and ZnONPs spiked in natural water. AgNPs and ZnONPs aggregated after spiking in natural water samples. AgNPs, silver nanoparticles; ZnONPs, zinc oxide nanoparticles; TEM, transmission electron microscopy.

3.3. Elemental Mapping of Natural Water

Furthermore, we performed elemental mapping to determine the elements present in natural water that reacted with AgNPs and ZnONPs. NCKU Lake, Zengwun Reservoir, and Erren River water samples were spiked with AgNPs and ZnONPs. We found that NCKU Lake, Zengwun Reservoir, and Erren River water precipitants contained abundant sulfur (S) and chlorine (Cl) and that Ag perfectly colocalized with them (Figure 3a–c). For ZnONP-spiked water samples, zinc was accompanied by magnesium (Mg) and potassium (K) in NCKU Lake water (Figure 3d). Furthermore, Zengwun Reservoir water samples contained calcium (Ca) and chlorine that colocalized with zinc (Figure 3e). Last, the Erren River water contained sulfur and calcium that overlapped with zinc (Figure 3f). Based on these results, we determined that both AgNPs and ZnONPs undergo complex chemical reactions with the components in natural water.

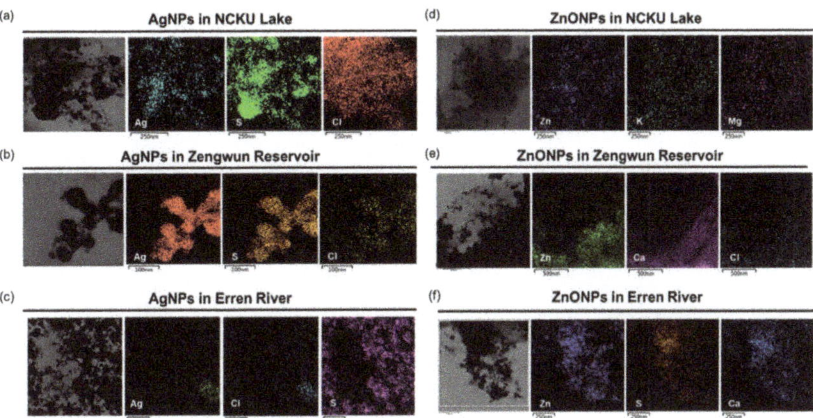

Figure 3. Elemental mapping images of AgNPs and ZnONPs in natural water. (**a–c**) Elemental mapping analysis of Ag spiked in NCKU Lake, Erren River, and Zengwun Reservoir water. (**d–f**) Elemental detection of ZnONPs spiked in natural water. AgNPs, silver nanoparticles; ZnONPs, zinc oxide nanoparticles; S, sulfur; Cl, chlorine; Mg, magnesium; K, potassium; Ca, calcium.

3.4. The Mortality of AgNPs and ZnONPs Spiked in Nature Water

First, we observed the survival of embryos after exposure to the three water samples to determine whether natural water exerted adverse effects on embryos. Compared with the positive control group (3,4-DCA), the original water samples that were not spiked with AgNPs and ZnONPs did not cause obvious mortality at 24, 48, and 96 hpf. Even in the sample from the polluted Erren River, the survival rate was still high (Figure S1a–d). When the original water samples were filtered through membranes with different pore sizes of 1 μm, 0.45 μm, 0.22 μm, and 0.1 μm, the filtrates also did not produce significant adverse effects. The survival rates of all groups were still greater than 80% (Figure S1a–d).

According to previous studies, natural substances, such as humic acid, in natural water mitigated the toxic effects of NPs [13]. We designed additional experiments to confirm whether the substances in natural water resulted in a reduction in toxicity. First, the larger substances in natural water were removed with a 0.45 μm filter in advance and then the samples were spiked with AgNPs and ZnONPs (Figure 4a,b). Second, the original natural water was spiked with AgNPs and ZnONPs first and then passed through a 0.45 μm filter (to remove deposited ENPs) (Figure 4a,b). Third, we spiked 1 and 10 μg/mL AgNPs and ZnONPs in MQ water, respectively, for comparison and observed a substantial increase in mortality (Figure 4c,d). Accordingly, ZnONPs spiked in the natural water samples from NCKU Lake, Zengwun Reservoir, and Erren River exerted minor toxic effects, even in the 10 μg/mL groups (Figure 4e,f). In terms of the deposited NP removal groups, the elimination of AgNPs and ZnONPs deposits in natural water reversed the mortality, especially in the 10 μg treatment group (Figure 4g–j). In summary, although AgNPs and ZnONPs spiked into natural water exhibited a dramatically reduced toxicity, they still exerted adverse effects on the ecological environment at higher concentrations.

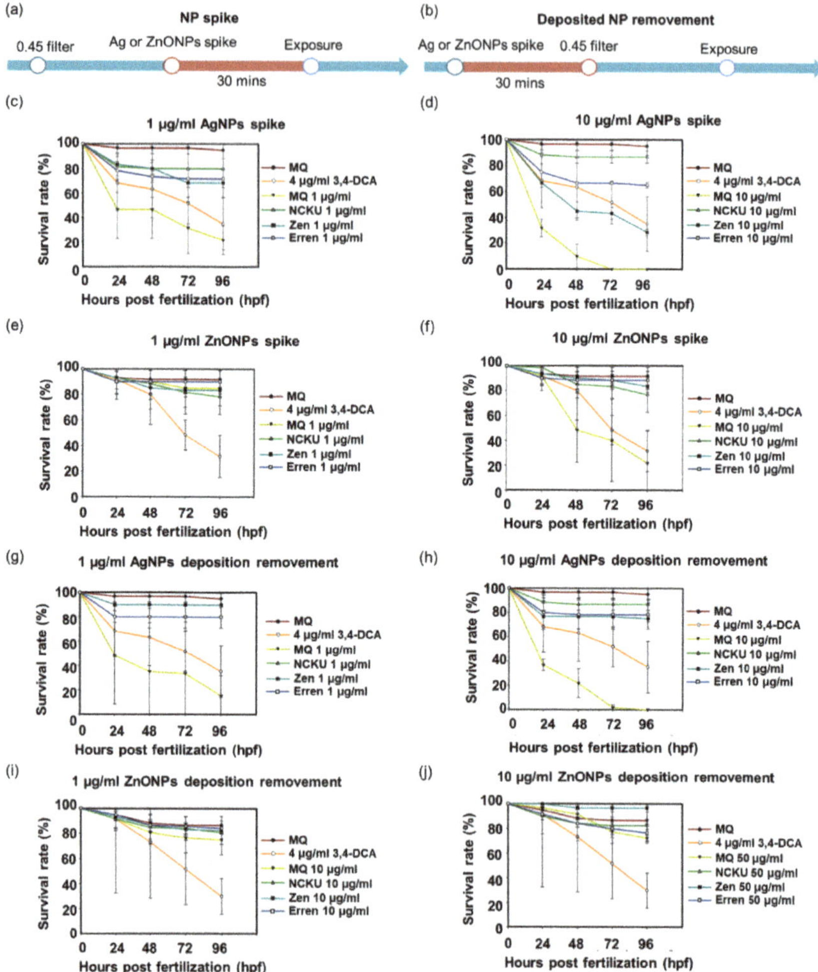

Figure 4. Lethality of AgNPs and ZnONPs spiked in natural water. The study design included two processes: (**a**) NP spiking and (**b**) deposited NP removal. The embryos were treated with (**c**) 1 μg/mL AgNPs, (**d**) 10 μg/mL AgNPs, (**e**) 1 μg/mL ZnONPs, and (**f**) 10 μg/mL ZnONP-spiked natural water. The mortality of embryos exposed to (**g**,**h**) 1 μg/mL and 10 μg/mL AgNP-spiked water, as well aI (**i**,**j**) 1 μg/mL and 10 μg/mL ZnONP-spiked natural water in which the aggregates were removed by a 0.45 μm filter. AgNPs, silver nanoparticles; ZnONPs, zinc oxide nanoparticles; 3,4-DCA, 3,4-dichloroaniline.

3.5. ZnONPs Caused Developmental Toxicity

We measured body length to explore the relationship between zebrafish embryo development and substances in the natural water environment. Interestingly, the filtrates of water samples from NCKU Lake, Zengwun Reservoir, and Erren River that were passed through 1 μm, 0.45 μm, 0.22 μm, and 0.1 μm filters did not alter the body length of the zebrafish compared with the control group (MQ water) (Figure S2). The body length of the zebrafish even increased after exposure to filtrates of natural water compared to the control group, but significant differences were not observed between the filtrate groups. In addition to the survival rate and body length, the hatching rate is also an indicator of developmental toxicity. When ZnONPs were spiked into the three different natural water samples and then passed through a 0.45 μm filter, we observed a significantly later hatching time at 96 hpf in embryos exposed to a higher concentration of 10 μg/mL (Figure 5b). Nonetheless, the embryos exposed to ZnONPs still hatched successfully without deformities. The AgNPs did not induce a delayed hatching time in this study (data not shown).

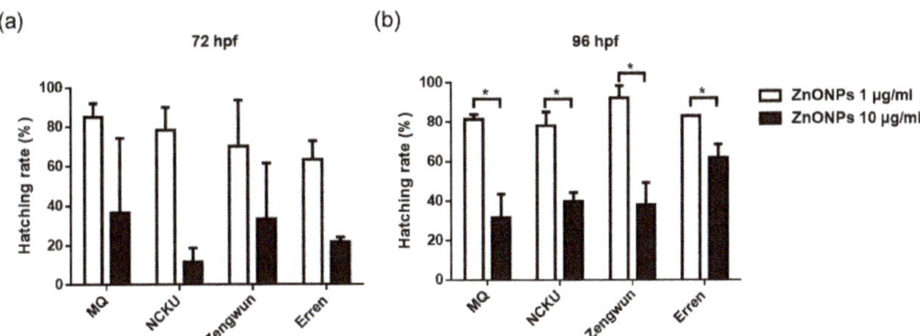

Figure 5. Developmental toxicity of ZnONPs. Hatching rate of zebrafish embryos exposed to ZnONPs. ZnONPs spiked in NCKU Lake, Zengwun Reservoir, and Erren River water. Hatching rate at (**a**) 72 hpf and (**b**) 96 hpf after exposure to different natural water samples. ZnONPs spiked in different water samples resulted in later hatching times in all groups. The significance between each different concentration groups were represented by * sign. (* $p < 0.05$). ZnONPs, zinc oxide nanoparticles; hpf, hours post-fertilization.

3.6. AgNPs and ZnONPs Induced Programmed Cell Death

Several studies have revealed that AgNPs and ZnONPs lead to cytotoxicity by inducing programmed cell death [8,14,15]. We applied the TUNEL assay to investigate the cytotoxic effects of AgNPs and ZnONPs spiked in deionized (MQ) or natural water at 72 hpf. DNase I served as a positive control group and resulted in a significant increase in the number of apoptotic cells (Figure 5a,b). Groups treated with AgNPs and ZnONPs spiked in the MQ water exhibited an increased number of apoptotic cells, and the signals were mainly located in the yolk sac (Figure 5a,b). Surprisingly, groups treated with both AgNPs and ZnONPs spiked in natural water also exhibited substantial cytotoxicity. These results provide evidence that ENPs in natural water exert cytotoxic effects by activating apoptosis.

3.7. AgNPs and ZnONPs Induced ROS Production

Excessive oxidative stress is an important cellular mechanism of ENPs and is recognized as an initial form of cell damage [8,14,16,17]. Next, we investigated whether AgNPs and ZnONPs spiked in different water samples would increase the ROS level in zebrafish embryos. We used dichloro-dihydro-fluorescein diacetate (DCFH-DA) dye to detect ROS generation in embryos exposed to these nanoparticles at 72 hpf. As expected, the H_2O_2 groups (positive control) displayed increased ROS generation (Figure 6a,b). Similarly, 10 μg/mL AgNPs and ZnONPs spiked in the MQ water increased ROS generation, and the signals were primarily observed in the intestine (Figure 6a,b). As mentioned above, we concluded that ENPs spiked into the natural water increase ROS level in zebrafish embryos.

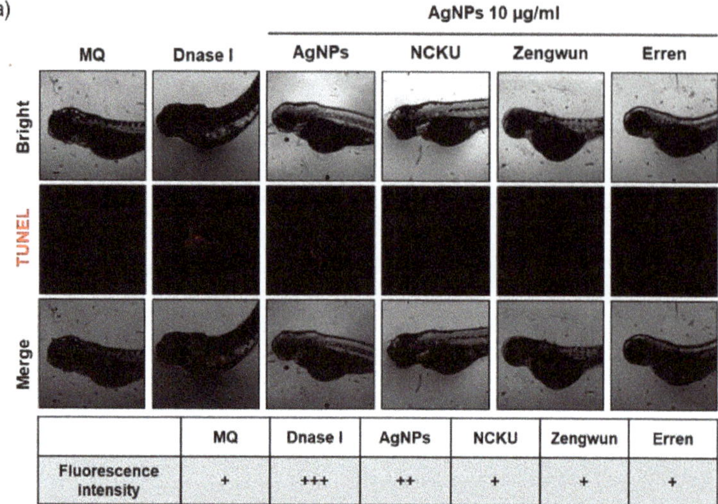

	MQ	Dnase I	AgNPs	NCKU	Zengwun	Erren
Fluorescence intensity	+	+++	++	+	+	+

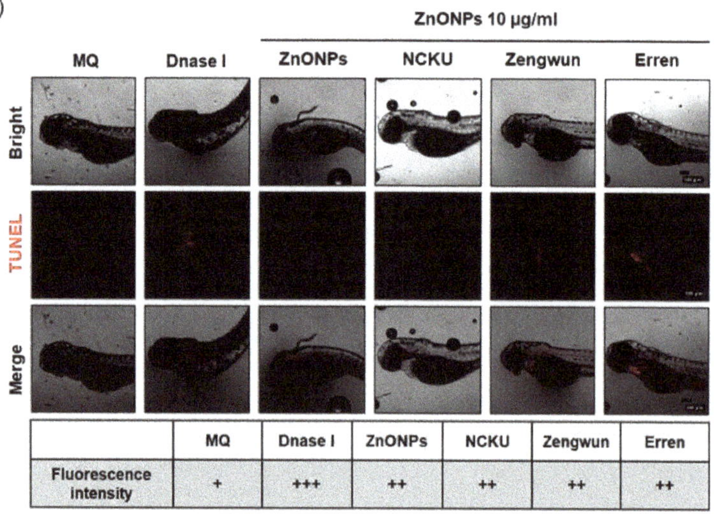

	MQ	Dnase I	ZnONPs	NCKU	Zengwun	Erren
Fluorescence intensity	+	+++	++	++	++	++

Figure 6. Cytotoxicity of AgNPs and ZnONPs. Zebrafish embryos were exposed to natural water after passing through a 0.45 μm filter and then spiked with NPs. After 72 hpf, zebrafish embryos were subjected to a TUNEL assay. Fluorescence images and intensity of (**a**) AgNPs and (**b**) ZnONPs. Apoptotic cells are represented by red fluorescence signals. The fluorescence intensity was assigned as + (weak), ++ (middle), and +++ (strong). Both AgNPs and ZnONPs spiked in natural water produced increased apoptosis signals in the 10 μg/mL NP groups. AgNPs, silver nanoparticles; ZnONPs, zinc oxide nanoparticles; hpf, hours post-fertilization.

3.8. AgNPs and ZnONPs Induced Autophagy

According to recent studies, AgNPs and ZnONPs cause dysfunctional autophagy in vitro [14,18]. Hence, we used the LysoTracker RED probe to evaluate lysosomal activity in zebrafish embryos incubated with AgNPs and ZnONPs spiked in different water samples. Treatment with 10 μg/mL AgNPs and ZnONPs in the MQ water activated autophagy, and the signals were detected in the yolk sac (Figure 7a,b). For natural water, AgNPs spiked in Zengwun Reservoir natural water apparently increased the activation of autophagy (Figure 7a). Some lysosomal activity was also observed in the other groups. Therefore, AgNPs and ZnONPs spiked in natural water induce autophagy.

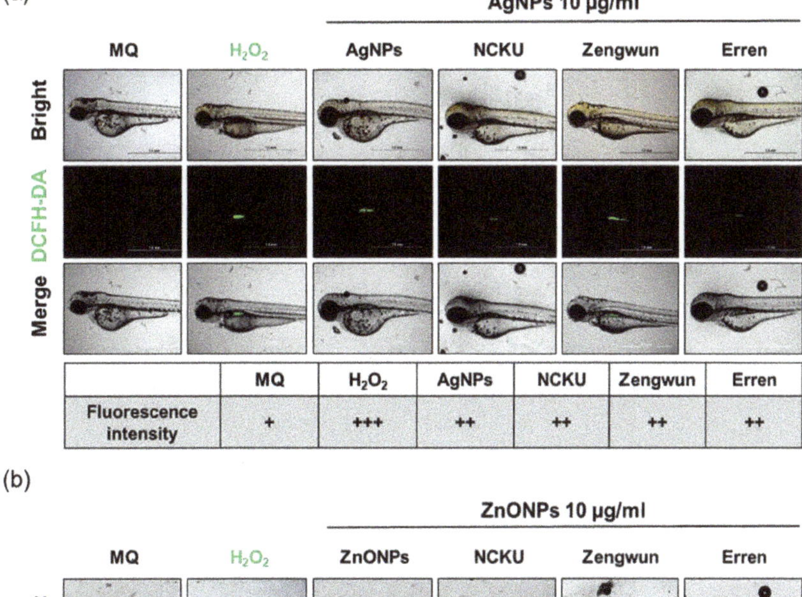

	MQ	H_2O_2	AgNPs	NCKU	Zengwun	Erren
Fluorescence intensity	+	+++	++	++	++	++

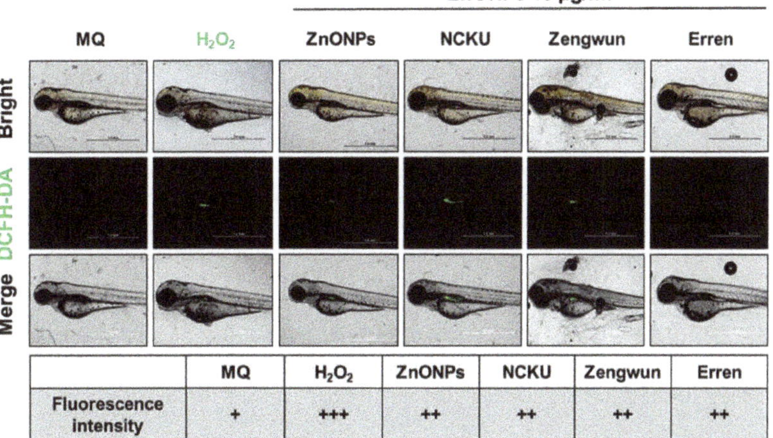

	MQ	H_2O_2	ZnONPs	NCKU	Zengwun	Erren
Fluorescence intensity	+	+++	++	++	++	++

Figure 7. ROS generation induced by AgNPs and ZnONPs. Zebrafish embryos were exposed to natural water after passing through a 0.45 μm filter and then spiked with (**a**) AgNPs and (**b**) ZnONPs. After 72 hpf, zebrafish embryos were stained with DCFH-DA (green). The fluorescence intensity was assigned as + (weak), ++ (middle), and +++ (strong). The H_2O_2-positive control group showed ROS generation. Both AgNPs and ZnONPs spiked in natural water resulted in increased ROS signals. AgNPs, silver nanoparticles; ZnONPs, zinc oxide nanoparticles; DCFH-DA, dichlorodihydrofluorescein diacetate; ROS, reactive oxygen species.

4. Discussion

Engineered nanomaterials and nanoparticles (ENPs) emission are estimated to be mainly derived from landfills (63–91%), with over 260,000–309,000 metric tons of global production in 2010, followed by release to soil (8–28%), water bodies (0.4–7%), and air (0.1–0.5%) [19]. In certain environmental compartments, ENPs may pose a relatively low risk, whereas organisms residing near ENP production plant outfalls or waste treatment plants may be at increased risk [20]. When ENPs enter terrestrial and aquatic systems, they may threaten the ecological environment and human health [21]. In general, the fate of ENPs in aquatic systems is mainly determined by three processes: heteroaggregation, dissolution, and sedimentation [22,23]. The factors that influence ENP behavior in the environment include size, surface coating materials and their changes (e.g., degradation or replacement by natural organic matter (NOM)), oxidation, dissolution, sulfidation, heteroaggregation, homoaggregation, and sedimentation/resuspension [24–26]. Other environmental factors include sunlight, the pH of the solution, inorganic salts, the interaction with surrounding metals, and dissolved NOM, which will interact with ENPs and lead to their transformation [27–31].

After AgNPs and ZnONPs were added to natural water, we observed that the color and particle size of the nanoparticles changed significantly (Figures 1 and 2). Thus, the nanoparticles may undergo chemical reactions with other components in the water environment. The dissolution of nanoparticles might be generally promoted after interacting with NOM. Two major surface transformation processes, oxidation and sulfidation, may occur on the surface of nanoparticles in the presence of NOM [32,33]. Therefore, NOM alters the toxicity of ENPs by changing suspension stabilization, the bioavailability of metal ions, electrostatic interactions and steric repulsion between nanoparticles and organisms, and induced reactive oxygen generation [33]. Bundschuh et al. noted that the phenomenon of co-occurring contaminants interacting with nanoparticles and indicated that nanoparticles serve as a sink for organic and inorganic co-contaminants in the water column [34]. Therefore, we conducted an elemental mapping analysis to investigate the composition of sediments of AgNPs or ZnONPs after their addition to natural water. Sulfur (S) and chlorine (CI) attached to AgNPs and ZnONPs, and iron (Fe) and phosphorous (P) attached on ZnONPs, which may in turn change the physicochemical properties of the nanoparticles due to the interaction between these molecules (Figure 3). AgNPs release silver only after they are oxidized by dissolved oxygen, and the released silver is readsorbed onto the surface of the nanoparticles or forms a secondary precipitate with complexing species (e.g., Cl^- and SO_4^{2-}) [35]. Sulfidation of AgNPs or ZnONPs frequently occurs under various environmental conditions and leads to the formation of core–shell Ag0–Ag2S structures or hollow Ag2S NPs. Sulfidation leads to nearly inert NP surfaces that alter their reactivity and toxicity [34]. The fate and stability of nanoparticles in both raw lake water and filtered lake water containing different NOM lead to different aggregation profiles [36]. The authors concluded that the use of pure NOM analogs may not accurately represent nanoparticles' interactions and fate in real natural systems [36]. Our experiment results suggested that the natural water may have mitigated the toxic effects of AgNPs and ZnONPs through nanoparticles aggregation and interaction with NOM, resulting in the formation of larger particles and sedimentation. Nonetheless, the underlying mechanisms of the interactions and relationships among nanoparticles and organic/inorganic substances in the ecosystem require further investigation.

With the advantages of rapid development and optical transparency, the zebrafish embryos are rapidly becoming an attractive vertebrate model species for screening ENPs [37]. Our current study showed a very high survival rate of zebrafish embryos exposed to three different original natural waters samples and their filtrates obtained after passing through different pore sizes of filter. Interestingly, all the above-mentioned natural water samples led to a longer body length of larva than the embryos exposed to MQ water (Figure S2). On the one hand, natural water samples may contain certain essential elements that enhance the development of the embryos. On the other hand, chemicals and/or ENP contamination of surface waters from rivers, lakes, and reservoirs in Taiwan may still be limited and promote the survival of zebrafish embryos. Currently, aquatic AgNP concentrations in fresh water are predicted to range from approximately a few pg/L to 10 ng/L between 2017 and 2050, which might be nontoxic to fish embryos [20]. Although the current ENP contamination may pose a relatively low risk to natural aquatic systems, the organisms living in the ecosystem near ENP production plants or waste treatment plants may be at higher risk. Therefore, we conducted acute zebrafish embryo toxicity assays by spiking AgNPs or ZnONPs into natural water samples. As shown above (Figures 6–8), both AgNPs and ZnONPs led to significant acute toxicity toward zebrafish embryos in a dose-dependent manner. The level of acute toxicity was relatively lower in the filtered natural water samples than in the MQ water samples, indicating that the interaction and transformation of these nanoparticles with the complex components in a water environment led to a reduced toxicity.

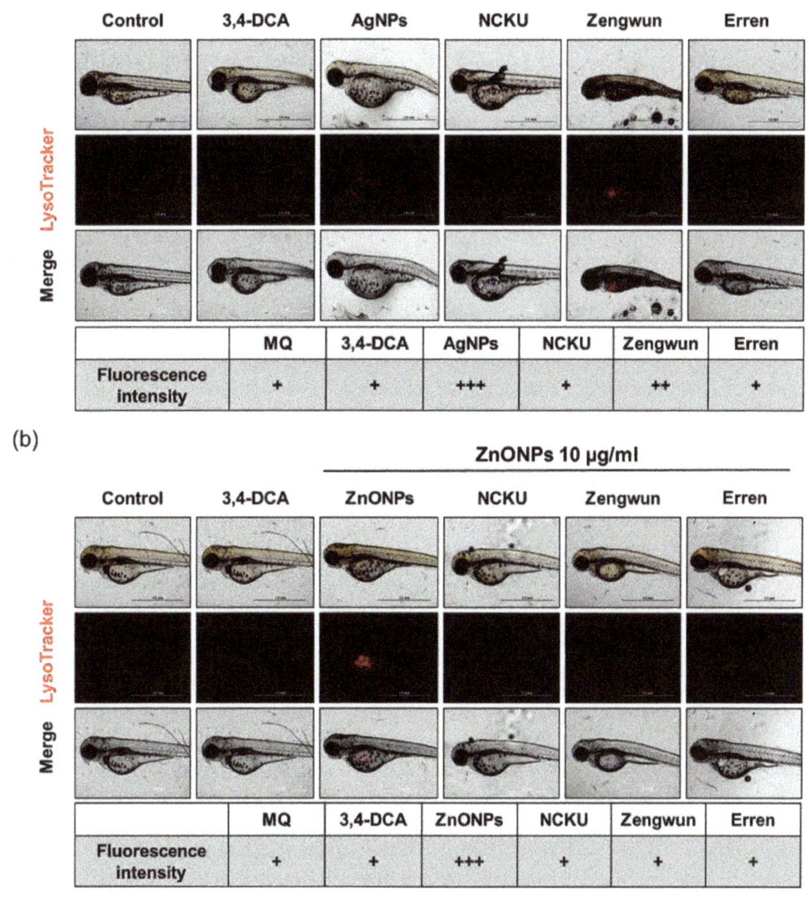

Figure 8. Autophagy induced by AgNPs and ZnONPs. Zebrafish embryos were exposed to natural water after passing through a 0.45 μm filter and then spiked with (**a**) AgNPs and (**b**) ZnONPs. After 72 hpf, zebrafish embryos were stained with LysoTracker RED. The fluorescence intensity was assigned as + (weak), ++ (middle), and +++ (strong). The 10 μg/mL MQ groups exhibited higher autophagy signals than groups treated with NPs spiked in natural water samples. AgNPs, silver nanoparticles; ZnONPs, zinc oxide nanoparticles; 3,4-DCA, 3,4-dichloroaniline; hpf, hours post-fertilization.

Zebrafish embryotoxicity tests have been indicated as a suitable approach for assessing the toxicity of both traditional chemicals and ENPs [38,39]. Nonetheless, the majority of the published studies were conducted in the laboratory with controlled standard water samples. Here, we aimed to reveal the potential toxic effects and mechanisms of AgNPs and ZnONPs on zebrafish embryos in natural water. One of the interesting findings is that ZnONPs, but not AgNPs, triggered a significant delay in embryo hatching (Figure 5). Consistent with our finding, Chen. et al. reported that exposure to ZnONPs suspensions and their respective centrifuged supernatants caused similar hatching delays, whereas the supernatants did not cause larval mortality or malformation. In addition, coexposure to N-acetylcysteine (NAC), a well-known antioxidant, did not alter the effects of ZnONPs on hatchability but rescued their behavioral effect [40]. Thus, the toxicity of ZnONPs may be due to a combination of the effects of dissolved Zn ions and particle-induced oxidative stress. Zinc is an essential transition metal in living organisms that plays an important role in the maintenance of protein structure and enzymatic function. However, excessive free Zn ions are toxic and may be bound by Zn-binding proteins, such as metalloproteins [41]. The dissolved Zn ions interfere with embryo hatching through a chelator-sensitive mechanism that involves the ligation of histidines in the metalloprotease ZHE1, which is responsible for degradation of the chorionic membrane [42,43]. The effects of ZnONPs on delaying hatching were attenuated in filtered natural water samples compared with MQ water, suggesting the slow release of Zn ions from ZnONPs and the interaction of dissolved Zn ions with the complex components in the water environment, which subsequently mitigate effects on embryonic development. Surface coating with different chemicals or NOM may influence the colloidal stability and solubility of ZnONPs or AgNPs and thereby modulate toxicity [44].

Both ENP-induced mortality and developmental toxicity seem to be related to oxidative stress. Excess ROS production may contribute to tissue damage and participate in signal transduction, the proliferative response, gene expression, and protein redox regulation [45]. ENP-induced oxidative stress was proposed as one of the initiators of the disruption of the mitochondrial membrane potential, the induction of ER stress, and cell death mediated by apoptosis and/or autophagy [9]. The mechanisms underlying ENP-induced toxicity have become one of the most frequently studied topics in toxicology during the last two decades. Our previous studies were the first to show that autophagy activated by AgNPs fails to trigger the lysosomal degradation pathway and leads to dysfunctional autophagy, which is relevant to the accelerated cellular pathogenesis of diseases [10,45,46]. More recently, we also prioritized the factors affecting the toxic potential of AgNPs, which included exposure dose/time, cell type, and the size and surface coating of AgNPs. Using an in silico decision tree-based knowledge discovery-in-databases process, the toxicity-related parameters are ranked as follows: exposure dose > cell type > particle size > exposure time $\geq$ surface coating [47]. AgNPs with larger particle sizes appeared to induce higher levels of autophagy during the earlier phase of both subcytotoxic and cytotoxic exposures in the in vitro cell culture models, whereas apoptosis, but not necrosis, accounted for the compromised cell survival over the same dosage range [47]. In addition, we determined the skin toxicity and the potential mechanisms of ZnONPs combined with UVB exposure and the preventive effect of a well-known antioxidant, pterostilbene. Exposure to both ZnONPs and UVB disrupts cellular autophagy, which in turn increases exosome release from cells. Application of the antioxidant pterostilbene reversed autophagy abnormalities by restoring normal autophagy flux and decreasing NLRP3 inflammasome-loaded exosome release through the attenuation of total ROS and mitochondrial ROS levels [11]. In general, autophagy is a cellular recycling pathway by which lysosomes degrade damaged organelles and/or proteins to maintain cellular homeostasis. However, ENPs have been proven to induce autophagic cell death in several cell types by interfering with autophagy flux and disrupting lysosomal function [48]. The leakage of lysosomal enzymes activates procaspases or damages the mitochondrial outer membrane to induce apoptosis. As shown in the present study, zebrafish embryos exposed to filtered natural water spiked with AgNPs or ZnONPs presented increased ROS levels, apoptosis, and lysosomal activity, an indicator of autophagy (Figures 7 and 8). To the best of our knowledge, the induction of autophagy in zebrafish embryos triggered by ENPs in natural water has seldom or never been reported previously. As human being and ecosystem exposure to ENPs is unavoidable, an in-depth understanding of ENP-modulated autophagy is required to assess their safety [48].

The existing literature on the embryotoxicity and teratogenicity of ENPs in zebrafish has been reported in a recent review article [39]. The interaction and bioaccumulation of ENPs in zebrafish embryos are associated with several toxic effects, such as delayed hatching, yolk sac alterations, circulatory changes, and musculoskeletal disorders. In addition, the toxic effects of ENPs on innate immunity in a zebrafish model have also been reported [49]. Most of the abovementioned toxic effects are related to dysregulated autophagy. Since autophagy is considered an early indicator of ENP interactions with cells and has been recognized as an important form of cell death in ENP-induced toxicity, creating an autophagy-related transgenic zebrafish line could be a good approach to monitor the ENP pollution in an ecosystem. Overall, our study revealed that AgNPs and ZnONPs spiked in natural water increased zebrafish embryo mortality at higher concentrations, delayed the hatching rate, and induced ROS production, autophagy, and apoptosis (Figure 9). Our current study focused on AgNPs and ZnONPs, which are widely used in several industries, and described their behavior, characteristics, embryotoxicity, and underlying mechanisms in natural aquatic systems. These results will enable the development of more relevant testing methods to predict the possible long-term ecotoxicity of ENPs and can be applied in the future for regulatory decision-making and risk assessments of ENPs.

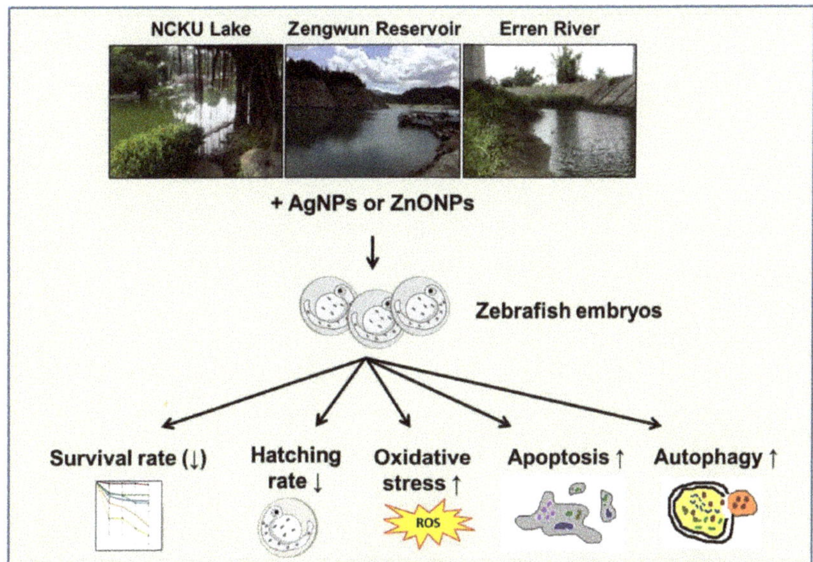

Figure 9. Illustration of the toxic mechanisms of AgNPs and ZnONPs spiked into natural water. The NP suspension was spiked in natural water obtained from NCKU Lake, Zengwun Reservoir, and Erren River and caused lethality and developmental toxicity in embryos. Mechanistically, AgNPs and ZnONPs spiked in natural water induced excessive ROS production, programmed cell death, and overactivated autophagy. AgNPs, silver nanoparticles; ZnONPs, zinc oxide nanoparticles.

5. Conclusions

Our research confirmed that nanoparticles of AgNPs and ZnONPs spiked in NCKU Lake, Zengwun Reservoir, and Erren River cause minor toxic effects. We speculated that AgNPs and ZnONPs spiked in the natural water had a stronger aggregation and changed their physicochemical properties by interacting with the surrounding environment, which consequently mitigated their toxic effects. However, AgNPs and ZnONPs spiked in the natural water increased the mortality of zebrafish embryos and delayed hatching time in higher concentrations, as well as induced ROS, autophagy, and apoptosis.

Supplementary Materials: The following supporting information can be downloaded at: https://www.mdpi.com/article/10.3390/nano12040717/s1, Figure S1: Survival rate of zebrafish embryos exposed to original water. (**a**) The survival rate of embryos treated with original water of NCKU Lake, Zengwun Reservoir, and Erren River. Survival rates of each natural water group were high. Four mg/L 3,4-dichloroaniline (3,4 DCA) was the positive control of acute embryo toxicity assay. Survival rate of zebrafish embryos exposed to original water after passing through 1 μm, 0.45 μm, 0.22 μm, and 0.1 μm filters of (**b**) NCKU Lake, (**c**) Zengwun Reservoir, and (**d**) Erren River at 0, 24, 48, 72, and 96 hpf. hpf, hours post-fertilization; Figure S2: Body length of zebrafish embryos in original water. Body length of zebrafish embryos exposed to original water after passing through 1 μm, 0.45 μm, 0.22 μm, and 0.1 μm filters with different natural water: (**a**) NCKU Lake, (**b**) Zengwun Reservoir, and (**c**) Erren River at 72 hpf. Body length was significantly longer than MQ water in each natural water (* $p \leq 0.05$). Body length of zebrafish embryos were measured by View 7 software. hpf, hours post-fertilization.

Author Contributions: Y.-L.L.: conceptualization, data curation, methodology, writing—original draft; Y.-S.S.: investigation, methodology; Z.-Y.C.: conceptualization, methodology, data curation, validation, visualization, writing—original draft; J.-Y.L.: formal analysis, investigation; F.-Y.C. and J.-Y.L.: formal analysis, investigation, Y.-H.W.: conceptualization, methodology, writing—review and editing; Y.-J.W.: conceptualization, writing—review and editing, funding acquisition, resources. All authors have read and agreed to the published version of the manuscript.

Funding: This study was supported by the Ministry of Science and Technology, Taiwan (MOST 109-2314-B-006-051-MY3), Toxic and Chemical Substances Bureau, Environmental Protection Administration, Executive Yuan, Taiwan (110A012) and the Environmental Analysis Laboratory, Environmental Protection Administration, Executive Yuan, Taiwan (Q106-P014).

Institutional Review Board Statement: Not applicable.

Informed Consent Statement: Not applicable.

Data Availability Statement: Not applicable.

Acknowledgments: We gratefully acknowledge the National Cheng Kung University Medical College Core Research Laboratory for administrative and technical support. We also thank the Taiwan Zebrafish Technology and Resource Center and the Center for Micro/Nano Science and Technology for technical help.

Conflicts of Interest: The authors declare no conflict of interest.

References

1. Chen, R.J.; Chen, Y.Y.; Liao, M.Y.; Lee, Y.H.; Chen, Z.Y.; Yan, S.J.; Yeh, Y.L.; Yang, L.X.; Lee, Y.L.; Wu, Y.H.; et al. The Current Understanding of Autophagy in Nanomaterial Toxicity and Its Implementation in Safety Assessment-Related Alternative Testing Strategies. *Int. J. Mol. Sci.* **2020**, *21*, 2387. [CrossRef] [PubMed]
2. Jahan, S.; Yusoff, I.B.; Alias, Y.B.; Bakar, A. Reviews of the toxicity behavior of five potential engineered nanomaterials (ENMs) into the aquatic ecosystem. *Toxicol. Rep.* **2017**, *4*, 211–220. [CrossRef] [PubMed]
3. Adeleye, A.S.; Keller, A.A.; Miller, R.J.; Lenihan, H.S. Persistence of commercial nanoscaled zero-valent iron (nZVI) and by-products. *J. Nanopart. Res.* **2013**, *15*, 1418. [CrossRef]
4. Gao, J.; Mahapatra, C.T.; Mapes, C.D.; Khlebnikova, M.; Wei, A.; Sepúlveda, M.S. Vascular toxicity of silver nanoparticles to developing zebrafish (*Danio rerio*). *Nanotoxicology* **2016**, *10*, 1363–1372. [CrossRef] [PubMed]
5. Ottofuelling, S.; Von der Kammer, F.; Hofmann, T. Commercial titanium dioxide nanoparticles in both natural and synthetic water: Comprehensive multidimensional testing and prediction of aggregation behavior. *Environ. Sci. Technol.* **2011**, *45*, 10045–10052. [CrossRef] [PubMed]
6. Espinasse, B.P.; Geitner, N.K.; Schierz, A.; Therezien, M.; Richardson, C.J.; Lowry, G.V.; Ferguson, L.; Wiesner, M.R. Comparative Persistence of Engineered Nanoparticles in a Complex Aquatic Ecosystem. *Environ. Sci. Technol.* **2018**, *52*, 4072–4078. [CrossRef]
7. Bind, V.; Kumar, A. Current issue on nanoparticle toxicity to aquatic organism. *MOJ Toxicol.* **2019**, *5*, 66–67. [CrossRef]
8. Chen, Z.Y.; Li, N.J.; Cheng, F.Y.; Hsueh, J.F.; Huang, C.C.; Lu, F.I.; Fu, T.F.; Yan, S.J.; Lee, Y.H.; Wang, Y.J. The Effect of the Chorion on Size-Dependent Acute Toxicity and Underlying Mechanisms of Amine-Modified Silver Nanoparticles in Zebrafish Embryos. *Int. J. Mol. Sci.* **2020**, *21*, 2864. [CrossRef]

9. Wu, Y.H.; Ho, S.Y.; Wang, B.J.; Wang, Y.J. The Recent Progress in Nanotoxicology and Nanosafety from the Point of View of Both Toxicology and Ecotoxicology. *Int. J. Mol. Sci.* **2020**, *21*, 4209. [CrossRef]
10. Mao, B.H.; Tsai, J.C.; Chen, C.W.; Yan, S.J.; Wang, Y.J. Mechanisms of silver nanoparticle-induced toxicity and important role of autophagy. *Nanotoxicology* **2016**, *10*, 1021–1040. [CrossRef]
11. Chen, Y.Y.; Lee, Y.H.; Wang, B.J.; Chen, R.J.; Wang, Y.J. Skin damage induced by zinc oxide nanoparticles combined with UVB is mediated by activating cell pyroptosis via the NLRP3 inflammasome-autophagy-exosomal pathway. *Part. Fibre Toxicol.* **2022**, *19*, 2. [CrossRef]
12. Abbas, Q.; Yousaf, B.; Fahad, A.; Ubaid Ali, M.; Munir, M.; El-Naggar, A.; Rinklebe, J.; Naushad, M. Transformation pathways and fate of engineered nanoparticles (ENPs) in distinct interactive environmental compartments: A review. *Environ. Int.* **2020**, *138*, 105646. [CrossRef] [PubMed]
13. Kansara, K.; Kumar, A.; Karakoti, A.S. Combination of humic acid and clay reduce the ecotoxic effect of TiO(2) NPs: A combined physico-chemical and genetic study using zebrafish embryo. *Sci. Total Environ.* **2020**, *698*, 134133. [CrossRef] [PubMed]
14. Mao, B.H.; Chen, Z.Y.; Wang, Y.J.; Yan, S.J. Silver nanoparticles have lethal and sublethal adverse effects on development and longevity by inducing ROS-mediated stress responses. *Sci. Rep.* **2018**, *8*, 2445. [CrossRef] [PubMed]
15. Attia, H.; Nounou, H.; Shalaby, M. Zinc Oxide Nanoparticles Induced Oxidative DNA Damage, Inflammation and Apoptosis in Rat's Brain after Oral Exposure. *Toxics* **2018**, *6*, 29. [CrossRef]
16. Kang, K.; Jung, H.; Lim, J.S. Cell Death by Polyvinylpyrrolidine-Coated Silver Nanoparticles is Mediated by ROS-Dependent Signaling. *Biomol. Ther.* **2012**, *20*, 399–405. [CrossRef]
17. Akter, M.; Sikder, M.T.; Rahman, M.M.; Ullah, A.; Hossain, K.F.B.; Banik, S.; Hosokawa, T.; Saito, T.; Kurasaki, M. A systematic review on silver nanoparticles-induced cytotoxicity: Physicochemical properties and perspectives. *J. Adv. Res.* **2018**, *9*, 1–16. [CrossRef]
18. Zhang, J.; Qin, X.; Wang, B.; Xu, G.; Qin, Z.; Wang, J.; Wu, L.; Ju, X.; Bose, D.D.; Qiu, F.; et al. Zinc oxide nanoparticles harness autophagy to induce cell death in lung epithelial cells. *Cell Death Dis.* **2017**, *8*, e2954. [CrossRef]
19. Keller, A.A.; McFerran, S.; Lazareva, A.; Suh, S. Global life cycle releases of engineered nanomaterials. *J. Nanopart. Res.* **2013**, *15*, 1692. [CrossRef]
20. Giese, B.; Klaessig, F.; Park, B.; Kaegi, R.; Steinfeldt, M.; Wigger, H.; von Gleich, A.; Gottschalk, F. Risks, Release and Concentrations of Engineered Nanomaterial in the Environment. *Sci. Rep.* **2018**, *8*, 1565. [CrossRef]
21. Bathi, J.R.; Moazeni, F.; Upadhyayula, V.K.K.; Chowdhury, I.; Palchoudhury, S.; Potts, G.E.; Gadhamshetty, V. Behavior of engineered nanoparticles in aquatic environmental samples: Current status and challenges. *Sci. Total Environ.* **2021**, *793*, 148560. [CrossRef] [PubMed]
22. Dale, A.L.; Casman, E.A.; Lowry, G.V.; Lead, J.R.; Viparelli, E.; Baalousha, M. Modeling nanomaterial environmental fate in aquatic systems. *Environ. Sci. Technol.* **2015**, *49*, 2587–2593. [CrossRef] [PubMed]
23. Xu, L.; Xu, M.; Wang, R.; Yin, Y.; Lynch, I.; Liu, S. The Crucial Role of Environmental Coronas in Determining the Biological Effects of Engineered Nanomaterials. *Small* **2020**, *16*, 2003691. [CrossRef] [PubMed]
24. Lowry, G.V.; Gregory, K.B.; Apte, S.C.; Lead, J.R. Transformations of Nanomaterials in the Environment. *Environ. Sci. Technol.* **2012**, *46*, 6893–6899. [CrossRef]
25. Batley, G.E.; Kirby, J.K.; McLaughlin, M.J. Fate and risks of nanomaterials in aquatic and terrestrial environments. *Acc. Chem. Res.* **2013**, *46*, 854–862. [CrossRef] [PubMed]
26. Natarajan, L.; Jenifer, M.A.; Mukherjee, A. Eco-corona formation on the nanomaterials in the aquatic systems lessens their toxic impact: A comprehensive review. *Environ. Res.* **2021**, *194*, 110669. [CrossRef]
27. Yu, S.J.; Yin, Y.G.; Chao, J.B.; Shen, M.H.; Liu, J.F. Highly Dynamic PVP-Coated Silver Nanoparticles in Aquatic Environments: Chemical and Morphology Change Induced by Oxidation of Ag0 and Reduction of Ag+. *Environ. Sci. Technol.* **2014**, *48*, 403–411. [CrossRef]
28. Han, Y.; Kim, D.; Hwang, G.; Lee, B.; Eom, I.; Kim, P.J.; Tong, M.; Kim, H. Aggregation and dissolution of ZnO nanoparticles synthesized by different methods: Influence of ionic strength and humic acid. *Colloids Surf. A Physicochem. Eng. Asp.* **2014**, *451*, 7–15. [CrossRef]
29. Cupi, D.; Hartmann, N.; Baun, A. The influence of natural organic matter and aging on suspension stability in guideline toxicity testing of Ag, ZnO, and TiO_2 nanoparticles with Daphnia magna. *Environ. Toxicol. Chem.* **2015**, *34*, 497–506. [CrossRef]
30. Peng, C.; Zhang, W.; Gao, H.; Li, Y.; Tong, X.; Li, K.; Zhu, X.; Wang, Y.; Chen, Y. Behavior and Potential Impacts of Metal-Based Engineered Nanoparticles in Aquatic Environments. *Nanomaterials* **2017**, *7*, 21. [CrossRef]
31. Lead, J.R.; Batley, G.E.; Alvarez, P.J.J.; Croteau, M.N.; Handy, R.D.; McLaughlin, M.J.; Judy, J.D.; Schirmer, K. Nanomaterials in the environment: Behavior, fate, bioavailability, and effects-An updated review. *Environ. Toxicol. Chem.* **2018**, *37*, 2029–2063. [CrossRef] [PubMed]
32. Wang, Z.; Zhang, L.; Zhao, J.; Xing, B. Environmental processes and toxicity of metallic nanoparticles in aquatic systems as affected by natural organic matter. *Environ. Sci. Nano* **2016**, *3*, 240–255. [CrossRef]
33. Zhao, J.; Wang, X.; Hoang, S.A.; Bolan, N.S.; Kirkham, M.B.; Liu, J.; Xia, X.; Li, Y. Silver nanoparticles in aquatic sediments: Occurrence, chemical transformations, toxicity, and analytical methods. *J. Hazard. Mater.* **2021**, *418*, 126368. [CrossRef] [PubMed]
34. Bundschuh, M.; Filser, J.; Lüderwald, S.; McKee, M.S.; Metreveli, G.; Schaumann, G.E.; Schulz, R.; Wagner, S. Nanoparticles in the environment: Where do we come from, where do we go to? *Environ. Sci. Eur.* **2018**, *30*, 6. [CrossRef] [PubMed]

35. Li, X.; Lenhart, J.J. Aggregation and Dissolution of Silver Nanoparticles in Natural Surface Water. *Environ. Sci. Technol.* **2012**, *46*, 5378–5386. [CrossRef]
36. Slomberg, D.L.; Ollivier, P.; Miche, H.; Angeletti, B.; Bruchet, A.; Philibert, M.; Brant, J.; Labille, J. Nanoparticle stability in lake water shaped by natural organic matter properties and presence of particulate matter. *Sci. Total Environ.* **2019**, *656*, 338–346. [CrossRef]
37. Dal, N.K.; Kocere, A.; Wohlmann, J.; Van Herck, S.; Bauer, T.A.; Resseguier, J.; Bagherifam, S.; Hyldmo, H.; Barz, M.; De Geest, B.G.; et al. Zebrafish Embryos Allow Prediction of Nanoparticle Circulation Times in Mice and Facilitate Quantification of Nanoparticle–Cell Interactions. *Small* **2020**, *16*, 1906719. [CrossRef]
38. Chakraborty, C.; Sharma, A.R.; Sharma, G.; Lee, S.S. Zebrafish: A complete animal model to enumerate the nanoparticle toxicity. *J. Nanobiotechnol.* **2016**, *14*, 1–13. [CrossRef]
39. Pereira, A.C.; Gomes, T.; Ferreira Machado, M.R.; Rocha, T.L. The zebrafish embryotoxicity test (ZET) for nanotoxicity assessment: From morphological to molecular approach. *Environ. Pollut.* **2019**, *252*, 1841–1853. [CrossRef]
40. Chen, T.H.; Lin, C.C.; Meng, P.J. Zinc oxide nanoparticles alter hatching and larval locomotor activity in zebrafish (*Danio rerio*). *J. Hazard. Mater.* **2014**, *277*, 134–140. [CrossRef]
41. Wu, W.; Bromberg, P.A.; Samet, J.M. Zinc ions as effectors of environmental oxidative lung injury. *Free Radic. Biol. Med.* **2013**, *65*, 57–69. [CrossRef] [PubMed]
42. Hansjosten, I.; Takamiya, M.; Rapp, J.; Reiner, L.; Fritsch-Decker, S.; Mattern, D.; Andraschko, S.; Anders, C.; Pace, G.; Dickmeis, T.; et al. Surface functionalisation-dependent adverse effects of metal nanoparticles and nanoplastics in zebrafish embryos. *Environ. Sci. Nano* **2021**, *9*, 375–392. [CrossRef]
43. Lin, S.; Zhao, Y.; Ji, Z.; Ear, J.; Chang, C.H.; Zhang, H.; Low-Kam, C.; Yamada, K.; Meng, H.; Wang, X.; et al. Zebrafish high-throughput screening to study the impact of dissolvable metal oxide nanoparticles on the hatching enzyme, ZHE1. *Small* **2013**, *9*, 1776–1785. [CrossRef] [PubMed]
44. Osborne, O.J.; Johnston, B.D.; Moger, J.; Balousha, M.; Lead, J.R.; Kudoh, T.; Tyler, C.R. Effects of particle size and coating on nanoscale Ag and TiO_2 exposure in zebrafish (*Danio rerio*) embryos. *Nanotoxicology* **2013**, *7*, 1315–1324. [CrossRef] [PubMed]
45. Lee, Y.H.; Cheng, F.Y.; Chiu, H.W.; Tsai, J.C.; Fang, C.Y.; Chen, C.W.; Wang, Y.J. Cytotoxicity, oxidative stress, apoptosis and the autophagic effects of silver nanoparticles in mouse embryonic fibroblasts. *Biomaterials* **2014**, *35*, 4706–4715. [CrossRef]
46. Lee, Y.H.; Fang, C.Y.; Chiu, H.W.; Cheng, F.Y.; Tsai, J.C.; Chen, C.W.; Wang, Y.J. Endoplasmic Reticulum Stress-Triggered Autophagy and Lysosomal Dysfunction Contribute to the Cytotoxicity of Amine-Modified Silver Nanoparticles in NIH 3T3 Cells. *J. Biomed. Nanotech* **2017**, *13*, 778–794. [CrossRef]
47. Mao, B.H.; Luo, Y.K.; Wang, B.J.; Cheng, F.Y.; Lee, Y.H.; Yan, S.J.; Wang, Y.J. Use of An In Silico Knowledge Discovery Approach To Translate Mechanistic Studies of Silver Nanoparticles-Induced Toxicity From In Vitro to In Vivo. *Part. Fibre Toxicol.* **2022**, *19*, 6. [CrossRef]
48. Guo, L.; He, N.; Zhao, Y.; Liu, T.; Deng, Y. Autophagy Modulated by Inorganic Nanomaterials. *Theranostics* **2020**, *10*, 3206–3222. [CrossRef]
49. Chen, R.J.; Huang, C.C.; Pranata, R.; Lee, Y.H.; Chen, Y.Y.; Wu, Y.H.; Wang, Y.J. Modulation of Innate Immune Toxicity by Silver Nanoparticle Exposure and the Preventive Effects of Pterostilbene. *Int. J. Mol. Sci.* **2021**, *22*, 2536. [CrossRef]

nanomaterials

MDPI

Article

Towards Standardization for Determining Dissolution Kinetics of Nanomaterials in Natural Aquatic Environments: Continuous Flow Dissolution of Ag Nanoparticles

Lucie Stetten, Aiga Mackevica, Nathalie Tepe, Thilo Hofmann and Frank von der Kammer *

Environmental Geosciences, Centre for Microbiology and Environmental Systems Science, University of Vienna, Althanstrasse 14, UZA II, 1090 Vienna, Austria; lucie.stetten@univie.ac.at (L.S.); aima@env.dtu.dk (A.M.); nathalie.tepe@univie.ac.at (N.T.); thilo.hofmann@univie.ac.at (T.H.)

* Correspondence: frank.von.der.kammer@univie.ac.at

Abstract: The dissolution of metal-based engineered nanomaterials (ENMs) in aquatic environments is an important mechanism governing the release of toxic dissolved metals. For the registration of ENMs at regulatory bodies such as REACH, their dissolution behavior must therefore be assessed using standardized experimental approaches. To date, there are no standardized procedures for dissolution testing of ENMs in environmentally relevant aquatic media, and the Organisation for Economic Co-operation and Development (OECD) strongly encourages their development into test guidelines. According to a survey of surface water hydrochemistry, we propose to use media with low concentrations of Ca^{2+} and Mg^{2+} for a better simulation of the ionic background of surface waters, at pH values representing acidic (5 < pH < 6) and near-neutral/alkaline (7 < pH < 8) waters. We evaluated a continuous flow setup adapted to expose small amounts of ENMs to aqueous media, to mimic ENMs in surface waters. For this purpose, silver nanoparticles (Ag NPs) were used as model for soluble metal-bearing ENMs. Ag NPs were deposited onto a 10 kg.mol^{-1} membrane through the injection of 500 μL of a 5 mg.L^{-1} or 20 mg.L^{-1} Ag NP dispersion, in order to expose only a few micrograms of Ag NPs to the aqueous media. The dissolution rate of Ag NPs in 10 mM $NaNO_3$ was more than two times higher for ~2 μg compared with ~8 μg of Ag NPs deposited onto the membrane, emphasizing the importance of evaluating the dissolution of ENMs at low concentrations in order to keep a realistic scenario. Dissolution rates of Ag NPs in artificial waters (2 mM $Ca(NO_3)_2$, 0.5 mM $MgSO_4$, 0–5 mM $NaHCO_3$) were also determined, proving the feasibility of the test using environmentally relevant media. In view of the current lack of harmonized methods, this work encourages the standardization of continuous flow dissolution methods toward OECD guidelines focused on natural aquatic environments, for systematic comparisons of nanomaterials and adapted risk assessments.

Keywords: engineered nanomaterials; flow-through dissolution testing; aquatic environments; OECD guidelines; environmental risk assessment

Citation: Stetten, L.; Mackevica, A.; Tepe, N.; Hofmann, T.; von der Kammer, F. Towards Standardization for Determining Dissolution Kinetics of Nanomaterials in Natural Aquatic Environments: Continuous Flow Dissolution of Ag Nanoparticles. *Nanomaterials* **2022**, *12*, 519. https://doi.org/10.3390/nano12030519

Academic Editor: Vivian Hsiu-Chuan Liao

Received: 19 December 2021
Accepted: 1 February 2022
Published: 2 February 2022

1. Introduction

The benefits related to the specific properties of engineered nanomaterials (ENMs) have led to an increase in their use and production, raising particular attention to their environmental behavior and toxicological impact. The wide disparity in estimates of industrial production of ENMs demonstrates an important uncertainty in the quantification of ENM discharges into the environment [1–3]. Nevertheless, early studies have predicted that both terrestrial and aquatic environments are likely to be exposed to ENMs [1,4]. In the case of European surface waters, modeled concentrations of nanoparticles (NPs) were found to exceed 0.18 ng.L^{-1} and 150 ng.L^{-1} for Ag and ZnO NPs, respectively, in the 10% most exposed rivers, likely corresponding to rivers located near large cities and downstream river systems [5]. ENMs can be transformed before entering the environment, for instance if they are released from wastewaters treatment plants [6]. Moreover, the discharge of pristine ENMs in natural environments should also be considered, since ENMs can be released through accidental events, weathering, or directly from the use of consumer products [7–10].

1.1. Dissolution of Nanoparticles in Aquatic Environments

Dissolution is emphasized as one of the most important mechanisms governing the fate of ENMs in aquatic environments. The oxidative dissolution of Ag NPs has been largely studied in aquatic media [11–18]. For example, Li and Lenhart, (2012) [18] showed that the dissolution of Ag NPs and the release of Ag^+ is significantly limited in natural river waters. It has been argued that aggregation of Ag NPs may hamper dissolution due to a substantial decrease in the specific surface area [19]. However, several studies found that dissolution of Ag NPs is dependent on the particle size rather than aggregation [14,20]. For regulatory testing of ENMs dissolution rates, this information is important since up to now it was seen as a requirement to bring ENMs into a stable dispersion for dissolution testing to avoid effects of agglomeration. Extrinsic parameters such as pH, ionic strength, inorganic/organic ligands, and temperature play also an important role. Gondikas et al. [21] demonstrated that dissolution of citrate- and PVP-capped Ag NPs increases in the presence of cysteine. In contrast, several studies showed that Ag NP dissolution is inhibited in the presence of humic and fulvic acids or in the presence of natural organic matter (NOM) with high sulfur and nitrogen content [17,22]. This points to a more complex process of ENMs dissolution than only the complexation of the released cations by NOM, lowering the free ion activity and thereby enhancing the apparent solubility and dissolution rate. Finally, pH is also an important factor controlling the dissolution of Ag NPs, being enhanced at low pH values [23–25].

For soluble ENMs such as Ag NPs, it has been established that the aqueous forms (e.g., dissolved ions or aqueous complexes) are to a large extent causing the toxic effects exerted by these ENMs [26–29]. Aqueous solubility and dissolution kinetics are thus essential for (eco-)toxicological hazard testing and risk assessment strategies and are a requested information for registration of ENMs at regulatory bodies such as the European Chemicals Agency (ECHA) [30]. Determining the dissolution rates of ENMs is particularly important in determining risk/hazard since the rate of release of toxic ions prior to interaction with ligands may be more important than equilibrium concentrations, and the lifetime of the particulate form will determine the exposure to this ENM. However, high degree of uncertainty remains when it comes to understanding dissolution kinetics of ENMs in aquatic environments, mostly due to the lack of studies using aqueous media representing the heterogeneity of the world's aquatic environments, and adapted experimental setups.

1.2. Experimental Methods to Study the Dissolution of Nanoparticles in Aquatic Environments

When it comes to understanding the fate of ENMs in aquatic environments, a limiting factor is the lack of standardized experimental procedures to study the specific endpoints (solubility, dissolution rate, chemical transformation, hetero-agglomeration). Standardized procedures would allow easier comparison between different ENMs and forms of the same ENM, enable grouping and read-across between nanoforms, as well as the production of more reliable data that can be implemented in geochemical models. International regulatory bodies, such as ECHA, the Organization for Economic Cooperation and Development (OECD) and the International Organization for Standardization (ISO), have argued that these points should be addressed [31]. To date, several methods have been applied for dissolution testing of ENMs [16]. However, there is no standardized OECD guideline for dissolution testing of ENMs other than the guidance in OECD GD 318 [32,33]. The OECD TG 105 [34] has been evaluated as not applicable for ENMs [35] and OECD GD 29 has been identified as being in principle applicable with some adjustments for determining solubility [36,37]. Nanomaterial-specific OECD test guidelines for dissolution rate testing in environmental media would thus be beneficial for generating data that are regulatory relevant and reliable [38].

Generally, dissolution-testing methods can be divided into static batch and continuous flow/flow-through approaches. Batch testing leads to a direct measure of the ENM solubility in a specific aqueous medium. Moreover, when using ultra-trace analytical techniques, it is possible to measure very low ion concentrations. However, it has several limitations, especially when dealing with particles that are rapidly dissolving. For instance, the ultra-centrifugation or centrifugal ultrafiltration of rapidly dissolving particles can lead to an overestimation of the dissolution, because of the dissolution of the particles during the centrifugation process. To the opposite, the dissolved fraction could be underestimated using filtration techniques if the dissolved ions are adsorbed to the ultrafiltration membrane or the walls of the filtration device. Furthermore, batch experiments are performed under static conditions (i.e., limited supply of exposure medium) and may lead to a change of the medium composition, due to ENMs dissolution and/or the re-precipitation of secondary species. The challenges associated with the batch systems can be overcome by using a flow-through/continuous flow setup. The continuous flow dissolution systems are considered to be more appropriate for dissolution rate measurements and present a more environmentally realistic system. It allows for working with high liquid-to-solid ratios (realistic high dilution of the ENM), and when using appropriate analytical instrumentation, it allows for the detection of low levels of dissolved species. Because of a constant and almost unlimited provision of the medium, the back reactions are limited and low electrolyte concentrations in the exposure medium can be used to mimic the chemical composition of natural waters. These results can finally serve as inputs into environmental fate modeling. The flow-through setup has been described in ISO TR 19057 [39]. For instance, Koltermann-Juelly et al. [40] studied the dissolution rates of 24 types of nanomaterials in phagolysosomal simulant fluids, and Bove et al. [31] investigated dissolution of Ag NPs in an in vitro assay simulating conditions which likely occur in human digestion. More recently, Keller et al. [41] applied continuous flow systems to study the dissolution of $BaSO_4$ in phagolysosomal and lung lining fluids. These studies demonstrated the advantage of using continuous flow rather than static incubations to investigate and compare the dissolution rates of ENMs. Nevertheless, when adapted to simulate ENMs dissolution in specific biological media, these studies applied relatively low flow rates and high loadings of ENMs, which resemble elevated solid-to-liquid ratios and would not be translatable to environmentally relevant conditions.

Here, we aim to provide guidance and recommendations for nanomaterials dissolution testing and dissolution rates determination under environmentally relevant conditions. Based on a survey of surface water hydrochemistry, we outline the environmental concentration ranges of key parameters to be considered in studying ENMs dissolution in natural aquatic environments. We present the proof of concept of a continuous flow setup suitable to investigate dissolution kinetics at low ENMs concentrations, relevant for natural environments. The results obtained on Ag NPs exposed to artificial waters demonstrate the suitability of the continuous flow dissolution method to determine dissolution rates of ENMs under non-equilibrium environmentally relevant conditions. This work will aid in the standardization of a continuous flow dissolution approach adapted to study ENMs dissolution kinetics in natural aquatic environments, and supports the ongoing development towards OECD dissolution guidelines.

2. Materials and Methods

2.1. Continuous Flow Testing

Our experimental setup consisted of one 10 kDa Hydrosart® regenerated cellulose membrane of 25 mm diameter (Sartorius Stedim Biotech GmbH, Goettingen, Germany) enclosed into a metal-free polyether ether ketone (PEEK) filter holder (Wyatt Technology Europe GmbH, Dernbach, Germany). After deposition of the NPs onto the membrane, a high-performance liquid chromatography (HPLC)-type piston pump (Postnova Analytics

GmbH, Landsberg am Lech, Germany) delivered the continuous flow of medium into the ultrafiltration cell at rates between 0.2 to a few mL.min^{-1} (Figure 1). The exposed solution was then collected at the outlet using an auto-sampler and analyzed for dissolved elemental concentrations. This setup allows the direct separation between the nanoparticulate and dissolved fractions, which are defined by the nominal cut-off of the ultrafiltration membrane, here equal to 10 kDa (10.000 g.mol^{-1}), which corresponds to a spherical particle of ~3 nm diameter. Nanoparticles are deposited onto the membrane filter through the injection of a NPs dispersion. The number of nanoparticles deposited at the surface of the membrane is controlled by the concentration of the NPs suspension and the volume injected. In contrast to previous continuous flow dissolution testing, where milligrams of NPs powder are loaded between two membranes, such a method allows to inject smaller amount of ENMs and to have a visible homogeneous coverage of ENMs at the surface of the membrane [10,40]. In particular, a PTFE tube was used as an injection loop for all experiments, in order to inject a volume of 500 µL of the NPs dispersion. This injection loop was connected to the flow-through setup with an injection valve (Figure 1). At the beginning of the experiment, the flow-through system was started in order to let the eluent solution flush the whole system. After a few minutes, the injection loop was connected to the system in order to inject the Ag NP dispersion.

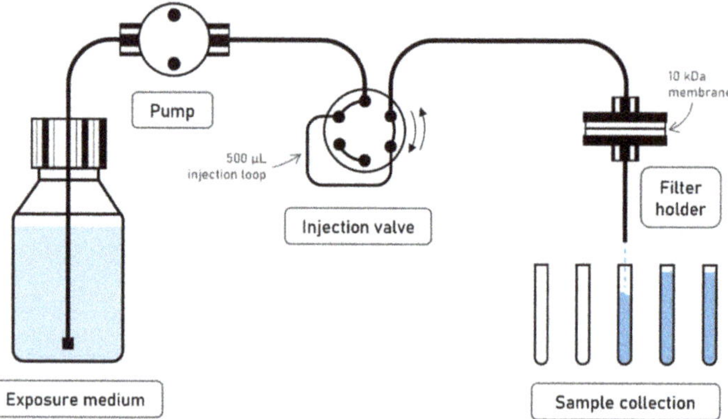

Figure 1. Representation of the setup used for the flow-through dissolution experiments.

2.2. Survey of Surface Water Hydrochemistry

In order to define a media composition more representative of surface aquatic environments, a survey on selected parameters known to influence the dissolution of ENMs was performed for several river systems. Specifically, pH, Ca^{2+}, Mg^{2+}, dissolved organic carbon (DOC), orthophosphates, and conductivity were chosen as key factors [21–23,42–44]. Data for the Danube River were extracted from the TransNational Monitoring Network (TNMN) dataset of the International Commission for the Protection of the Danube River (ICPDR) database [45]. For the Rhine river, data were extracted from the FGG Rhein database for two measuring stations, Karlsruhe and Bad Honnef, located along the Rhine main tributary [46]. Data for the Elbe River were obtained from the specialized information system (FIS) of the FGG Elbe database which have been collected at important measuring stations in the area of the Elbe catchment within the national measuring programs [47]. For these databases, values from 2015 to 2017 were extracted, corresponding to specific sampling locations and times [45–47]. In addition, data of the same parameters were extracted from the Forum of European Geological Surveys (FOREGS) [48] and the European Environment Agency (EEA) [49] databases in order to represent a large and global pool of European surface waters. From these data sets, minimum and maximum values, and median, first (Q1), and third (Q3) quartiles were calculated and plotted.

2.3. Experimental Conditions

Dissolution tests were performed using spherical 80 nm citrate-coated Ag NPs (NanoXact^{TM}) obtained from Nanocomposix, San Diego, CA, USA (JRD0035). In order to assess the feasibility of the flow-through dissolution method, a set of experiments were performed at pH 5 with 10 mM $NaNO_3$ (Merck, Darmstadt, Germany) as exposure medium. These experiments were performed in order (1) to define optimal experimental parameters for the test and (2) to investigate the influence of extrinsic parameters (i.e., injection velocity, flow rate and ENMs loading). Table 1 summarizes the experimental conditions tested for the dissolution experiments. Ag_0.2mL/min_2.2µg_$NaNO_3$ and Ag_0.2mL/min_8.2µg_$NaNO_3$ were performed with 2.2 and 8.2 µg of Ag NPs loaded onto the filter, corresponding to 2.99 % and 0.91 % of the filter area covered by Ag NPs, respectively. For these experiments, the injection flow rate was set to 1 mL.min^{-1} during 1 h. After injection, the flow rate was reduced to 0.2 mL.min^{-1} until the end of the experiment. Ag_0.5mL/min_8.2µg_$NaNO_3$ was performed with 8.2 µg of Ag NPs loaded onto the membrane at a flow rate of 0.5 mL.min^{-1}, constant during all the time of the experiment. Ag_0.2mL/min_2.2µg_$NaNO_3$, Ag_0.2mL/min_8.2µg_$NaNO_3$ and Ag_0.5mL/min_8.2µg_$NaNO_3$ were performed in duplicates. All other experiments were performed with an Ag NPs loading of 8.2 µg and a constant flow rate of 0.5 mL.min^{-1}. For the latter, 2 mM $Ca(NO_3)_2$ and 0.5 mM $MgSO_4$ solutions (Merck, Darmstadt, Germany) were used in order to have a Ca^{2+}/Mg^{2+} ratio of 4:1, as recommended by the OECD guideline 318 [33]. To simulate acidic surface waters (ASW), Ag_0.5mL/min_8.2µg_ASW was performed at pH 5. The pH of the eluent was adjusted using 0.2 M HNO_3. To simulate near-neutral surface waters (NSW), Ag_0.5mL/min_8.2µg_NSW was performed at pH 7.5 by adding 5 mM $NaHCO_3^-$ to the $Ca(NO_3)_2$-$MgSO_4$ solution, acting as pH buffer. As part of this work, no DOC and orthophosphate were included, the aim being to show a proof of concept of the continuous flow dissolution setup instead of discussing the impact of ligands on the dissolution kinetic of Ag NPs.

Table 1. Experimental parameters and media composition used to perform the flow-through experiments.

	Ag_0.2mL/min _2.2µg_ $NaNO_3$	**Ag_0.2mL/min _8.2µg_ $NaNO_3$**	**Ag_0.5mL/min _8.2µg_ $NaNO_3$**	**Ag_0.5mL/min _8.2µg_ ASW**	**Ag_0.5mL/min _8.2µg_ NSW**
Setup parameters					
[Ag NPs][1] [mg.L^{-1}]	4.3	16.4	16.4	16.4	16.4
Volume injected [µL]	500	500	500	500	500
Ag NPs loading [µg]	2.2	8.2	8.2	8.2	8.2
Injection time [h]	1	1	-	-	-
Injection flow rate [mL.min^{-1}]	1	1	-	-	-
Flow rate experiment [mL.min^{-1}]	0.2	0.2	0.5	0.5	0.5
Exposure media					
pH	5	5	5	5	7.5
$NaNO_3$ [mM]	10	10	10	-	-
$Ca(NO_3)_2$ [mM]	-	-	-	2	2
$MgSO_4$ [mM]	-			0.5	0.5
HCO_3^- [mM]	-	-	-	-	5

[1] the Ag NPs concentration injected was determined by acid digestion of 500 µL of the 20 mg.L^{-1} Ag NPs stock solution. For Ag_0.2mL/min_2.2µg_$NaNO_3$, the volume of the 20 mg.L^{-1} Ag NPs stock solution used to prepare the 5 mg.L^{-1} solution was acid digested and Ag NPs concentration was thus assume from the dilution factor.

2.4. Samples Measurement and Data Treatment

For all experiments, samples were taken at 5, 10, 15, or 20 min intervals directly into 15 mL PP vials, acidified with 20 µL of 65%HNO_3 (AnalaR Normapur®, VWR, Austria), and analyzed by inductively coupled plasma-mass spectrometry (Agilent 7900 ICP-MS, Agilent Technologies, Tokyo, Japan) for dissolved Ag concentrations. The ICP-MS was equipped with a Ni cone, a MicroMist nebulizer and a scott-type double pass spray chamber. It was operated on standard mode, under continuous Ar gas flow (UHP, Nebulizer gas flow rate at 0.8 $L.min^{-1}$; Dilution gas flow rate of 0.4 $L.min^{-1}$). The two stable isotopes ^{107}Ag and ^{109}Ag were measured for higher accuracy, and Rhodium (Rh, solution of 10 $\mu g.L^{-1}$) was used as internal standard. The ICP-MS was calibrated with dissolved Ag standards ranging from 5 $ng.L^{-1}$ to 50 $\mu g.L^{-1}$ prepared from a single-element Ag standard (1000 $\mu g.mL^{-1}$; CGAG1−125 ml, Inorganic Ventures, Christiansburg, VA, USA) diluted with 2% HNO_3. The acid blanks for all measurement times show an averaged limit of detection (LOD; 3× standard deviation + mean) of 0.05 $\mu g.mL^{-1}$and an averaged limit of quantification (LOQ; 10× standard deviation + mean) of 0.12 $\mu g.mL^{-1}$. Measurements were carried out in triplicates.

Experiment Ag_0.2mL/min_2.2µg_NaNO3 was performed for 5 h. Ag_0.2mL/min_8.2µg_NaNO3, Ag_0.5mL/min_8.2µg_NaNO3, Ag_0.5mL/min_8.2µg_ASW and Ag_0.5mL/min_8.2µg_NSW were performed for 8 h. For each experiment, apparent dissolution rates k were calculated from the outflow concentrations using Equation (1):

$$k = [Ag]_{outlet} \times F/SA \quad (1)$$

where k in $\mu g.s^{-1}.m^{-2}$, $[Ag]_{outlet}$ the Ag concentration measured at the outlet in $\mu g.mL^{-1}$, F the flow rate of the eluent in $mL.s^{-1}$, and SA the combined surface area of the NPs deposited on the membrane in m^2. SA was determined considering the surface area of an 80 nm diameter spherical NP, NP_{SA} and the total number of NPs deposited onto the membrane, $NPs_{membrane}$ ($SA = NP_{SA} \times NPs_{membrane}$).

3. Results and Discussion

3.1. Hydrochemical Conditions to Investigate ENMs Dissolution in Natural Aquatic Media

Studies on solubility and dissolution kinetics are usually performed using deionized water or simple background electrolyte solutions (e.g., NaCl, $NaNO_3$ or $Ca(NO_3)_2$) for specific pH values [15,16]. However, the lack of complexity in the exposure media is not suited to mimic ENMs dissolution in natural waters, which depends on an interplay between intrinsic properties and extrinsic environmental parameters, unique to each environment. Few variations are observed within one river system, such as the Danube and Rhine rivers, and to a larger extent for the Elbe River. In contrast, the FOREGS database, issued from the Geochemical Atlas of Europe [48], presents a wide range of values (Figure 2) covering the diversity of various catchments on the continent. The FOREGS database is, to date, the most relevant database to define media composition, mimicking a large pool of natural surface waters, from alkaline rivers such as Danube and Rhine (Figure 2) to more acidic waters mostly present in base-poor buffering capacity regions and/or organic-rich acid buffering regions. However, it does not provide data on all chemical components, such as orthophosphates. For this study, orthophosphate concentrations that are representative of a pool of rivers were obtained from the European Environment Agency (EEA) database "Waterbase-Water Quality" [49].

As recommended by OECD GD318 for the testing of nanomaterials dispersion stability [32], exposure media harboring low concentrations of the major ions reported in surfaces waters, Ca^{2+} and Mg^{2+}, would allow a better simulation of the ionic background of surface waters [48]. In addition to standard tests performed using 10 mM $Na(NO_3)$, we thus propose to use 2 mM of $Ca(NO_3)_2$ and 0.5 mM of $MgSO_4$ to mimic a more realistic scenario of natural waters. Such values are based on Ca^{2+} and Mg^{2+} concentrations reported by the FOREGS database, with average concentration of 1.4 mM and 0.5 mM for Ca^{2+} and Mg^{2+}, respectively [48]. To simulate alkaline surface waters (pH 7–8), 2 to 5 mM HCO_3^- should be added to the exposure media.

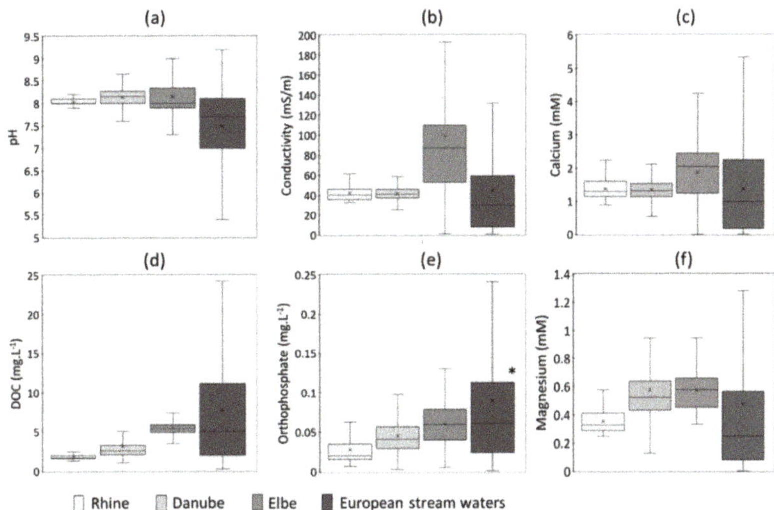

Figure 2. Boxplots illustrating the range values of (**a**) pH, (**b**) conductivity, (**c**) calcium concentration, (**d**) DOC, (**e**) orthophosphate and (**f**) magnesium concentrations in specific and a pool of surface waters. Data from 2015 to 2017 were obtained from The International Commission for the Protection of the Danube River (ICPDR) [45], the River Basin Communities of Rhine [46] and the Elbe Data Information System (FIS) of the River Basin Community [47] for the Danube, Rhine, and Elbe rivers, respectively. Data from the Rhine River were obtained at the locations Bad Honnef and Karlsruhe. The FOREGS (Forum of European Geological Surveys) Geochemical database [48] and the European Environment Agency (EEA) database were used to represent European stream waters. * Data obtained from the EEA database "Waterbase-Water Quality" [49].

For all databases, phosphate concentrations are significant, with values ranging between 0.001 to 0.190 mg.L^{-1} (Figure 2). Regarding the expected concentrations of ENMs into aquatic environments, for example a maximum of 150 ng.L^{-1} for ZnO NPs [5], phosphate can thus play an important role by initiating transformation or formation of lower soluble metal-phosphate coatings. Similarly, NOM in surface waters (Figure 2) might play an important role, enhancing or inhibiting dissolution of ENMs [21,22,50]. Thus, for a more realistic scenario, test medium should also include NOM and inorganic phosphate. We should, however, point out the importance of separating two pivotal processes in environmental aquatic media: dissolution and chemical transformation. Testing using a complex aquatic chemistry (i.e., sulfide, Cl^-, PO_4^{3-}, HCO_3^-, DOC) might trigger ENM transformations into less soluble phases [51–53] and affect the direct evaluation of the solubility and dissolution rate. Determination of the dissolution rates in a medium that represents natural conditions but does not induce other reactions is then necessary. In this regard, precipitation of secondary species must be preliminarily predicted using thermodynamic modeling.

3.2. Dissolution of Ag NPs in Simple Background Electrolyte: Validation of the Continuous Flow System for Low Ag NPs Loadings

For all experiments performed at pH 5 with 10 mM $NaNO_3$ as the eluent, higher dissolved Ag concentrations were measured at the beginning of the exposure experiments (Figure 3). During the first minutes of the exposure, the dissolution behavior of the Ag NPs might be governed by the positioning of the particles in the filter holder. Since the particles are not immediately fixed onto the membrane, they may remain dispersed in the filter holder's dead volume above the membrane, increasing the contact time with the eluent and consequently the Ag concentration. Similarly, the increase in Ag concentrations after 60 min exposure for the experiment Ag_0.2mL/min_2.2µg_$NaNO_3$ and Ag_0.2mL/min_8.2µg_$NaNO_3$ correspond to an increase in contact time with the eluant due to the decrease in the flow rate from 1 mL.min^{-1} to 0.2 mL.min^{-1}. The presence of dissolved Ag^+ in the injected Ag suspension and sorbed Ag^+ at the surface of the NPs could also explain higher Ag concentrations at the beginning of the experiments.

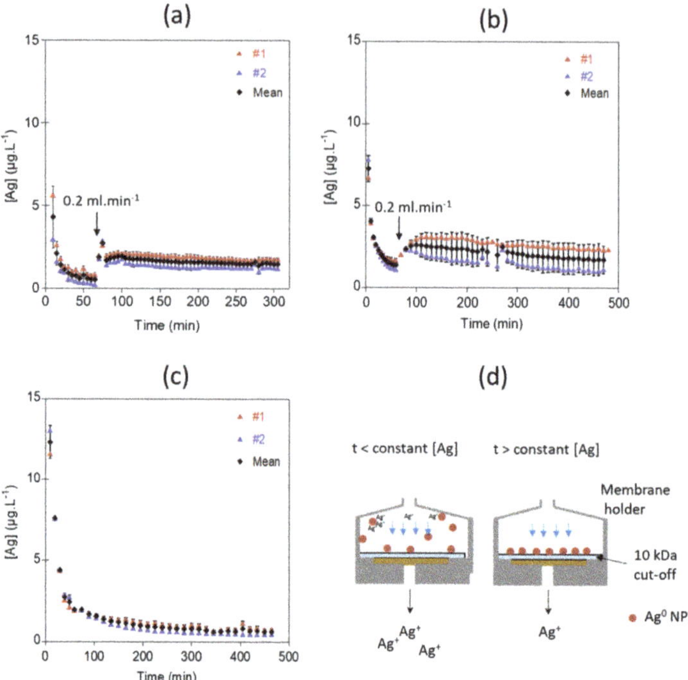

Figure 3. Ag concentrations measured at the outlet of the flow-through setup for (**a**) 500 µL of a 4.3 mg.L^{-1} Ag NPs suspension injected with a flow rate of 1 mL.min^{-1} for 1 h and set to 0.2 mL.min^{-1} for the rest of the experiment, Ag_0.2mL/min_2.2µg_$NaNO_3$ experiment (**b**) for 500 µL of a 16.4 mg.L^{-1} Ag NPs suspension injected with a flow rate of 1 mL.min^{-1} for 1 h and set to 0.2 mL.min^{-1} for the rest of the experiment, Ag_0.2mL/min_8.2µg_$NaNO_3$ experiment and (**c**) for 500 µL of a 16.4 mg.L^{-1} Ag NPs suspension injected with a flow rate set to 0.5 mL.min^{-1} during all the time of the experiment, Ag_0.5mL/min_8.2µg_$NaNO_3$ experiment. (**d**) Conceptual illustration of the behavior of Ag NPs in the filter holder during the flow-through experiment.

Once the particles are deposited onto the membrane and are exposed to a continuous and unchanged flow rate, outflow Ag concentration becomes stable (Figure 3). The apparent dissolution rates calculated from the Ag concentrations (Figure 4) reached a steady state after 225 min for Ag_0.2mL/min_2.2µg_$NaNO_3$, 400 min for Ag_0.2mL/min_8.2µg_$NaNO_3$, and 345 min for Ag_0.5mL/min_8.2µg_$NaNO_3$. The average Ag concentrations and averaged dissolution rates were calculated on the steady state range. Experiments performed with 8.2 µg of Ag NPs loaded onto the membrane at two different flow rates show similar averaged dissolution rates (k = 0.10 ± 0.002 µg.m^{-2}.s^{-1} and k = 0.093 ± 0.009 µg.m^{-2}.s^{-1} for Ag_0.2mL/min_8.2µg_$NaNO_3$, and Ag_0.5mL/min_8.2µg_$NaNO_3$, respectively). Nevertheless, for the experiment performed at a flow rate of 0.2 mL.min^{-1}, the averaged Ag concentration at the outlet was 2.7 times higher, due to a longer contact time of the eluent with Ag NPs (Figures 3 and 4d). Theoretically, lower flow rates would allow more contact time between the eluent and the surface of the NPs, leading to higher Ag concentrations in the reaction zone and result in a larger diffusion boundary layer, both limiting the dissolution process. The study of Keller et al. [54] illustrates well the effect of the flow-rate on the dissolution kinetics of nanomaterials such as $BaSO_4$ NPs, by performing long term experiments with higher NPs loadings and flow rate ramping between 0.1 and 3.0 mL.h^{-1}. Dissolution rates are not much different between the experiments Ag_0.2mL/min_8.2μg_$NaNO_3$, and Ag_0.5mL/min_8.2μg_$NaNO_3$. However, it is likely that at higher flow rates a more pronounced difference will be observed. However, for highly soluble ENMs such as CuO and ZnO NPs, the influence of the flow rate on dissolution might be reduced [54].

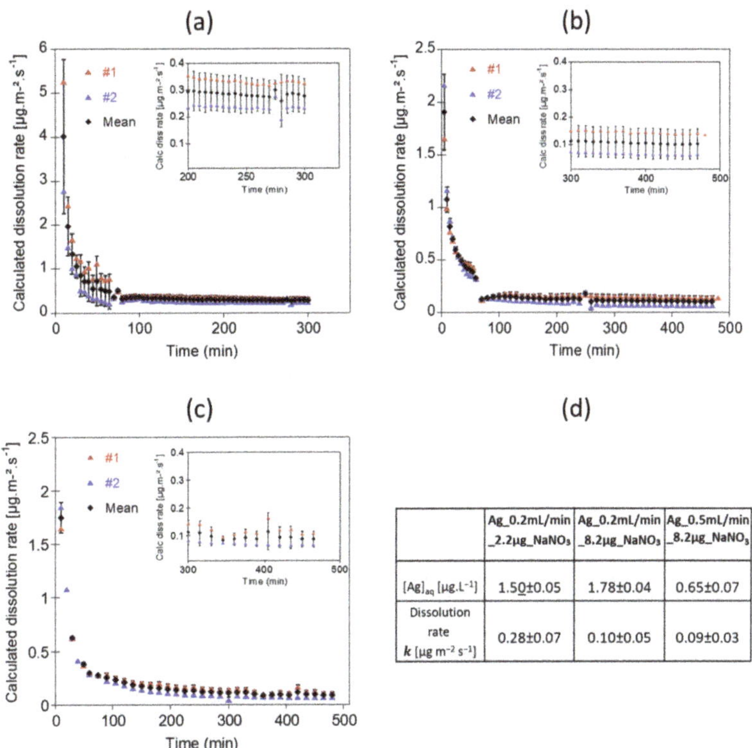

	Ag_0.2mL/min _2.2μg_$NaNO_3$	Ag_0.2mL/min _8.2μg_$NaNO_3$	Ag_0.5mL/min _8.2μg_$NaNO_3$
$[Ag]_{aq}$ [μg.L^{-1}]	1.50±0.05	1.78±0.04	0.65±0.07
Dissolution rate ***k*** [μg m^{-2} s^{-1}]	0.28±0.07	0.10±0.05	0.09±0.03

Figure 4. Dissolution rates of 80 nm Ag NPs in 10 mM $NaNO_3$ at pH 5, for different particles loadings and flow rates. (**a**) Dissolution rates obtained for the Ag_0.2mL/min_2.2μg_$NaNO_3$ experiment corresponding to 2.2 μg Ag NPs exposed at a flow rate of 1 mL.min^{-1} for 1 h and 0.2 mL.min^{-1} for the rest of the experiment. (**b**) Dissolution rates obtained for Ag_0.2mL/min_8.2μg_$NaNO_3$ experiment corresponding to 8.2 μg Ag NPs exposed at a flow rate of 1 mL.min^{-1} for 1 h and 0.2 mL.min^{-1} for the rest of the experiment. (**c**) Dissolution rates obtained for Ag_0.5 mL/min_8.2μg_$NaNO_3$ experiment corresponding to 8.2 μg Ag NPs exposed at a flow rate of 0.5 mL.min^{-1}. (**d**) Table showing the average [Ag] concentrations and the averaged dissolution rates calculated on the steady state ranges.

Ag NPs loading appear to influence significantly the dissolution rate of Ag NPs when applying the same 0.2 mL.min^{-1} flow (experiments Ag_0.2mL/min_2.2μg_$NaNO_3$ and Ag_0.2mL/min_8.2μg_$NaNO_3$, Figure 4d). The dissolution rate was more than two times higher for the experiment performed with 2.2 μg of Ag NPs (Ag_0.2mL/min_2.2μg_$NaNO_3$, k = 0.28 ± 0.07 μg.m^{-2}.s^{-1}) compared with the experiment performed with 8.2 μg of Ag NPs (Ag_0.2mL/min_8.2μg_$NaNO_3$, k = 0.10 ± 0.05 μg.m^{-2}.s^{-1}). A high particle loading increases the dissolved ion concentration in the vicinity of the particle surfaces, limiting their dissolution. To avoid local saturation around the nanoparticles, a higher flow rate might be used. Nevertheless, the dissolution rates determined at a flow rate of 0.5 and

0.2 $mL.min^{-1}$ are similar (k = 0.09 ± 0.03 compared with k = 0.10 ± 0.05 $\mu g.m^{-2}.s^{-1}$, respectively, both at 8.2 μg loading and with $NaNO_3$). This indicates that the dissolution rate of Ag NPs depends on the initial Ag NPs concentration. Concentration-dependent dissolution of Ag NPs was already reported by several studies, showing higher dissolution rates for lower Ag NPs concentrations [55,56]. Keller et al. [54] also highlighted the influence of the initial NPs loading on the dissolution rate for $BaSO_4$, CuO, ZnO and TiO_2 NPs. Such findings emphasize the importance to investigate ENMs dissolution at environmentally relevant ENMs concentrations. Indeed, performing flow through dissolution experiments using high initial NPs loading may result in wrong assessments of the dissolution rate of an ENM in natural aquatic systems. We may also hypothesize that a larger amount of Ag NPs injected does not result in the deposition of Ag NPs as a monolayer.

The disparity between the dissolution rates calculated for each experiment performed using the same Ag NPs and the same exposure medium demonstrated that the dissolution kinetic of Ag NPs is dependent on the initial NPs concentration. The flow rate is also an important parameter to adjust for the success of the test and an appropriate representation of the system to mimic. A low flow rate allows for longer interaction between the medium and the ENMs, which would lead to a more reliable measurement of the dissolved fraction at the outlet. Suitable for ENMs with low solubility, it may also result in local saturation and underestimation of the dissolution rate. Experimental parameters (i.e., flow rate, exposure time and ENMs loading) need to be adjusted to the environmental conditions we want to mimic and to the solubility property of the studied ENM, keeping the feasibility of the test. Based on the results obtained through these tests, we recommend performing continuous flow dissolution experiments at a flow rate between 0.5 and 1 $mL.min^{-1}$, sufficient to reach a steady-state after 5–6 h of exposure, and allowing robust determination of the outlet dissolved concentrations while reducing potential saturation effects.

3.3. Dissolution Rate of Ag NPs in Artificial Surface Waters

Continuous flow experiments were also performed using artificial waters intended to mimic acidic streams (2 mM $Ca(NO_3)_2$ and 0.5 mM $MgSO_4$ solution at pH 5) and near-neutral/alkaline surface waters (2 mM $Ca(NO_3)_2$, 0.5 mM $MgSO_4$ and 5mM $NaHCO_3^-$ solution at pH 7.5). In both experiments, the steady-state plateau was reached after 400 min (Figure 5a,b). The averaged dissolution rate of Ag NPs calculated for the near-neutral artificial water exposure experiment was relatively low (Ag_0.5mL/min_8.2μg_NSW, k = 0.08 $\mu g.m^{-2}.s^{-1}$). For this latter experiment, the averaged concentration of Ag measured in the eluant ([Ag] = 0.56 ± 0.01 $\mu g.L^{-1}$) was higher than the LOD (0.05 $\mu g.L^{-1}$) and LOQ (0.12 $\mu g.L^{-1}$) values determined for the ICP-MS. However, for the lower Ag NP dissolution rate, lower Ag concentrations at the outlet might impact the relevance and significance of the test, for example, if dissolved Ag concentrations are too low to be accurately measured by ICP-MS (Figure 5c). In such case, lower dilution factors and/or a lower flow rate might thus be suitable.

At a lower pH, the averaged Ag dissolution rate was higher (Ag_0.5mL/min_8.2μg_ASW, k = 0.13 $\mu g.m^{-2}.s^{-1}$) than at pH 7.5 (Ag_0.5mL/min_8.2μg_NSW, k = 0.08 $\mu g.m^{-2}.s^{-1}$). This trend is consistent with previous work showing an increase in Ag NPs dissolution in acidic media [23]. Using the classic batch experiment, Mitrano et al. [57] investigated the dissolution rate of 100 nm and 60 nm spherical citrate-coated Ag NPs in artificial and natural media. They reported dissolution rate values as logr = −11.72 and −12.23 $mol.cm^{-2}.s^{-1}$ in deionized water (pH = 6.7), logr = −12.14 and −12.71 $mol.cm^{-2}.s^{-1}$ in a natural surface water (pH = 7.3) and logr = −12.46 $mol.cm^{-2}.s^{-1}$ in deionized water containing 2 $mg.L^{-1}$ of DOC (pH = 4.8). Such values are close to the values obtained in this study, with log k ranging between −12.58 and −13.13 $mol.cm^{-2}.s^{-1}$ for Ag_0.2mL/min_2.2μg_$NaNO_3$ and Ag_0.5mL/min_8.2μg_NSW, respectively.

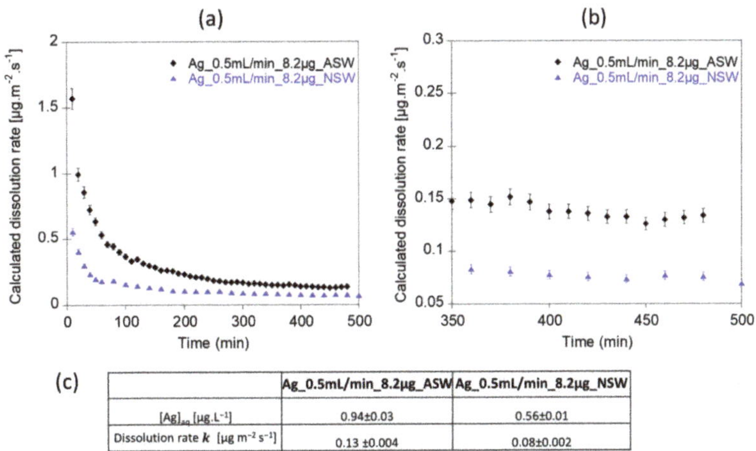

	Ag_0.5mL/min_8.2µg_ASW	Ag_0.5mL/min_8.2µg_NSW
$[Ag]_{aq}$ [µg.L^{-1}]	0.94±0.03	0.56±0.01
Dissolution rate k [µg m^{-2} s^{-1}]	0.13 ±0.004	0.08±0.002

Figure 5. (**a**,**b**) Dissolution rates of 80 nm Ag NPs in environmental aqueous media. (**c**) Table showing the average [Ag] concentrations and the averaged dissolution rates calculated on the steady state ranges. Ag_0.5mL/min_8.2µg_ASW was performed at pH 5 with 2 mM $Ca(NO_3)_2$ and 0.5 mM $MgSO_4$ as exposure medium. Ag_0.5mL/min_8.2µg_NSW was performed at pH 7.5 with 2 mM $Ca(NO_3)_2$, 0.5 mM $MgSO_4$ and 5 mM of $NaHCO_3^-$ as exposure medium. Error bars correspond to the incertitude of Ag concentrations measured by ICP-MS.

4. Conclusions

It is encouraged to develop standardized experimental procedures into OECD test guidelines to evaluate the dissolution rate of nanomaterials in natural aquatic environments [32]. Flow-through/continuous flow dissolution experiments are well suited to study dissolution kinetics of NPs and to systematically investigate NPs dissolution rates under environmentally relevant conditions. The main advantages are (1) a continuous supply of exposure media, (2) the possibility to expose small amounts of NPs (low solid/liquid ratio), and (3) the instantaneous separation between the dissolved and the solid fractions. For ENMs of very high or very low solubility, the flow-through testing can be easily adjusted according to the fast depletion of the material in the one, and too small outlet concentrations in the other case (i.e., NPs loading, flow rate). In contrast to recent studies investigating dissolution of ENMs in physiological fluids using a continuous flow system [40,41,54], our experimental approach uses the injection of ENMs dispersions which allows to load significantly lower amount of ENMs, relevant to mimic realistic natural environmental settings [5,58]. Our results demonstrated the feasibility to run flow-through dissolution tests for ENMs loadings down to the microgram range using environmentally relevant exposure media. The influence of the flow rate on the thickness of the diffusional boundary layer, the dissolution rate [54], and the local saturation at the vicinity of the nanoparticles should be further investigated. In particular, local saturation at the vicinity of the nanoparticles may occur in natural settings, for example, in water-saturated soils and sediments where pore waters may have longer residence time. In such specific aquatic compartments, it is relevant to determine various dissolution rates from long term experiments with various flow rates. In addition, establishing a clear recommendation of media compositions that would represent a wide range of natural waters, covering the dissolution-relevant species, would bring the tests closer to environmental realism. Here, it must be differentiated between standard testing for the comparison of materials regarding only their dissolution behavior, and tests, which aim to observe realistic environmental behavior including possible ENMs transformations. In the first case, a controlled pH and a simple inert background electrolyte would be suitable, whereas in the latter case, a more complex water chemistry including phosphate, sulfate, chloride, sulfide, and NOM (and others) would be chosen. In summary, there is a growing need for standardizing and implementing continuous flow dissolution testing in test guidelines to address a more realistic scenario of ENMs in natural aquatic systems. This would improve reproducibility and comparability of results for the wide diversity of pristine and transformed ENMs and allow their registration with international regulatory bodies.

Author Contributions: Conceptualization, F.v.d.K.; investigation, L.S., A.M. and N.T.; writing—original draft preparation, L.S.; writing—review and editing, A.M., N.T., F.v.d.K., L.S. and T.H.; funding acquisition, F.v.d.K. and T.H. All authors have read and agreed to the published version of the manuscript.

Funding: This work was part of the NanoFASE and Gov4Nano projects funded by the European Union's Horizon 2020 research and innovation programme under the grant agreements 646002 and 814401.

Acknowledgments: The authors thank Wolfgang Obermaier for his support on ICP-MS measurements.

Conflicts of Interest: The authors declare no conflict of interest.

References

1. Giese, B.; Klaessig, F.; Park, B.; Kaegi, R.; Steinfeldt, M.; Wigger, H.; von Gleich, A.; Gottschalk, F. Risks, release and concentrations of engineered nanomaterial in the environment. *Sci. Rep.* **2018**, *8*, 1565. [CrossRef]
2. Pulit-Prociak, J.; Banach, M. Silver nanoparticles—A material of the future ... ? *Open Chem.* **2016**, *14*, 76–91. [CrossRef]
3. Piccinno, F.; Gottschalk, F.; Seeger, S.; Nowack, B. Industrial production quantities and uses of ten engineered nanomaterials in Europe and the world. *J. Nanoparticle Res.* **2012**, *14*, 1109. [CrossRef]
4. Bundschuh, M.; Filser, J.; Lüderwald, S.; McKee, M.S.; Metreveli, G.; Schaumann, G.E.; Schulz, R.; Wagner, S. Nanoparticles in the environment: Where do we come from, where do we go to? *Environ. Sci. Eur.* **2018**, *30*, 6. [CrossRef] [PubMed]
5. Dumont, E.; Johnson, A.C.; Keller, V.D.J.; Williams, R.J. Nano silver and nano zinc-oxide in surface waters—Exposure estimation for Europe at high spatial and temporal resolution. *Environ. Pollut.* **2015**, *196*, 341–349. [CrossRef]
6. Kaegi, R.; Voegelin, A.; Sinnet, B.; Zuleeg, S.; Hagendorfer, H.; Burkhardt, M.; Siegrist, H. Behavior of metallic silver nanoparticles in a pilot wastewater treatment plant. *Environ. Sci. Technol.* **2011**, *45*, 3902–3908. [CrossRef]
7. Mitrano, D.M.; Nowack, B. The need for a life-cycle based aging paradigm for nanomaterials: Importance of real-world test systems to identify realistic particle transformations. *Nanotechnology* **2017**, *28*, 072001. [CrossRef]
8. Kaegi, R.; Ulrich, A.; Sinnet, B.; Vonbank, R.; Wichser, A.; Zuleeg, S.; Simmler, H.; Brunner, S.; Vonmont, H.; Burkhardt, M.; et al. Synthetic TiO2nanoparticle emission from exterior facades into the aquatic environment. *Environ. Pollut.* **2008**, *156*, 233–239. [CrossRef]
9. Benn, T.M.; Westerhoff, P. Nanoparticle silver released into water from commercially available sock fabrics. *Environ. Sci. Technol.* **2008**, *42*, 4133–4139. [CrossRef]
10. Gondikas, A.P.; von der Kammer, F.; Reed, R.B.; Wagner, S.; Ranville, J.F.; Hofmann, T. Release of TiO_2 nanoparticles from sunscreens into surface waters: A one-year survey at the Old Danube Recreational Lake. *Environ. Sci. Technol.* **2014**, *48*, 5415–5422. [CrossRef]
11. Zhang, W.; Yao, Y.; Sullivan, N.; Chen, Y. Modeling the primary size effects of citrate-coated silver nanoparticles on their ion release kinetics. *Environ. Sci. Technol.* **2011**, *45*, 4422–4428. [CrossRef]
12. Ho, C.-M.; Yau, S.K.W.; Lok, C.N.; So, M.H.; Che, C.M. Oxidative dissolution of silver nanoparticles by biologically relevant oxidants: A kinetic and mechanistic study. *Chem.–Asian J.* **2010**, *5*, 285–293. [CrossRef] [PubMed]
13. Dobias, J.; Bernier-Latmani, R. Silver release from silver nanoparticles in natural waters. *Environ. Sci. Technol.* **2013**, *47*, 4140–4146. [CrossRef] [PubMed]
14. Peretyazhko, T.S.; Zhang, Q.; Colvin, V.L. Size-controlled dissolution of silver nanoparticles at neutral and acidic pH conditions: Kinetics and size changes. *Environ. Sci. Technol.* **2014**, *48*, 11954–11961. [CrossRef] [PubMed]
15. Borm, P.; Klaessig, F.C.; Landry, T.D.; Moudgil, B.; Pauluhn, J.; Thomas, K.; Trottier, R.; Wood, S. Research strategies for safety evaluation of nanomaterials, part V: Role of dissolution in biological fate and effects of nanoscale particles. *Toxicol. Sci.* **2006**, *90*, 23–32. [CrossRef]
16. Misra, S.K.; Dybowska, A.; Berhanu, D.; Luoma, S.N.; Valsami-Jones, E. The complexity of nanoparticle dissolution and its importance in nanotoxicological studies. *Sci. Total Environ.* **2012**, *438*, 225–232. [CrossRef]
17. Liu, J.Y.; Hurt, R.H. Ion release kinetics and particle persistence in aqueous nano-silver colloids. *Environ. Sci. Technol.* **2010**, *44*, 2169–2175. [CrossRef]
18. Li, X.; Lenhart, J.J. Aggregation and dissolution of silver nanoparticles in natural surface water. *Environ. Sci. Technol.* **2012**, *46*, 5378–5386. [CrossRef]
19. Skjolding, L.M.; Sørensen, S.N.; Hartmann, N.B.; Hjorth, R.; Hansen, S.F.; Baun, A. Aquatic ecotoxicity testing of nanoparticles—The quest to disclose nanoparticle effects. *Angew. Chem. Int. Ed. Engl.* **2016**, *55*, 15224–15239. [CrossRef]

20. Kent, R.D.; Vikesland, P.J. Controlled evaluation of silver nanoparticle dissolution using atomic force microscopy. *Environ. Sci. Technol.* **2012**, *46*, 6977–6984. [CrossRef] [PubMed]
21. Gondikas, A.P.; Morris, A.; Reinsch, B.C.; Marinakos, S.M.; Lowry, G.V.; Hsu-Kim, H. Cysteine-induced modifications of zero-valent silver nanomaterials: Implications for particle surface chemistry, aggregation, dissolution, and silver speciation. *Environ. Sci. Technol.* **2012**, *46*, 7037–7045. [CrossRef] [PubMed]
22. Gunsolus, I.L.; Mousavi, M.P.S.; Hussein, K.; Bühlmann, P.; Haynes, C.L. Effects of humic and fulvic acids on silver nanoparticle stability, dissolution, and toxicity. *Environ. Sci. Technol.* **2015**, *49*, 8078–8086. [CrossRef]
23. Fernando, I.; Zhou, Y. Impact of pH on the stability, dissolution and aggregation kinetics of silver nanoparticles. *Chemosphere* **2019**, *216*, 297–305. [CrossRef] [PubMed]
24. Odzak, N.; Kistler, D.; Behra, R.; Sigg, L. Dissolution of metal and metal oxide nanoparticles under natural freshwater conditions. *Environ. Chem.* **2014**, *12*, 138–148. [CrossRef]
25. Molleman, B.; Hiemstra, T. Time, pH, and size dependency of silver nanoparticle dissolution: The road to equilibrium. *Environ. Sci. Nano* **2017**, *4*, 1314–1327. [CrossRef]
26. Jo, H.J.; Choi, J.W.; Lee, S.H.; Hong, S.W. Acute toxicity of Ag and CuO nanoparticle suspensions against *Daphnia magna*: The importance of their dissolved fraction varying with preparation methods. *J. Hazard. Mater.* **2012**, *227–228*, 301–308. [CrossRef] [PubMed]
27. Hoheisel, S.M.; Diamond, S.; Mount, D. Comparison of nanosilver and ionic silver toxicity in *Daphnia magna* and *Pimephales promelas*. *Environ. Toxicol. Chem.* **2012**, *31*, 2557–2563. [CrossRef]
28. Notter, D.A.; Mitrano, D.M.; Nowack, B. Are nanosized or dissolved metals more toxic in the environment? A meta-analysis. *Environ. Toxicol. Chem.* **2014**, *33*, 2733–2739. [CrossRef]
29. Levard, C.; Hotze, E.M.; Lowry, G.V.; Brown, G.E. Environmental transformations of silver nanoparticles: Impact on stability and toxicity. *Environ. Sci. Technol.* **2012**, *46*, 6900–6914. [CrossRef]
30. European Union, Commission Regulation (EU) 2018/1881 of 3 December 2018 Amending Regulation (EC) No 1907/2006 of the European Parliament and of the Council on the Registration, Evaluation, Authorisation and Restriction of Chemicals (REACH) as Regards Annexes I, III, VI, VII, VIII, IX, X, XI, and XII to Address Nanoforms of Substances. Commission Regulation (EU) 2018/1881. 2018. Available online: http://data.europa.eu/eli/reg/2018/1881/oj (accessed on 15 November 2021).
31. Bove, P.; Malvindi, M.A.; Kote, S.S.; Bertorelli, R.; Summa, M.; Sabella, S. Dissolution test for risk assessment of nanoparticles: A pilot study. *Nanoscale* **2017**, *9*, 6315–6326. [CrossRef]
32. OECD GD 3018, Guidance Document for the Testing of Dissolution and Dispersion Stability of Nanomaterials and the Use of the Data for Further Environmental Testing and Assessment Strategies. 2020. Available online: https://www.oecd.org/chemicalsafety/testing/series-testing-assessment-publications-number.htm (accessed on 13 November 2021).
33. *OECD TG 318, Test Guideline 318: Dispersion Stability of Nanomaterials in Simulated Environmental Media*; OECD Test Guidelines for Testing Chemicals Organisation for Economic Co-Operation and Development: Paris, France, 2017. Available online: http://www.oecd-ilibrary.org/environment/test-no-318-dispersion-stability-of-nanomaterials-in-simulated-environmental-media_9789264284142-en (accessed on 13 November 2021).
34. *OECD TG 105, TG 105 Water Solubility. OECD Test Guideline for the Testing of Chemicals*; Organisation for Economic Co-Operation and Development (OECD): Paris, France, 1995. Available online: https://www.oecd-ilibrary.org/environment/test-no$-$105-water-solubility_9789264069589-en (accessed on 13 November 2021).
35. Baun, A.; Sayre, P.; Steinhäuser, K.G.; Rose, J. Regulatory relevant and reliable methods and data for determining the environmental fate of manufactured nanomaterials. *NanoImpact* **2017**, *8*, 1–10. [CrossRef]
36. *OECD GD 29, OECD Series on Testing and Assessment, No. 29. Guidance Document on Transformation/Dissolution of Metals and Metal Compounds in Aqueous Media*; Organisation for Economic Co-Operation and Development (OECD): Paris, France, 2001. Available online: http://www.oecd.org/env/ehs/testing/seriesontestingandassessmenttestingforenvironmentalfate.htm (accessed on 13 November 2021).
37. Wasmuth, C.; Rüdel, H.; Düring, R.A.; Klawonn, T. Assessing the suitability of the OECD 29 guidance document to investigate the transformation and dissolution of silver nanoparticles in aqueous media. *Chemosphere* **2016**, *144*, 2018–2023. [CrossRef]
38. Hankin, S.M.; Peters, S.A.K.; Poland, C.A.; Hansen, S.F.; Holmqvist, J.; Ross, B.L.; Varet, J.; Aitken, R.J. *Specific Advice on Fulfilling Information Requirements for Nanomaterials under REACH (RIP-oN 2)—Final Project Report*; European Commission: Brussels, Belgium, 2011.
39. International Organisation for Standardisation. *ISO/TR 19057 Nanotechnologies—Use and Application of Acellular In Vitro Tests and Methodologies to Assess Nanomaterial Biodurability*; ISO: Geneva, Switzerland, 2017.
40. Koltermann-Jülly, J.; Keller, J.G.; Vennemann, A.; Werle, K.; Müller, P.; Ma-Hock, L.; Landsiedel, R.; Wiemann, M.; Wohlleben, W. Abiotic dissolution rates of 24 (nano) forms of 6 substances compared to macrophage-assisted dissolution and in vivo pulmonary clearance: Grouping by biodissolution and transformation. *NanoImpact* **2018**, *12*, 29–41. [CrossRef]
41. Keller, J.G.; Graham, U.M.; Koltermann-Jülly, J.; Gelein, R.; Ma-Hock, L.; Landsiedel, R.; Wiemann, M.; Oberdörster, G.; Elder, A.; Wohlleben, W. Predicting dissolution and transformation of inhaled nanoparticles in the lung using abiotic flow cells: The case of barium sulfate. *Sci. Rep.* **2020**, *10*, 458. [CrossRef] [PubMed]
42. Li, L.; Schuster, M. Influence of phosphate and solution pH on the mobility of ZnO nanoparticles in saturated sand. *Sci. Total Environ.* **2014**, *472*, 971–978. [CrossRef]

43. Lv, J.; Zhang, S.; Luo, L.; Han, W.; Zhang, J.; Yang, K.; Christie, P. Dissolution and microstructural transformation of ZnO nanoparticles under the influence of phosphate. *Environ. Sci. Technol.* **2012**, *46*, 7215–7221. [CrossRef]
44. Conway, J.R.; Adeleye, A.S.; Gardea-Torresdey, J.; Keller, A.A. Aggregation, dissolution, and transformation of copper nanoparticles in natural waters. *Environ. Sci. Technol.* **2015**, *49*, 2749–2756. [CrossRef] [PubMed]
45. Water Quality in the Danube River Basin—2017, ICPDR—International Commission for the Protection of the Danube River. Editor: Igor Liska, ICPDR Secretariat. Danube River Basin Water Quality Database. Available online: https://www.icpdr.org/wq-db/ (accessed on 21 February 2021).
46. FGG Rhein. Flussgebietsgemeinschaft Rhein (River Basin Community Rhine)—Hydrologische Datenbank HYDABA bei der Bundesanstalt für Gewässerkunde (BfG). Available online: http://www.fgg-rhein.de/servlet/is/4254/ (accessed on 21 February 2021).
47. FGG Elbe. Flussgebietsgemeinschaft Elbe (River Basin Community Elbe) Database FIS. Available online: http://176.28.42.206/FisFggElbe/content/start/ZurStartseite.action;jssionid=0DCA632198EC95EA6F8B51A066F5C040 (accessed on 21 February 2021).
48. Salminen, R.; Batista, M.J.; Bidovec, M.; Demetriades, A.; De Vivo, B.; De Vos, W.; Duris, M.; Gilucis, A.; Gregorauskiene, V.; Halamic, J.; et al. *Geochemical Atlas of Europe. Part 1—Background Information, Methodology and Maps*; Geological Survey of Finland: Espoo, Finland, 2005; p. 526.
49. European Environment Agency (EEA), Nutrients in freshwater in Europe (CSI 020/WAT 003). Data from Waterbase-Water Quality. Available online: https://www.eea.europa.eu/data-and-maps/indicators/nutrients-in-freshwater/nutrients-in-freshwater-assessment-published-9 (accessed on 21 February 2021).
50. Yu, S.; Liu, J.; Yin, Y.; Shen, M. Interactions between engineered nanoparticles and dissolved organic matter: A review on mechanisms and environmental effects. *J. Environ. Sci.* **2018**, *63*, 198–217. [CrossRef]
51. Khaksar, M.; Jolley, D.F.; Sekine, R.; Vasilev, K.; Johannessen, B.; Donner, E.; Lombi, E. In situ chemical transformations of silver nanoparticles along the water–sediment continuum. *Environ. Sci. Technol.* **2015**, *49*, 318–325. [CrossRef]
52. Sivry, Y.; Gelabert, A.; Cordier, L.; Ferrari, R.; Lazar, H.; Juillot, F.; Menguy, N.; Benedetti, M.F. Behavior and fate of industrial zinc oxide nanoparticles in a carbonate-rich river water. *Chemosphere* **2014**, *95*, 519–526. [CrossRef] [PubMed]
53. Reed, R.B.; Ladner, D.A.; Higgins, C.P.; Westerhoff, P.; Ranville, J.F. Solubility of nano-zinc oxide in environmentally and biologically important matrices. *Environ. Toxicol. Chem.* **2012**, *31*, 93–99. [CrossRef] [PubMed]
54. Keller, J.; Peijnenburg, W.; Werle, K.; Landsiedel, R.; Wohlleben, W. Understanding dissolution rates via continuous flow systems with physiologically relevant metal ion saturation in lysosome. *Nanomaterials* **2020**, *10*, 311. [CrossRef] [PubMed]
55. Sikder, M.; Lead, J.R.; Chandler, G.T.; Baalousha, M. A rapid approach for measuring silver nanoparticle concentration and dissolution in seawater by UV–Vis. *Sci. Total Environ.* **2018**, *618*, 597–607. [CrossRef] [PubMed]
56. Merrifield, R.C.; Stephan, C.; Lead, J. Determining the concentration dependent transformations of Ag Nanoparticles in complex media: Using SP-ICP-MS and Au≅Ag core–shell nanoparticles as tracers. *Environ. Sci. Technol.* **2017**, *51*, 3206–3213. [CrossRef]
57. Mitrano, D.M.; Ranville, J.F.; Bednar, A.; Kazor, K.; Hering, A.S.; Higgins, C.P. Tracking dissolution of silver nanoparticles at environmentally relevant concentrations in laboratory, natural, and processed waters using single particle ICP-MS (SpICP-MS). *Environ. Sci. Nano* **2014**, *1*, 248–259. [CrossRef]
58. Gottschalk, F.; Sun, T.; Nowack, B. Environmental concentrations of engineered nanomaterials: Review of modeling and analytical studies. *Environ. Pollut.* **2013**, *181*, 287–300. [CrossRef]

MDPI

Article

Assessment of Nanopollution from Commercial Products in Water Environments

Raisibe Florence Lehutso [1,2] and Melusi Thwala [1,3,4,*]

1 Water Centre, Council for Scientific and Industrial Research, Pretoria 0001, South Africa; flehutso@csir.co.za
2 Department of Chemical Sciences, University of Johannesburg, Johannesburg 2028, South Africa
3 Department of Environmental Health, Nelson Mandela University, Gqeberha 6019, South Africa
4 Centre for Environmental Management, University of the Free State, Bloemfontein 9031, South Africa
* Correspondence: mthwala@csir.co.za

Abstract: The use of nano-enabled products (NEPs) can release engineered nanomaterials (ENMs) into water resources, and the increasing commercialisation of NEPs raises the environmental exposure potential. The current study investigated the release of ENMs and their characteristics from six commercial products (sunscreens, body creams, sanitiser, and socks) containing nTiO_2, nAg, and nZnO. ENMs were released in aqueous media from all investigated NEPs and were associated with ions (Ag^+ and Zn^{2+}) and coating agents (Si and Al). NEPs generally released elongated (7–9 × 66–70 nm) and angular (21–80 × 25–79 nm) nTiO_2, near-spherical (12–49 nm) and angular nAg (21–76 × 29–77 nm), and angular nZnO (32–36 × 32–40 nm). NEPs released varying ENMs' total concentrations (*ca* 0.4–95%) of total Ti, Ag, Ag^+, Zn, and Zn^{2+} relative to the initial amount of ENMs added in NEPs, influenced by the nature of the product and recipient water quality. The findings confirmed the use of the examined NEPs as sources of nanopollution in water resources, and the physicochemical properties of the nanopollutants were determined. Exposure assessment data from real-life sources are highly valuable for enriching the robust environmental risk assessment of nanotechnology.

Keywords: nanopollution; nano-enabled products; engineered nanomaterials; physicochemical properties; aquatic environments

Citation: Lehutso, R.F.; Thwala, M. Assessment of Nanopollution from Commercial Products in Water Environments . *Nanomaterials* **2021**, *11*, 2537. https://doi.org/10.3390/nano11102537

Academic Editor: Vivian Hsiu-Chuan Liao

Received: 24 August 2021
Accepted: 18 September 2021
Published: 28 September 2021

1. Introduction

The global commercialisation of nano-enabled products (NEPs) is growing rapidly year on year [1], and it is estimated to grow from USD 39.2 billion in 2016 to over USD 125 billion by 2024 [2]. Approximately 5000 NEPs were identified in various global inventories between 2015 and 2021, belonging to six product categories, namely: health and fitness, electronics and computers, home and garden, appliances, automotive, and food beverages [3–7]. These inventories are generally dominated by health and fitness NEPs, such as sunscreens, personal care products, and clothing products [3–7], which exhibit medium to high probability of emitting engineered nanomaterials (ENMs) into the environment during use, especially water resources (i.e., environmental exposure) [1,6].

Increasing the production and use of NEPs consequently raises the probability of proportional ENMs' release into aquatic environments; therefore, NEPs are potential sources of daily nanopollution [6,8]. For instance, the release of some commonly applied ENMs in NEPs such as silver (nAg) and zinc oxide (nZnO) into surface water is estimated at approximately 4.9–1700 t/annually [9]. Elsewhere, it was estimated that 50–95% of ENMs (nAg and nTiO_2) are released into water resources along the life cycle of NEPs [10]. Furthermore, environmental concentrations of ENMs in water systems differ from estimates from in silico studies [11–15]. For example, Ag and Ti' predicted environmental concentrations (PECs) are reported, respectively, as 0.7–16 μg/L [16–19], and 0.014–2.2 μg/L [18–20], while measured environmental concentrations (MECs) were quantified at 0.03–19.7 μg/L (Ag) [21–23],

0.67–150 μg/L (Ti) [12–14,24–26]. Continuous release of ENMs leads to concentrations of nanopollutants reaching levels that can be hazardous in water resources [27].

In order to address concerns related to nanopollution, a considerable proportion of studies have been undertaken on pristine ENMs [28], but the generated data cannot be directly transferred to ENMs released from NEPs (product-released ENMs) due to differences in physicochemical properties [29]. Differences in physicochemical properties are due to (i) the manipulation of pristine ENMs during preparation for incorporation into NEP, (ii) association of product-released-ENMs with other product components, and (iii) the influence of the NEPs life cycle [28–30]. For example, before incorporation into NEPs (e.g., cosmetics), the surface of $nTiO_2$ are commonly modified with coating agents such as aluminium hydroxide ($Al(OH)_3$) or polydimethylsiloxane (PDMS) to facilitate dispersion in the matrix of NEPs, to prevent/reduce photooxidation and generation of reactive oxygen species (ROS) [31]. The behaviour of such functionalised $nTiO_2$ does not resemble pristine ENMs counterparts. Similarly, during the use of NEPs, some of the physicochemical properties of ENMs can be altered after exposure of NEPs (i.e., clothing) to environmental stressors such as ultraviolet and physical forms, the use of fabrics, and how they are washed [32]. For instance, nano-enabled socks that were used released ca 50–100 nm compared to 1–2 nm counterparts released by unused socks [33]. Furthermore, the washing method may also influence the amount and properties of ENMs released from fabrics [34,35].

In that context, scientists focused on examining the environmental risk arising from product-released ENMs, and studies on product-released ENMs and other articles (nanocomposites) have grown from 96 in 2017 [36] to approximately 120 in 2021. The studies illustrate that the concentration of product-released ENMs varies considerably (0.01–35%), and so does the size (<100–385 nm) and other physicochemical properties [36]. Due to the low sample mass/volume attainable after sample preparation and the limited analytical equipment capability to analyse ENMs in complex matrices, fewer studies have optimally characterised product-released ENMs [36,37]. As such, there is a considerable knowledge gap on the exposure characteristics of product-released ENMs, and consequently, robust and realistic risk assessment of product-released ENMs in the environment remains to be established [38,39].

In order to establish and address environmentally realistic risks of product-released ENMs, exposure assessment data need to be strengthened at the various stages of the life cycle of NEPs (production, usage, and end of life) [36,40]. The current study examined the release and exposure characteristics of product-released ENMs from a wide array of NEPs that exhibit a medium to high nanopollution potential toward water resources [6]. The NEPs samples were from the category of health and fitness products: sunscreens, hand sanitiser, body cream, and socks samples. The health and fitness category, specifically personal care products, has been shown to dominate NEPs markets worldwide [3–7]. The selection of NEPs was further influenced by, but not limited to, include a few chemical identities of ENMs, the physicochemical properties of the applied ENMs, and the location of the ENMs within the product; all of which influence the environmental exposure potential of ENMs [6]. By considering the current data gap regarding the environmental risk associated with the use of NEPs, the current study sought to enrich the data on the physicochemical properties of product-released ENMs as an essential component to advance global efforts to determine the probable risk of nanopollutants in aquatic environments.

2. Materials and Methods

Six NEPs, namely three sunscreens, SUN1 ($nTiO_2$ + nZnO), SUN2–3 ($nTiO_2$); body cream CA1 ($nTiO_2$ + nAg); sanitiser; SAN1 (nAg); and socks SK1 ($nTiO_2$ + nAg) were purchased from South African retailers. The physicochemical properties of the ENMs incorporated in the six NEPs varied and were previously reported [41]. Briefly, SUN1 contained elongated $nTiO_2$, and angular nZnO particles sized 14 × 62 nm and 35 × 38 nm, respectively. SUN2 and SUN3 contained angular-shaped $nTiO_2$ with a size range of

20–28 × 27–32 nm and 20–28 × 27–32 nm, respectively. Near-spherical nAg particles with the size range of 22–37 nm were observed in SAN1. CA1 and SK1, respectively, contained $nTiO_2$ particles sized at 8 × 53 nm (elongated) and 32–203 × 48–135 nm (angular), the NEPs also contained near-spherical nAg ranging 18–28 nm. The particles were negatively charged, and the $nTiO_2$ phase was rutile or anatase. The total amounts of ENMs differed between NEPs [41].

2.1. Procedures for ENMs Release

The procedures used to investigate the release of ENMs from NEPs differed and are described in Sections 2.1.1 and 2.1.2. In all instances, the procedures were adapted to simulate conditions (but not fully replicate actual life cycle stages) that promote ENMs release from the NEPs. However, some stages applicable during the use of current NEPs (e.g., application of sunscreen or sanitiser to the skin, prior wearing of socks) were eliminated as the primary focus of the study was the analytical determination of the potential for ENMs release. Furthermore, simple (or standard) aqueous media were preferred to avoid physicochemical complexation, which occurs when using complex media and uncharacterised commercial detergents [42]; however, release evaluations in complex media with greater environmental realism are the cornerstone of nanotechnology risk assessment and are recommended for future studies as analytical capabilities and access advances. All investigations were carried out in triplicate.

2.1.1. Release of ENMs from Suncreen 1–3 (SUN1–3) and Body Cream 1 (CA1)

The release of ENMs from SUN1–3 and CA1 followed a slightly modified protocol of Botta et al. [43]. Briefly, 2 g of SUN1–3 and CA1 were aged in 180 mL of release media for 48 h at 25 °C under darkness (plastic beakers were capped and covered with heavy-duty aluminum foil) and under illumination: −6000 lux (uncapped transparent plastic beakers). The SUN1–3 ENMs were released in Milli-Q water (18 MΩ·cm), freshwater, seawater or swimming pool water (S1.1), while the CA1 ENMs were released using Milli-Q water only; the properties of the release media are given in S1.1. ENMs were released by agitating the suspension at 400 rpm for 48 h, and the sample volume was maintained by continuously adding aqueous media throughout the 48 h. Agitation was stopped after 48 h, and the samples were allowed to settle for another 48 h, a step that caused sedimentation that resulted in two phases (surface suspension and sediment); the overall duration of ENMs release was 96 h. The two phases were separated by sampling 150 mL of stable suspension; the sediments were not disturbed during sampling. The stable suspensions were prepared for product-released ENMs analysis (Section 2.2).

2.1.2. Release of ENMs from Sock1 (SK1) and Sanitiser1 (SAN1)

The release of ENMs from SK1 was undertaken by adapting previously developed methods [44,45]. Briefly, areas (spots/regions) of the sock material marketed and experimentally confirmed to be incorporated with nAg and $nTiO_2$ [41] were cut from SK1 samples and transferred into 1 L glass bottles, washed with 200 mL of the release media (Milli-Q water, tap water, and sodium dodecyl sulfate as a detergent). The detergent media was prepared in two ways: (i) sodium dodecyl sulfate 1, which was prepared in Milli-Q water, while (ii) sodium dodecyl sulfate 2 was prepared in tap water. The samples were washed by shaking at 350 rpm at 40 °C for 12 h (2 washes). After the final washing cycle, the fabrics were removed from the washing water, and the samples were prepared for analysis (Section 2.2).

For SAN1, ENMs were released following a slightly modified method of Benn et al. [46] and Mackevica et al. [47]. Briefly, 1 mg/L of the sample was prepared in Milli-Q water and ENMs released by agitating at 350 rpm at 40 °C for 24 h. After ENMs' release, samples were prepared for analysis (given in Section 2.2).

SUN1 (nZnO), CA1 (nAg), SAN1 (nAg), and SK1 (nAg) were incorporated with ENMs that are relatively soluble. The released ions were recovered from the release media through sequential filtration. The samples were sequentially filtered using Amicon® Ultra-15 30 K centrifugal filters (30000 MWCO, Merck, South Africa), followed by further centrifugation using Amicon® Ultra-15 3 K centrifugal filter devices (3000 MWCO, Merck, South Africa) for 30 min at 10,000 rpm for each filtration step. The released ions (filtrates from the 3 K centrifugal filter device) were quantified (Section 2.2.3).

2.2. Physicochemical Properties of Product-Released ENMs

2.2.1. Electron Microscopy

Images of product-released ENMs were obtained using a JEOL-JEM 2100 high-resolution transmission electron microscope coupled to energy-dispersive X-ray spectroscopy (HR-TEM-EDX) (Tokyo, Japan) fitted with a LaB6 filament operated at 200 kV. A Cu grid with a holey carbon film was dipped in the sample solution and air-dried for 12 h, followed by TEM-EDX analysis. Multiple images were captured at different spots on the grid to measure the product-released ENMs' size (minimum particle set at 50) using the ImageJ software.

2.2.2. Surface Charge of Product-Released ENMs

A Zetasizer Nano ZS (Malvern Instruments, Worcestershire, United Kingdom) was used to determine the zeta (ζ) potential of product-released ENMs in the release media, which measured the physicochemical properties reported in Table S1.

2.2.3. Elemental Analysis

Elemental analysis of product-released ENMs was performed using inductively coupled plasma mass spectrometry (ICP-MS, Icap Q, Thermo Fisher Scientific, Waltham, United States of America). For total Ti, Zn, Ag, and Si analysis, samples were predigested following a modified MARS 6 Method Note Compendium [48]. Product-released ENMs samples were transferred into digestion vessels, 5.0 mL of nitric acid (HNO_3) (70%, Merck, Johannesburg, South Africa) was added, followed by swirling the vessel and leaving it open for approximately 10 min. After 10 min, 2.0 mL of hydrogen peroxide (H_2O_2) (37%, Merck, Johannesburg, South Africa) and hydrofluoric acid (HF) (49%, Merck, Johannesburg, South Africa) were added to samples containing Ag and Ti, respectively. The microwave digestion program followed the cosmetic and textile heating program highlighted in the MARS 6 Method Note Compendium [48]. All product-released ENMs' digests were filtered using a 0.45 µm filter syringe (Merck, Johannesburg, South Africa) and prepared for ICP-MS analysis, monitoring ^{66}Zn, ^{48}Ti, ^{107}Ag, ^{28}Si, ^{27}Al, and ^{45}Sc (internal). The performance of the digestion method was evaluated by digesting both bulk (Zn, Ti and Ag, Anatech instruments, Johannesburg, South Africa) and $nTiO_2$ (Tavo commercial nanocomposite, Merck, Johannesburg, South Africa), Ag (bare and aminated, nanoComposix, San Diego, United States of America) and ZnO (Z-cote, a commercial nanocomposite, BASF, Johannesburg, South Africa). The recovered filtrates were acidified (5% using HNO_3) and directly analysed for the dissolved ions. Appropriate sample dilutions were performed prior to analysis.

2.3. Data Analysis

Statistical analysis and drawing of graphs were performed using GraphPad Prism8 version 8.4.3 for Windows (GraphPad Software, La Jolla, San Diego, United States of America). Student's t-test and two-way ANOVA with Tukey's HSD post hoc test were applied to examine differences between samples, with significance tested at $\alpha = 0.05$.

3. Results and Discussion

3.1. Characterisation of Product-Released ENMs

The release methods used in the current study successfully released ENMs from all NEPs as confirmed in all exposure media variants (Sections 3.1.1–3.1.3).

3.1.1. Sunscreen 1–3 (SUN1–3) Product-Released ENMs

Product-released ENMs in different release media (Milli-Q, freshwater, swimming pool water, and seawater) obtained under light conditions are provided in Figure 1 (TEM images) and Figure S1 (elemental profile). TEM images and elemental profiles of product-released ENMs obtained under dark conditions are given in Figures S2 and S3, respectively. Variation of illumination conditions and release media did not influence the morphology of product-released ENMs. SUN1-released $nTiO_2$ were elongated, while nZnO was angular in shape. SUN2–3-released $nTiO_2$ were angular in shape, shapes that were previously reported [41].

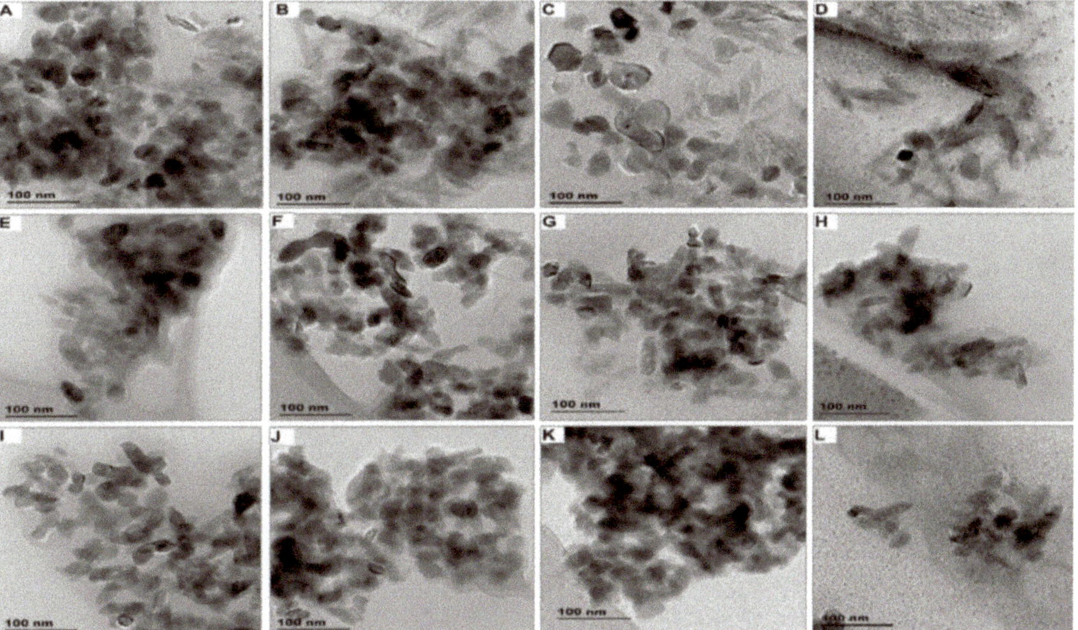

Figure 1. TEM images of product-released ENMs obtained under light conditions for SUN1 detected in Milli-Q water (**A**), freshwater (**B**), swimming pool water (**C**), seawater (**D**), SUN2 detected in Milli-Q water (**E**), freshwater (**F**), swimming pool water (**G**), seawater (**H**) and SUN3 detected in Milli-Q water (**I**), freshwater (**J**), swimming pool water (**K**), seawater (**L**).

Product-released ENMs were still predominantly associated with aluminium (Al) and silicon (Si) (Figures S1 and S3), indicative of remnants of coating agents [49–51], either intact on the surface of product-released ENMs or in the release media. The intensities of the Al and Si peak varied between the release media and the type of sunscreen (Figures S1 and S3), demonstrating that the release media or exposure conditions affected the ENMs coating agents differently.

The findings confirmed that product-released ENMs were not released in naked forms (pristine ENMs), supporting previous reports that product-released ENMs are commonly released associated the matrix of NEPs (transformed state) [14,42]. For example, in SUN1, Si appeared to have been predominantly released into the media, while Al partially remained adsorbed in product-released $nTiO_2$ (Figure S4). In SUN2 and SUN3, Si remained mainly attached to product-released $nTiO_2$ (Figure S5). The desorption of the coating agents from product-released ENMs surface has implications for the exposure potential of ENMs in aquatic environments, as they influence their reactivity [52], bioavailability and toxicity to aquatic organisms [53–55]. The findings partly illustrated that the environmental exposure characteristics arising from the use of NEPs could not be accurately established from studies using pristine ENMs, which leads to the need for refinement and standardisation of ENMs release protocols to improve exposure assessment data.

Similar to the shape of the released ENMs, the sizes (width × length) were also unaffected by both illumination and release media (Table 1 and Figure S6). The average size of SUN1-released nZnO was 32–36 × 32–40 nm, while product-released $nTiO_2$ were 7–9 × 66–70 nm. The SUN2-released ENMs sizes were 27–30 × 33–37 nm, while SUN3-released ENMs were 21–22 × 25–28 nm. The average size of SUN2-released ENMs was in agreement with previously reported sizes [41]. The distribution of the SUN1-released nZnO particles (W × L) in Milli-Q water and freshwater were comparable, and the distribution densities were similar to the ENMs incorporated in SUN1. SUN1-released nZnO distributions (W × L) in seawater and swimming pool water were similar; the distribution (upper and lower quartiles and violin density) was comparable to the ENMs found in SUN1. The particle distributions of SUN1-released $nTiO_2$ (W × L) were similar in all media. The upper and lower quartiles of the SUN1-released $nTiO_2$ (W × L) distribution slightly varied; the ENMs in SUN1 but were comparable in violin density [41]. In all release media, the product-released $nTiO_2$ distribution profiles of SUN2 and SUN3 were also similar and generally comparable to ENMs in the respective SUNs. While the distribution profiles in all SUN's were comparable to ENMs in NEPs, few exceptions, especially on violin shape/structure and quartiles, were observed.

Table 1. Particle shape and the average size of product-released ENMs in different release water media. [a] and [b] are the sizes of product-released ENMs obtained under light and dark conditions, respectively.

Release Media			Milli-Q Water	Freshwater	Swimming Pool Water	Seawater
Sample	**ENMs Type**	**ENMs Shape**	**Size (nm)**	**Size (nm)**	**Size (nm)**	**Size (nm)**
SUN1	ZnO [a]	Angular	34 ± 6 × 37 ± 7	35 ± 9 × 32 ± 8	38 ± 7 × 39 ± 5	36 ± 4 × 40 ± 7
	ZnO [b]	Angular	37 ± 9 × 39 ± 8	36 ± 4 × 37 ± 4	37 ± 9 × 39 ± 9	43 ± 6 × 42 ± 9
	TiO_2 [a]	Elongated	7 ± 2 × 66 ± 6	9 ± 2 × 66 ± 7	9 ± 3 × 67 ± 9	8 ± 2 × 70 ± 7
	TiO_2 [b]	Elongated	10 ± 2 × 68 ± 6	9 ± 3 × 64 ± 7	9 ± 3 × 64 ± 5	9 ± 2 × 67 ± 8
SUN2	TiO_2 [a]	Angular	30 ± 4 × 33 ± 7	30 ± 4 × 34 ± 5	27 ± 6 × 35 ± 6	29 ± 4 × 37 ± 4
	TiO_2 [b]	Angular	31 ± 7 × 36 ± 7	29 ± 6 × 32 ± 8	31 ± 4 × 35 ± 8	32 ± 4 × 38 ± 8
SUN3	TiO_2 [a]	Angular	21 ± 5 × 25 ± 5	22 ± 4 × 25 ± 5	22 ± 4 × 28 ± 5	21 ± 6 × 26 ± 5
	TiO_2 [b]	Angular	22 ± 5 × 28 ± 6	27 ± 4 × 25 ± 7	26 ± 5 × 26 ± 6	23 ± 5 × 26 ± 6

Product-released ENMs of all sunscreens were negatively charged, illustrating that, as expected, illumination did not influence the surface charge (under light: Figure 2 and dark: Figure S7). Although all product-released ENMs were negatively charged, the stability of product-released ENMs varied between the release media. Relatively high ζ potentials (negative or positive, a minimum value of 22 mV) are considered electrically stable, while lower ζ potentials are less stable and can lead to rapid agglomeration of nanoparticles [56]. All sunscreen-released ENMs were stable in Milli-Q and freshwater and unstable in seawater and swimming pool water. The difference in the stability of sunscreen-released ENMs is well corroborated with the TEM-EDX results (Figures S1 and S3), where the coating agents of product-released ENMs were affected differently by the different release media. ENMs are functionalised with coating agents to improve stability [57]; therefore, alteration of the ENMs coating agents will directly affect the stability of ENMs and their fate in aquatic systems [58,59].

The findings of the current study were comparable to previous reports. For example, the size range of elongated product-released $nTiO_2$ obtained in the current study was comparable to the range (10 × 139 nm) of sunscreen-released $nTiO_2$ (elongated) previously reported [42,43]. The negative surface potential of sunscreen-released ENMs was also previously reported [43,60].

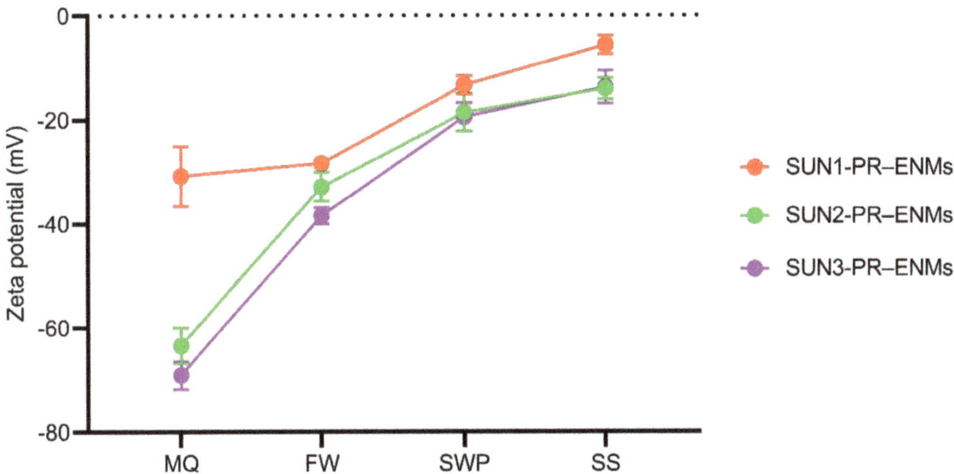

Figure 2. Zeta potential of product-released ENMs (PR–ENMs) obtained under light conditions in different release media of Milli-Q water (MQ), freshwater (FW), swimming pool water (SWP), and seawater (SS).

3.1.2. Sanitiser 1 (SAN1) and Body Cream (CA1) Product-Released ENMs

SAN1 and CA1 ENMs were successfully released into the respective media, as shown in Figure 3 and Figure S8 (size distribution of product-released ENMs). SAN1-released nAg were near-spherical and averaged 10 ± 2 and 23 ± 4 nm, indicating distinct size classes. The SAN1-released nAg generated two distribution profiles that differed in the upper quartiles; one of the profiles was comparable to the ENMs in SAN1 [41]. The other distribution differs from the ENMs profile in SAN1 on width, indicating possible agglomeration. The SAN1-released nAg ζ potentials were determined to be −32.5 ± 2.1 mV.

Binary CA1-released $nTiO_2$ and CA1-released nAg were detected under both illumination conditions (Figure 4 and Figure S9), the Si peak of the coating agents was also detected. The CA1-released $nTiO_2$ were elongated in shape and had an average size of $8 \pm 3 \times 60 \pm 13$ nm (under light) and $9 \pm 3 \times 66 \pm 9$ nm (under dark), indicating that the size was not affected by variation in illumination. Near-spherical CA1-released nAg were detected in three distinct average sizes of 12 ± 3, 27 ± 7, and 49 ± 9 nm under light conditions, relative to 10 ± 3, 28 ± 8, and 54 ± 8 nm under dark conditions, indicating that illumination variation did not affect ENMs sizes. The distribution and the violin density of CA1-released nAg obtained under light and dark conditions were similar. The distribution density of the CA1-released nAg and ENMs was comparable, but differed in the upper quartiles, indicating possible particle transformation. Similarly, the distribution of CA1 released $nTiO_2$ was comparable, except in the lower quartiles of CA1 released $nTiO_2$ obtained under dark conditions. CA1-released ENMs obtained under light and dark conditions were negatively charged at −23.6 ± 1.3 and −22.8 ± 1.2 mV, respectively.

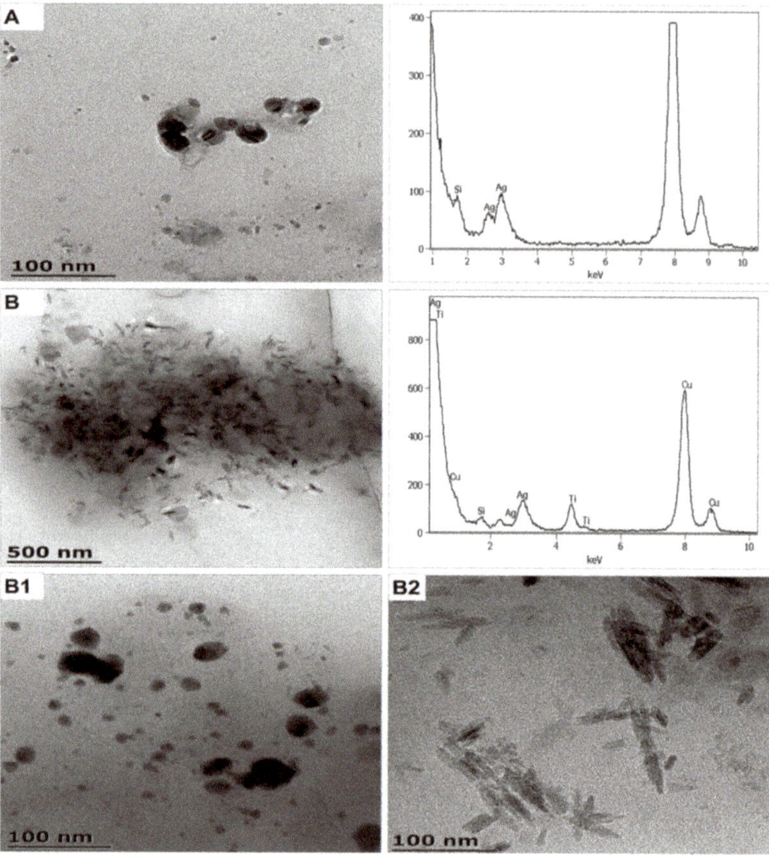

Figure 3. TEM-EDX illustrating SAN1-released ENMs (**A**) and binary CA1-released ENMs obtained under light conditions (**B**). (**B1**,**B2**) are higher magnification of image B showing product-released nAg and product-released $nTiO_2$, respectively.

The presence of different product-released nAg size classes indicated that the ENMs were transformed during release, since the primary size of the ENMs incorporated in the NEPs averaged 21.7 ± 6 (CA1) and 22 ± 7 nm and 37 ± 4 nm (SAN1) [41]. In aquatic environments, pristine nAg are susceptible to undergo various transformations [61], including oxidative dissolution and reformation of Ag particles, leading to the formation of particles of different sizes [62,63]. Peretyazhko et al. [64] found that after the dissolution of pristine nAg, the size of the particles increased due to Ostwald ripening. In the case of particle size decrease, some studies attributed the reduction to the dissolution of nAg, followed by the reduction-driven formation of smaller nAg [65,66]. Furthermore, it was shown that in the absence of environmental factors such as ultraviolet radiation and environmental ligands, a simple dilution of concentrated nAg suspensions and colloidal Ag-based products such as SAN1 can cause particle destabilisation leading to the formation of agglomerates and the reduction in particle size [67,68]. The change in the particle size of nAg incorporated into body cream and mouth spray in artificial sweat and saliva was previously reported [69]; product-released ENMs experienced significant growth in size from 5 to 25 nm to 10 to 800 nm.

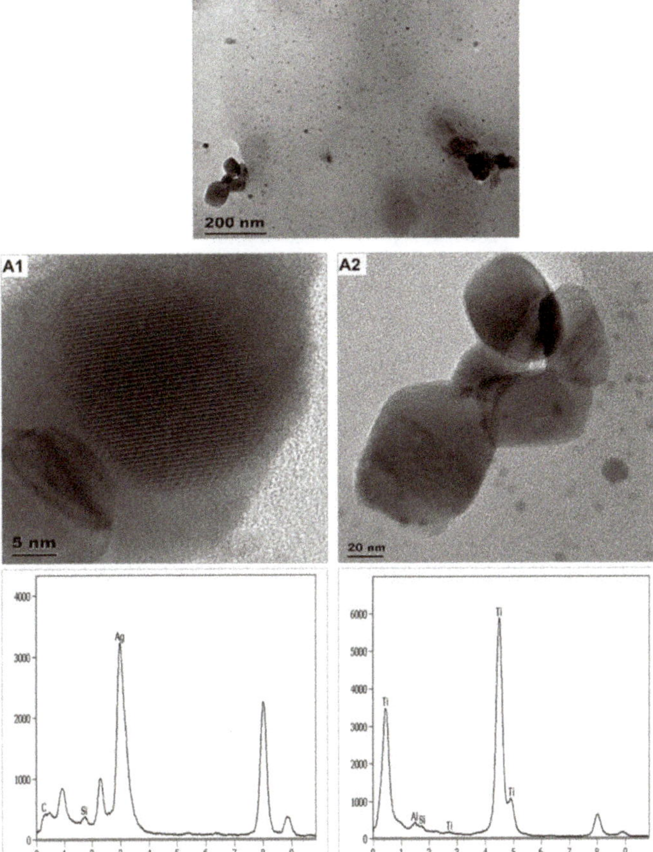

Figure 4. Images and respective spectra obtained from TEM-EDX characterisation of SK1-released ENMs (**A**). Images (**A1**,**A2**) are high magnification of image A, specifically showing near-spherical SK1-released nAg and angular SK1-released $nTiO_2$ particles, respectively.

Environmental exposure to product-released ENMs in aquatic environments has been reported mainly from commercial clothing [70,71], personal care products (toothbrushes, toothpaste, face masks, shampoo, and detergents) [46,47], and paints [72]. The sizes of personal care-released nAg and paint-released nAg were 42–500 nm [46,47] and <15–100 nm [72], respectively. Similar to most release studies, the product-released nAg were still embedded in the NEPs' matrix. Herein, the SAN1-released nAg did not appear to be embedded in the product matrix and were individually isolated or agglomerated; such findings further illustrate that ENMs release potential is influenced by their *loci* in products and product formulation. CA1-released ENMs were often visualised to be encircled by a layer that could not be accurately identified, whether being components of the NEP's matrix or ENMs coating agents; however, it is worth noting that Si was detected in the sample by EDX (Figure 4). The physicochemical state at which the product-released ENMs were detected in aqueous environments was predominantly related to the matrix of NEPs. For example, SAN1 was a clear liquid suspension with a viscosity comparable to water, while CA1 was a semi-solid cream made up of organic compounds.

3.1.3. Socks 1 (SK1) Product-Released ENMs

Washing SK1 released binary ENMs ($nTiO_2$ and nAg) (Figure 4). SK1-released nAg were near-spherical and angular in shape and averaged 8 ± 4 nm and 21–76 × 29–77 nm, respectively. The angular particles were smaller and rapidly agglomerated. SK1-released $nTiO_2$ were angular and averaged 80 ± 25 × 79 ± 29 nm (Figure S10). The distribution of SK1-released $nTiO_2$ and the spherically shaped SK1-released nAg and the respective ENMs in SK1 are comparable. The profile of angular/irregular shaped SK1-released nAg and nAg in SK1 slightly differs, an expected observation since nAg size was affected by the ashing procedure [41].

SK1-released ENMs were coated with Si, and Al, and the coating agents were found to be intact on some SK1-released ENMs (Figure S11). Similar to the previous product-released ENMs (in the preceding sections), the SK1-released ENMs' surface was negatively charged (−33.0 ± 2.1 mV). The current findings are in agreement with previous reports, whereby product-released nAg (20–40 nm) and product-released $nTiO_2$ (60–350 nm) were detected after washing nano-enhanced textiles [70,71].

It is worth mentioning that considerable analytical challenges were initially experienced during the characterisation of SK1-released ENMs. First, SK1-released ENMs were not detected (TEM-EDX) without a pre-enrichment step, especially for SK1-released nAg. After sample enrichment, small particles (~4–6 nm) were imaged but could not be identified because the EDX beam rapidly destroyed them. Finally, the washing detergent introduced a thick layer that concealed the SK1-released ENMs underneath (Figure S12). To improve TEM-EDX characterisation, the number of SK1 units washed concurrently was increased; for this part, the release media was limited to Milli-Q water. Increasing the number of SK1 samples washed simultaneously and concentrating the sample through centrifugation improved TEM-EDX characterisation and enabled SK1-released ENMs particle size quantification.

Overall, the characterisation of product-released ENMs showed that all NEPs investigated in the present study are potential nanopollution sources for water resources. The shapes of the respective product-released ENMs were similar to the ENMs incorporated into the respective NEPs, whose physicochemical properties were previously reported [41]. The sizes of SUNs-released ENMs were comparable to the sizes determined in the NEPs [41]. However, in the case of CA1, SAN1, SK1, the product-released ENMs sizes were slightly different from the ENMs incorporated into the NEPs [41], especially for nAg, where the transformation occurred in terms of the change in particle size (increase and decrease). The physical properties of product-released ENMs are crucial in understanding the behaviour, fate and effects of nanopollutants in aquatic environments, where several studies have already reported their presence in real environmental samples [13,14,73–75].

3.2. Elemental Quantification of Product-Released ENMs

The digestion, analysis, and recovery method of nano- and bulk reference standards were within the acceptable ranges of (75–107%) Ti, (72–97%) Ag, (74–98%) Zn, (70–91%) Al and (70–87%) Si.

3.2.1. Sunscreen 1–3 (SUN1–3) Product-Released ENMs

The total concentration of Ti, Zn, and Zn^{2+} released relative to the initial amount of ENMs added to the sunscreens varied and ranged in general between 0.4 and 8% (w/w) (Figure 5). SUN1–3 released Ti at different extents; in most exposure scenarios, SUN3 > SUN2 > SUN1. In addition to Ti release, SUN1 simultaneously released Zn and Zn^{2+} in the range of 0.67–5.7% (w/w) and 0.5–3.0% (w/w), respectively (Figure 6). The amounts of Zn and Zn^{2+} in the respective product-released ENMs release media were mostly different (Figure 6). Indicative that SUN1 generally releases Zn in particulate and ionic forms.

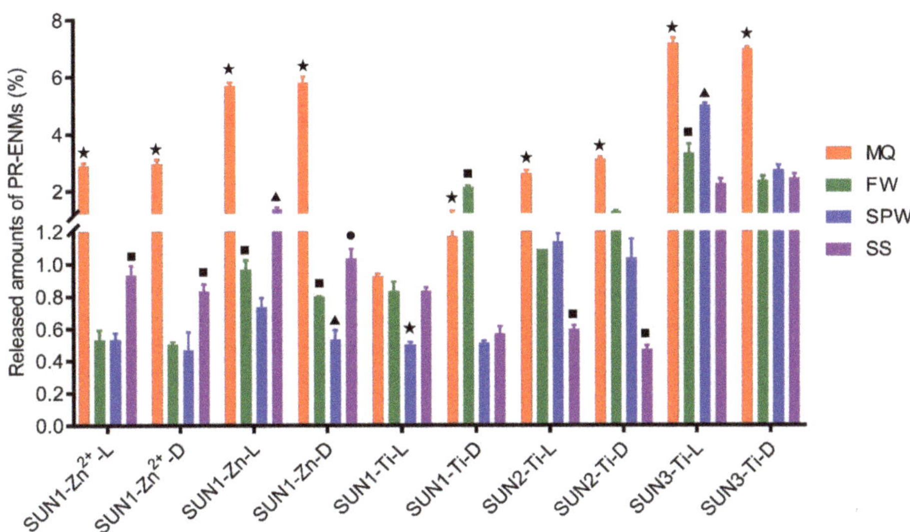

Figure 5. The amount of Zn^{2+}, Zn, and Ti released from SUN1–3 in different release media (Milli-Q water (MQ), freshwater (FW), swimming pool water (SPW), and seawater (SS) under light (L) and dark (D) conditions.The differing of symbols (★ ■ ▲) on top of error bars indicates statistical difference ($p < 0.05$) between the release media treatments.

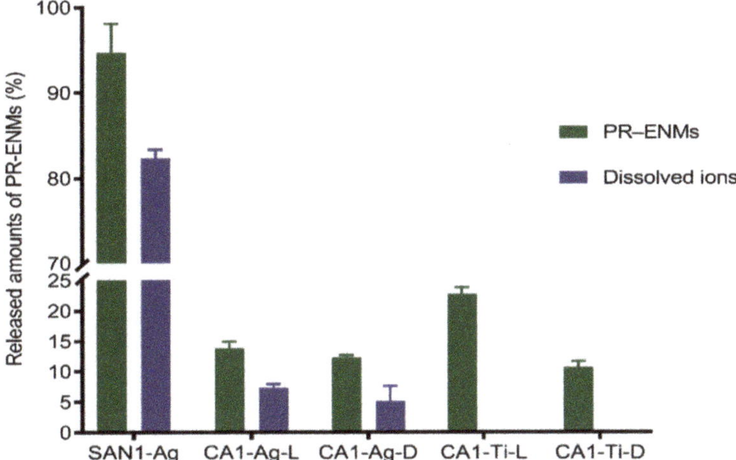

Figure 6. The amounts of SAN1 and CA1-released nAg and released Ag ions; L and D denote light and dark conditions, respectively.

The amounts of Ti, Zn, and Zn^{2+} released from sunscreens were influenced by nature of the NEPs formulation (the initial amount present in the NEP matrix and the product matrix) and simulated environmental conditions (water chemistry and variation in illumination). The influence of the initial amount present in the NEPs was observed between SUN2 and SUN3 (being of the same brand). SUN3, which contained more $nTiO_2$ [1.6% (*w*/*w*)] compared to SUN2 [0.95% (*w*/*w*)] [41], released relatively higher amounts of Ti ($p = 0.0001$–0.01). A further comparison of the amount of Ti released by SUN1–3 showed that the NEPs matrix also influenced Ti release. Although SUN1 contained relatively more $nTiO_2$ [4.31% (*w*/*w*)] than SUN2–3 [<3% (*w*/*w*)] [41], the total amounts of Ti released from

SUN1 were lower than SUN2–3 (Figure 6)—probably an influence of the formulation of the product on the release of ENMs.

In terms of environmental conditions, the amounts of Ti, Zn, and Zn^{2+} were mainly influenced by water chemistry rather than illumination variations; illumination rarely influenced the amounts released. In descending order, the amount of Ti released from SUN1 under light and dark conditions was Milli-Q water $\geq$ freshwater $\geq$ seawater > swimming pool water and freshwater > Milli-Q water > Seawater $\geq$ swimming pool water, respectively. In the case of Zn, the trend of the amounts released under light and dark conditions, in descending order, was Milli-Q water > freshwater> seawater > swimming pool water and Milli-Q water > seawater > freshwater > swimming pool water, respectively. The amounts of Zn^{2+} followed a descending order of Milli-Q water > seawater > freshwater $\geq$ swimming pool water for both illuminations. The Ti amount trends (descending order) of SUN2 and SUN3 were Milli-Q water > freshwater $\geq$ swimming pool water > seawater and Milli-Q water > swimming pool water > freshwater $\geq$ seawater for both illuminations, respectively.

The ionic strength of the release media probably enhanced the agglomeration and sedimentation rate, thus probably causing the differences in the amount released. The release media influenced the dispersion of the sunscreens in the media was different; for example, in Milli-Q water and freshwater, the sunscreens dispersed thoroughly and turned into a homogeneous milky solution, while in other cases, the sunscreen matrix fragmented and formed flocculates. The difference in the dispersion and sedimentation of the NEPs matrix in the release media has implications for ENMs' exposure dynamics in the aqueous phase, as the two (uniform mixture and flocculates) will have different sedimentation rates; flocs sediment faster due to gravity [76]. Different ENMs sedimentation rates were reported in different water chemistries and are influenced by ionic strength, ionic species, and dissolved oxygen [77].

Overall, the current findings illustrated the varying nanopollution characteristics arising from sunscreen NEPs in different water quality environments and that the degree of nanopollution depends on both the NEPs' matrix properties and recipient resource water quality. Furthermore, the results showed that the product-released ENMs will pollute not only the aqueous phase of aquatic environments but also sediments, in addition to adsorption to abiotic and biotic entities. The sedimentation rate influenced the concentrations detected in the suspension, a factor that will be at play in real water bodies as driven by the velocity of the water and other characteristics. Investigations of ENMs sedimentation were carried out on pristine ENMs, and it was found [78] that 50% and 70% of $nTiO_2$ and nZnO were found to sediment within the first 24 h and continued to slowly sediment for the next 2 to 14 days in natural water, respectively. Similarly, the study by Botta et al. [43] showed that a significant proportion of sunscreen-released $nTiO_2$ in seawater aggregated and sedimented. The rate of sedimentation influences the exposure dynamics of benthic organisms. Beyond the release stage, the behaviour of product-released ENMs in aquatic environments and the effects on benthic organisms are not well understood and warrant detailed attention. As such, at more robust levels, ENMs exposure assessment must consider aquatic resource characteristics.

3.2.2. Sanitiser 1 (SAN1) and Body Cream (CA1) Product-Released ENMs

The amount of Ag, Ag^+, and Ti released from SAN1 and CA1 varied (Figure 6). SAN1 released considerably higher amounts of Ag than CA1 ($p = 0.001$); the characteristics of the NEPs matrix probably caused the observed difference—further illustrating the influential role of the NEPs matrix in the potential for exposure to ENMs. Both SAN1 and CA1 released Ag in particulate and dissolved forms. The amount of Ag and Ag^+ in the respective release media of SAN1 ($p = 0.0002$) and CA1 ($p = 0.003$–0.005) varied, indicating the coexistence of particulate and ionic Ag. CA1 further released Ti amounts higher and comparable to Ag under light ($p = 0.02$) and dark ($p = 0.056$) conditions.

Some NEPs containing nAg were classified as having medium to high exposure potential to water resources [6,7,79]. The studies reported that toothbrushes released nAg (5.9–626 ng/L) [47], paints released nAg (30%) [72], and plush toy exterior fur released nAg (<1–35%) [80]. It is estimated that Ag can be released from products in the range of 25–100% in wastewater treatment plants [81]. In most cases, the NEPs release Ag in the particulate or ionic form at varying degrees [47,80]. The form and extent of Ag release from nAg are complex because speciation is influenced by various factors, such as particle size, coating agents, and release media characteristics [82–86].

The dissolution of nAg from both NEPs may be due to the small-sized particles incorporated in SAN1 (10–37 nm) and CA1 (13–44 nm) and the change in particle size of product-released nAg (as observed by the detection of particles of different sizes) may have contributed to the degree of dissolution observed in the exposures of SAN1 and CA1. Nanoparticle size reduction was previously reported to result in increased dissolution due to increased surface area [87,88].

3.2.3. Socks 1 (SK1)-Released ENMs

SK1 released 0.004–0.100 mg/L and 2.66–5.98 mg/L of total Ag and Ti, respectively (Figure 7). The Ag and Ti concentrations released from SK1 were not normalised back to the initial concentration incorporated into the NEPs because the Ag and Ti present in different SK1 materials were inconsistent. Incorporation of ENMs of different properties by manufacturers has recently been reported [89]. As illustrated in Figure 7, the amounts quantified in the different wash cycles varied between the release media and fractions. The amounts quantified for particulate fractions in Milli-Q water were not different (p = 0.22–0.67). In the case of tap water, the difference was only observed in the >0.45 μm fraction, where a higher Ag concentration was determined in the second cycle. The amount released from sodium dodecyl sulfate 1 and sodium dodecyl sulfate 2 also varied; relatively large amounts were detected in the first cycle for >0.45 μm (p = 0.001–0.03) and <0.45 μm (0.06–0.21). For Ag^+, only sodium dodecyl sulfate 2 released higher amounts in the first cycle (p = 0.01). Released Ag^+ was detected in comparable amounts between cycles in Milli-Q water, tap water, and sodium dodecyl sulfate 1 (p = 0.27–0.99). As shown in Figure 7, sodium dodecyl sulfate 1 release media mainly affected Ag forms, compared to tap water and especially Milli-Q water.

In cases where Ag amounts varied in different fractions, more Ag was detected in >0.45 μm, while <3 kDa was comparable or lower than <0.45 μm. Contrary to Ag and Ag^+, the amounts of Ti were comparable in the first and second wash cycles for all wash media, except for sodium dodecyl sulfate 1, where the amounts of Ti were higher in the first cycle. As shown in Figure 7, the amounts of Ti in the different fractions were comparable, except for tap water, sodium dodecyl sulfate 1 and sodium dodecyl sulfate 2, where the highest amounts were quantified in fractions> 0.45 m and fractions >0.45 μm and <0.45 μm fractions, respectively. Overall, nAg release was more affected by the simulated washing conditions than $nTiO_2$ incorporated in SK1.

The environmental exposure of particulates and ions of Ag [33–35,44,90,91] and Ti [71,92], respectively, released from commercial textile products enabled with nAg or $nTiO_2$ have been investigated, although, in some instances, there were differences between studies. The differences were mainly caused by the assessment of different clothing materials, the type of ENMs nanocomposite, the initial amounts of ENMs added to the NEPs, ENMs incorporation methods, ENMs location within the NEPs, and the chemistry of the release media.

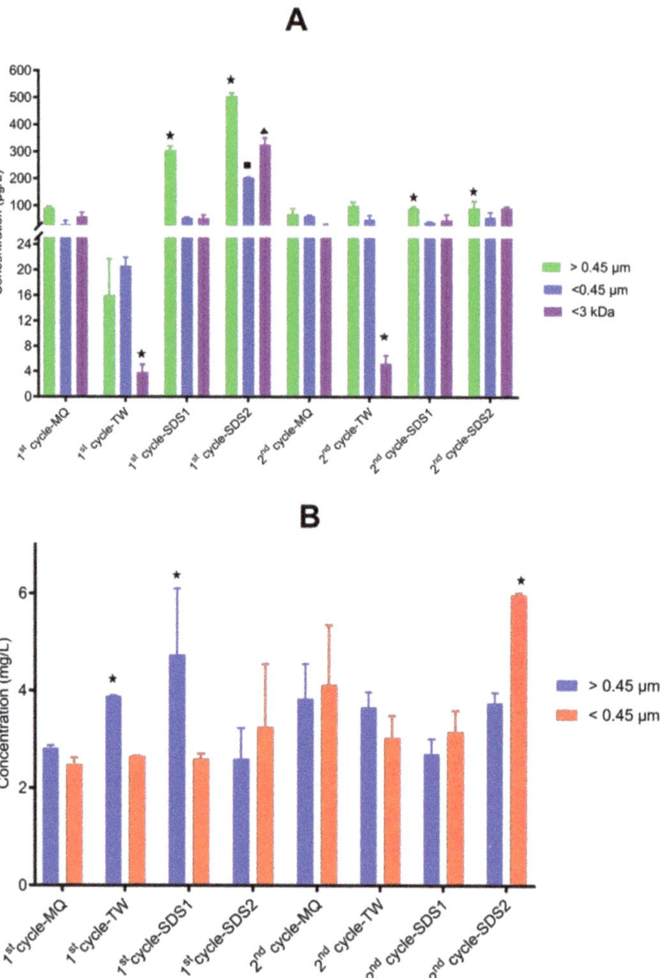

Figure 7. Total amounts of Ag (**A**) and Ti (**B**) released from SK1 in Milli-Q water (MQ), tap water (TW), sodium dodecyl sulfate 1 (SDS1), and sodium dodecyl sulfate (SDS2) in two wash cycles. The symbol (★ ■) on top of error bars denote statistical difference ($p < 0.05$) between the released Ag and Ti fractions per wash cycle.

Nonetheless, the current study correlated with some previous reports on the high amount of Ag detected mainly in the first wash cycle [14,75,77,80] and the detection of higher amounts of particulate Ag [44]. Thus, elevated ENMs release from fabric NEPs can be expected from initial washes after purchase. Overall, the amounts released in the current study were in agreement with the previous reporting of 0.32–38.5 mg/L for Ag [34,91,93] and 5 mg/L for Ti [71]. Exposure assessments of commercial TiO_2 nano-enabled textiles compared to nAg-enabled textiles are currently scarce. The high market penetration of nAg functional textiles and the primary function of nAg (antimicrobial properties) could be the reason behind the difference in the number of studies undertaken.

3.2.4. Release of ENMs Coating

ENMs coating agents in SUN1–3 (Figures S4 and S5), CA1 (Figure S13) and SK1 (Figure S8) were somewhat desorbed from the surface of ENMs, and the components of the coating agent (Si and Al) were released into the respective media. The Si and Al coating agents of the released ENMs were determined in <3 kDa filtrate to avoid coating agents still attached to the product-released ENMs. Although TEM-EDX analysis showed that Si and Al were coated on the surface of ENMs, the presence of these elements as part of the other matrix of NEPs may exist; for all NEPs, manufacturers neither declared the element nor the quantity. Because of uncertainties, for this exercise, the Si and Al are assumed to originate from ENMs coatings and the overall NEPs matrix.

The findings on the extent to which ENMs coating agents were released were recorded in Milli-Q water release media. The amount of Si released varied between the NEPs (Table 2).

Table 2. The concentration of Si released from SUN1–3, CA1, and SK1.

Sample Name	Release Amount of Si% (*w*/*w*)	
	Light Condition	**Dark Condition**
SUN1	10.1 ± 1.4	11.2 ± 2.5
SUN2	5.4 ± 0.9	5.7 ± 0.8
SUN3	3.7 ± 0.4	3.4 ± 0.4
CA1	23 ± 2	19 ± 6

Si was detected in the respective release media (SUN1–3 and CA1) in the descending order of CA1 > SUN1 > SUN3 $\geq$ SUN2 under both illumination conditions. The SUN1 ENMs, which were coated with Si and Al, released Si in large amounts compared to Al; Al amounts in the product-released ENMs media of SUN1 were below the detection limit (LOD = 10 μg/L). Elemental concentrations were consistent with the EDX observations, where Si desorbed and released into the product-released ENMs release media, while Al was still partially attached to the product-released ENMs (Figure S4). Similarly, the low amounts of Si detected in SUN2 and SUN3 corroborated the earlier findings of TEM-EDX (Figure S5), where Si was observed to be still attached to product-released ENMs. In the case of CA1, which released the highest amounts of Si, TEM-EDX analysis (Figure S13) showed that most of the Si disassociated from product-released ENMs. For SK1, the amounts of Si and Al were determined to be 10 and 4.76 mg/L, respectively, also confirming the release observed with TEM-EDX (Figure S11).

Although ENMs incorporated into NEPs are well known to be enclosed with coating agents [94] and have been shown to be altered during the aqueous ageing of nanocomposites (used in cosmetics) and released into aquatic environments [52,95–97]; the amount of the coating agent components released from NEPs is often not reported. Until recently, nanocomposites intended for NEPs formulations, such as sunscreens, were evaluated [97]; 1.5–2% (*w*/*w*) of Si was released into ultrapure water, while higher amounts of 88–98% (*w*/*w*) were simulated freshwater and seawater. It is imperative that the amounts of released coating agents are quantified and considered when the risks of product-released ENMs are evaluated, as it is currently unclear whether the components of the coating agent influence the product-released ENMs to what extent, and therefore future studies should evaluate their association.

4. Conclusions

The study successfully illustrated the nanopollution of water media during the simulated use of NEPs by characterising product-released ENMs from a wide range of products. The product-released $nTiO_2$ were elongated (7–9 × 66–70 nm) or angular (21–80 × 25–79 nm) in shape; product-released nAg were near-spherical (12–49 nm) or angular (21–76 × 29–77 nm) and product-released nZnO were angular (32–36 × 32–40 nm) in shape. The ENMs release rate was determined to be *ca* 0.4–95% relative to the initial amount of ENMs added to NEPs. The extent and characteristics of product-released ENMs were influenced by receiving water quality, ENMs *loci* in the product, and the formulation of the product matrix, while illumination variation essentially did not exert influence. Predominantly, the product-released ENMs were released in association with coating agents (Si and Al) and ionic forms. Considering the influential role the surface coating exerts on the behaviour and toxicity of ENMs in water resources, we highly recommend the reporting of the presence and characteristics of coating agents on product-released ENMs since it is currently not standard practice.

SUN1, CA1 and SK1 released binary ENMs. Typically, there is currently limited information on the environmental implications of ENMs mixtures, more so for product-released ENMs; hence, we encourage more studies to unravel the exposure and hazard dynamics of product-released ENMs mixtures.

Nanopollution is an emerging environmental health issue that is yet to be clearly quantified. Nevertheless, proactive mitigation measures can reduce environmental exposure, for instance, the reduction in ENMs quantity in NEPs (safety-by-design principle), since this study demonstrated that the NEPs sample caused nanopollution. In low- and middle-income countries, such as South Africa, where the current study was carried out, there must be accelerated efforts to estimate the size of the NEPs market to refine the extent of nanopollution, as developed regions have advanced in that aspect.

Supplementary Materials: The following are available online at https://www.mdpi.com/2079-4991/11/10/2537/s1, S1.1 Properties of the release media of SUN1–3, Table S1. Average physicochemical properties of release media before and after ENMs release, Figure S1. The EDX spectra of product-released ENMs obtained under light conditions for SUN1 detected in milli-Q water (A), freshwater (B), swimming pool water (C), seawater (D); SUN2 detected in milli-Q water (E), freshwater (F), swimming pool water (G), seawater (H) and SUN3 detected in milli-Q water (I), freshwater (J), swimming pool water (K), seawater (L), Figure S2. TEM images of product-released ENMs obtained under dark conditions for SUN1 detected in milli-Q water (A), freshwater (B), swimming pool water (C), seawater (D); SUN2 detected in milli-Q water (E), freshwater (F), swimming pool water (G), seawater (H) and SUN3 detected in milli-Q water (I), freshwater (J), swimming pool water (K), seawater (L), Figure S3. Corresponding EDX images of product-released ENMs obtained under dark conditions for SUN1 detected in milli-Q water (A), freshwater (B), swimming pool water (C), seawater (D); SUN2 detected in milli-Q water (E), freshwater (F), swimming pool water (G), seawater (H) and SUN3 detected in milli-Q water (I), freshwater (J), swimming pool water (K), seawater (L), Figure S4. EDX elemental mapping showing adsorption and desorption of ENMs coating agents (Si and Al) on SUN1-released ENMs, Figure S5. EDX elemental mapping showing adsorption and desorption of ENMs coating agents (Si) on SUN2 (A)- and SUN3 (B)-released ENMs, Figure S6A. Violin plot showing particle distribution of SUN1(A)-, SUN2 (B)-, and SUN3 (C)-released ENMs obtained under light conditions. The upper and lower quartiles are highlighted by a solid line, while the dotted line indicates the median. The denser the violin shape, the higher the number of particle size in that region, Figure S6B. Violin plot showing particle distribution of SUN1(A)-, SUN2 (B)-, and SUN3 (C)-released ENMs obtained under dark conditions. The upper and lower quartiles are highlighted by a solid line, while the dotted line indicates the median. The denser the violin shape, the higher the number of particle size in that region, Figure S7. Zeta potential of SUN1–3-released ENMs obtained under dark conditions in different release media of milli-Q water (MQ), freshwater (FW), swimming pool water (SPW), and seawater (SS), Figure S8. Violin plot showing particle distribution of CA1-released ENMs (obtained under light and dark conditions) and SAN1-released ENMs. The upper and lower quartiles are highlighted by a solid line, while the dotted line indicates the median. The denser the violin shape, the higher the number of particle size in that region, Figure S9. TEM-EDX image showing of CA1 product-released nAg and product-released $nTiO_2$ obtained under dark conditions, Figure S10. Violin plot showing particle distribution SK1-released ENMs. The upper and lower quartiles are highlighted by a solid line, while the dotted line indicates the median. The denser the violin shape, the higher the number of particle size in that region, Figure S11. Elemental mapping of binary SK1-released ENMs identified as product-released $nTiO_2$ (yellow) and product-released nAg (red). The images further show evidence of SK1-released $nTiO_2$ particles partially still coated with Si and Al, Figure S12. TEM images showing the thick layer introduced by washing SK1 with sodium dodecyl sulfate release media, Figure S13. EDX elemental mapping illustrating Si desorbed from CA1-released ENMs.

Author Contributions: Conceptualisation, M.T. and R.F.L.; methodology, R.F.L. and M.T.; software, R.F.L.; validation, M.T.; formal analysis, R.F.L.; investigation, R.F.L.; resources, M.T.; data curation, R.F.L.; writing—original draft preparation, R.F.L.; writing—review and editing, R.F.L. and M.T.; visualisation, R.F.L. and M.T.; supervision, M.T.; project administration, M.T.; funding acquisition, M.T. All authors have read and agreed to the published version of the manuscript.

Funding: This research was funded by the South African Department of Science and Technology (DST) under the Nanotechnology Health, Safety and Environment Risk Research Platform (grant number: 0085/2015).

Institutional Review Board Statement: Not applicable.

Informed Consent Statement: Not applicable.

Data Availability Statement: The data presented in this study are available on a reasonable request from the corresponding author.

Acknowledgments: This work was supported by the South African Department of Science and Technology (DST) under the Nanotechnology Health, Safety and Environment Risk Research Platform (grant number: 0085/2015). We further wish to thank Jérôme Rose, Melanie Auffan, and Danielle Slomberg funded by Gov4Nano H2020 European Commission project (grant number: 814401), the SERENADE project (Labex Serenade programme (grant number. ANR-11-LABX-0064), and the "Investissements d'Avenir" programme of the French National Research Agency (ANR) through the A*MIDEX project (grant number: ANR-11-IDEX-0001-02) for hosting and mentoring L.R.F.

Conflicts of Interest: The authors declare no conflict of interest.

References

1. Foss Hansen, S.; Heggelund, L.R.; Revilla Besora, P.; Mackevica, A.; Boldrin, A.; Baun, A.; Hansen, S.F.; Heggelund, L.R.; Revilla Besora, P.; Mackevica, A.; et al. Nanoproducts—What is actually available to European consumers? *Environ. Sci. Nano* **2016**, *3*, 169–180. [CrossRef]
2. BCC Research Global Nanotechnology Market (by Component and Applications), Funding & Investment, Patent Analysis and 27 Companies Profile & Recent Developments—Forecast to 2024. Available online: https://www.bccresearch.com/market-research/nanotechnology (accessed on 7 October 2020).
3. Vance, M.E.; Kuiken, T.; Vejerano, E.P.; McGinnis, S.P.; Hochella, M.F.; Rejeski, D.; Hull, M.S.; Hull, M.S.; Hull, D.R.; Rejeski, D.; et al. Nanotechnology in the real world: Redeveloping the nanomaterial consumer products inventory. *Beilstein J. Nanotechnol.* **2015**, *6*, 1769–1780. [CrossRef]
4. PEN Inventory Finds Increase in Consumer Products Containing Nanoscale Materials. News Archive. Nanotechnology Project. Available online: http://www.nanotechproject.org/news/archive/9242/ (accessed on 4 June 2019).
5. The Nanodatabase. The Nanodatabase is developed by the DTU Environment, the Danish Ecological Council and Danish Consumer Council. Available online: http://nanodb.dk/en/about-us/ (accessed on 21 July 2017).
6. Moeta, P.J.; Wesley-Smith, J.; Maity, A.; Thwala, M. Nano-enabled products in South Africa and the assessment of environmental exposure potential for engineered nanomaterials. *SN Appl. Sci.* **2019**, *1*, 577. [CrossRef]
7. Zhang, Y.; Leu, Y.R.; Aitken, R.J.; Riediker, M. Inventory of engineered nanoparticle-containing consumer products available in the singapore retail market and likelihood of release into the aquatic environment. *Int. J. Environ. Res. Public Health* **2015**, *12*, 8717–8743. [CrossRef] [PubMed]
8. Biswas, J.K.; Sarkar, D. Nanopollution in the Aquatic Environment and Ecotoxicity: No Nano Issue! *Curr. Pollut. Rep.* **2019**, *5*, 4–7. [CrossRef]
9. Rajkovic, S.; Bornhöft, N.A.; van der Weijden, R.; Nowack, B.; Adam, V. Dynamic probabilistic material flow analysis of engineered nanomaterials in European waste treatment systems. *Waste Manag.* **2020**, *113*, 118–131. [CrossRef] [PubMed]
10. Adam, V.; Caballero-Guzman, A.; Nowack, B. Considering the forms of released engineered nanomaterials in probabilistic material flow analysis. *Environ. Pollut.* **2018**, *243*, 17–27. [CrossRef]
11. Reed, R.B.; Martin, D.P.; Bednar, A.J.; Montaño, M.D.; Westerhoff, P.; Ranville, J.F. Multi-day diurnal measurements of Ti-containing nanoparticle and organic sunscreen chemical release during recreational use of a natural surface water. *Environ. Sci. Nano* **2017**, *4*, 69–77. [CrossRef]
12. Wang, J.; Nabi, M.M.; Mohanty, S.K.; Afrooz, A.N.; Cantando, E.; Aich, N.; Baalousha, M. Detection and quantification of engineered particles in urban runoff. *Chemosphere* **2020**, *248*, 126070. [CrossRef] [PubMed]

13. Slomberg, D.L.; Auffan, M.; Guéniche, N.; Angeletti, B.; Campos, A.; Borschneck, D.; Aguerre-Chariol, O.; Rose, J. Anthropogenic Release and Distribution of Titanium Dioxide Particles in a River Downstream of a Nanomaterial Manufacturer Industrial Site. *Front. Environ. Sci.* **2020**, *8*, 76. [CrossRef]
14. Gondikas, A.P.; von der Kammer, F.; Reed, R.B.; Wagner, S.; Ranville, J.F.; Hofmann, T.; Von Der Kammer, F.; Reed, R.B.; Wagner, S.; Ranville, J.F.; et al. Release of TiO_2 nanoparticles from sunscreens into surface waters: A one-year survey at the old danube recreational lake. *Environ. Sci. Technol.* **2014**, *48*, 5415–5422. [CrossRef] [PubMed]
15. Shi, X.; Li, Z.; Chen, W.; Qiang, L.; Xia, J.; Chen, M.; Zhu, L.; Alvarez, P.J.J. Fate of TiO_2 nanoparticles entering sewage treatment plants and bioaccumulation in fish in the receiving streams. *NanoImpact* **2016**, *3–4*, 96–103. [CrossRef]
16. Mueller, N.C.; Nowack, B. Exposure Modeling of Engineered Nanoparticles in the Environment. *Environ. Sci. Technol.* **2008**, *42*, 63. [CrossRef] [PubMed]
17. Gottschalk, F.; Sonderer, T.; Scholz, R.W.; Nowack, B. Modeled Environmental Concentrations of Engineered Nanomaterials (TiO_2, ZnO, Ag, CNT, Fullerenes) for Different Regions. *Environ. Sci. Technol.* **2009**, *43*, 9216–9222. [CrossRef]
18. Coll, C.; Notter, D.; Gottschalk, F.; Sun, T.; Som, C.; Nowack, B. Probabilistic environmental risk assessment of five nanomaterials (nano-TiO_2, nano-Ag, nano-ZnO, CNT, and fullerenes). *Nanotoxicology* **2016**, *10*, 436–444. [CrossRef]
19. Musee, N. Simulated environmental risk estimation of engineered nanomaterials: A case of cosmetics in Johannesburg City. *Hum. Exp. Toxicol.* **2011**, *30*, 1181–1195. [CrossRef]
20. Sun, T.Y.; Bornhöft, N.A.; Hungerbühler, K.; Nowack, B. Dynamic Probabilistic Modeling of Environmental Emissions of Engineered Nanomaterials. *Environ. Sci. Technol.* **2016**, *50*, 4701–4711. [CrossRef]
21. Peters, R.J.B.B.; van Bemmel, G.; Milani, N.B.L.L.; den Hertog, G.C.T.T.; Undas, A.K.; van der Lee, M.; Bouwmeester, H. Detection of nanoparticles in Dutch surface waters. *Sci. Total Environ.* **2018**, *621*, 210–218. [CrossRef]
22. Aznar, R.; Barahona, F.; Geiss, O.; Ponti, J.; Luis, T.J.; Barrero-Moreno, J. Quantification and size characterisation of silver nanoparticles in environmental aqueous samples and consumer products by single particleICPMS. *Talanta* **2017**, *175*, 200–208. [CrossRef]
23. Li, L.; Stoiber, M.; Wimmer, A.; Xu, Z.; Lindenblatt, C.; Helmreich, B.; Schuster, M. To What Extent Can Full-Scale Wastewater Treatment Plant Effluent Influence the Occurrence of Silver-Based Nanoparticles in Surface Waters? *Environ. Sci. Technol.* **2016**, *50*, 6327–6333. [CrossRef] [PubMed]
24. Venkatesan, A.K.; Reed, R.B.; Lee, S.; Bi, X.; Hanigan, D.; Yang, Y.; Ranville, J.F.; Herckes, P.; Westerhoff, P. Detection and Sizing of Ti-Containing Particles in Recreational Waters Using Single Particle ICP-MS. *Bull. Environ. Contam. Toxicol.* **2018**, *100*, 120–126. [CrossRef]
25. Wu, S.; Zhang, S.; Gong, Y.; Shi, L.; Zhou, B. Identification and quantification of titanium nanoparticles in surface water: A case study in Lake Taihu, China. *J. Hazard. Mater.* **2020**, *382*, 121045. [CrossRef]
26. Loosli, F.; Wang, J.; Rothenberg, S.; Bizimis, M.; Winkler, C.; Borovinskaya, O.; Flamigni, L.; Baalousha, M. Sewage spills are a major source of titanium dioxide engineered (nano)-particle release into the environment. *Environ. Sci. Nano* **2019**, *6*, 763–777. [CrossRef]
27. Garner, K.L.; Suh, S.; Keller, A.A. Assessing the Risk of Engineered Nanomaterials in the Environment: Development and Application of the nanoFate Model. *Environ. Sci. Technol.* **2017**, *51*, 5541–5551. [CrossRef]
28. Lehutso, R.F.; Tancu, Y.; Maity, A.; Thwala, M. Aquatic toxicity of transformed and product-released engineered nanomaterials: An overview of the current state of knowledge. *Process Saf. Environ. Prot.* **2020**, *138*, 39–56. [CrossRef]
29. Mitrano, D.M.; Motellier, S.; Clavaguera, S.; Nowack, B. Review of nanomaterial aging and transformations through the life cycle of nano-enhanced products. *Environ. Int.* **2015**, *77*, 132–147. [CrossRef] [PubMed]
30. Mitrano, D.M.; Nowack, B. The need for a life-cycle based aging paradigm for nanomaterials: Importance of real-world test systems to identify realistic particle transformations. *Nanotechnology* **2017**, *28*, 072001. [CrossRef]
31. Auffan, M.; Pedeutour, M.; Rose, J.; Masion, A.; Ziarelli, F.; Borschneck, D.; Chaneac, C.; Botta, C.; Chaurand, P.; Labille, J.; et al. Structural degradation at the surface of a TiO_2-based nanomaterial used in cosmetics. *Environ. Sci. Technol.* **2010**, *44*, 2689–2694. [CrossRef] [PubMed]
32. Mitrano, D.M.; Lombi, E.; Dasilva, Y.A.R.; Nowack, B.; Arroyo, Y.; Dasilva, R.; Nowack, B. Unraveling the Complexity in the Aging of Nanoenhanced Textiles: A Comprehensive Sequential Study on the Effects of Sunlight and Washing on Silver Nanoparticles. *Environ. Sci. Technol.* **2016**, *50*, 5790–5799. [CrossRef] [PubMed]
33. Gagnon, V.; Button, M.; Boparai, H.K.; Nearing, M.; O'Carroll, D.M.; Weber, K.P.; Nearing, M.; Weber, K.P.; Boparai, H.K.; Button, M.; et al. Influence of realistic wearing on the morphology and release of silver nanomaterials from textiles. *Environ. Sci. Nano* **2018**, *6*, 411–424. [CrossRef]
34. Lorenz, C.; Windler, L.; von Goetz, N.; Lehmann, R.P.P.; Schuppler, M.; Hungerbühler, K.; Heuberger, M.; Nowack, B. Characterization of silver release from commercially available functional (nano)textiles. *Chemosphere* **2012**, *89*, 817–824. [CrossRef]
35. Lombi, E.; Donner, E.; Scheckel, K.G.; Sekine, R.; Lorenz, C.; Von Goetz, N.; Nowack, B. Silver speciation and release in commercial antimicrobial textiles as influenced by washing. *Chemosphere* **2014**, *111*, 352–358. [CrossRef]
36. Koivisto, A.J.; Jensen, A.C.Ø.; Kling, K.I.; Nørgaard, A.; Brinch, A.; Christensen, F.; Jensen, K.A. Quantitative material releases from products and articles containing manufactured nanomaterials: Towards a release library. *NanoImpact* **2017**, *5*, 119–132. [CrossRef]

37. Nowack, B.; Mitrano, D.M. Procedures for the production and use of synthetically aged and product released nanomaterials for further environmental and ecotoxicity testing. *NanoImpact* **2018**, *10*, 70–80. [CrossRef]
38. Hristozov, D.R.; Gottardo, S.; Critto, A.; Marcomini, A. Risk assessment of engineered nanomaterials: A review of available data and approaches from a regulatory perspective. *Nanotoxicology* **2012**, *6*, 880–898. [CrossRef] [PubMed]
39. Jacobs, R.; Meesters, J.A.J.; ter Braak, C.J.F.; van de Meent, D.; van der Voet, H. Combining exposure and effect modeling into an integrated probabilistic environmental risk assessment for nanoparticles. *Environ. Toxicol. Chem.* **2016**, *35*, 2958–2967. [CrossRef]
40. Salieri, B.; Turner, D.A.; Nowack, B.; Hischier, R. Life cycle assessment of manufactured nanomaterials: Where are we? *NanoImpact* **2018**, *10*, 108–120. [CrossRef]
41. Lehutso, R.F.; Tancu, Y.; Maity, A.; Thwala, M. Molecules Characterisation of Engineered Nanomaterials in Nano-Enabled Products Exhibiting Priority Environmental Exposure. *Molecules* **2020**, *26*, 1370. [CrossRef]
42. Nthwane, Y.B.; Tancu, Y.; Maity, A.; Thwala, M. Characterisation of titanium oxide nanomaterials in sunscreens obtained by extraction and release exposure scenarios. *SN Appl. Sci.* **2019**, *1*, 312. [CrossRef]
43. Botta, C.; Labille, J.; Auffan, M.; Borschneck, D.; Miche, H.H.; Cabié, M.; Masion, A.; Rose, J.; Bottero, J.-Y.Y. TiO_2-based nanoparticles released in water from commercialized sunscreens in a life-cycle perspective: Structures and quantities. *Environ. Pollut.* **2011**, *159*, 1543–1550. [CrossRef]
44. Limpiteeprakan, P.; Babel, S.; Lohwacharin, J.; Takizawa, S. Release of silver nanoparticles from fabrics during the course of sequential washing. *Environ. Sci. Pollut. Res.* **2016**, *23*, 22810–22818. [CrossRef] [PubMed]
45. Mitrano, D.M.; Arroyo Rojas Dasilva, Y.; Nowack, B. Effect of Variations of Washing Solution Chemistry on Nanomaterial Physicochemical Changes in the Laundry Cycle. *Environ. Sci. Technol.* **2015**, *49*, 9665–9673. [CrossRef] [PubMed]
46. Benn, T.; Cavanagh, B.; Hristovski, K.; Posner, J.D.; Westerhoff, P. The Release of Nanosilver from Consumer Products Used in the Home. *J. Environ. Qual.* **2010**, *39*, 1875. [CrossRef] [PubMed]
47. Mackevica, A.; Olsson, M.E.; Hansen, S.F. The release of silver nanoparticles from commercial toothbrushes. *J. Hazard. Mater.* **2017**, *322*, 270–275. [CrossRef]
48. CEM MARS Microwave Digestion—Sample Preparation using Microwave Digestion. Available online: http://cem.com/microwave-digestion (accessed on 16 July 2018).
49. Lewicka, Z.A.; Benedetto, A.F.; Benoit, D.N.; Yu, W.W.; Fortner, J.D.; Colvin, V.L. The structure, composition, and dimensions of TiO_2 and ZnO nanomaterials in commercial sunscreens. *J. Nanopart. Res.* **2011**, *13*, 3607–3617. [CrossRef]
50. Philippe, A.; Košík, J.; Welle, A.; Guigner, J.-M.; Clemens, O.; Schaumann, G.E. Extraction and characterization methods for titanium dioxide nanoparticles from commercialized sunscreens. *Environ. Sci. Nano* **2018**, *5*, 191–202. [CrossRef]
51. Lu, P.-J.J.; Huang, S.-C.C.; Chen, Y.-P.P.; Chiueh, L.-C.C.; Shih, D.Y.-C.C. Analysis of titanium dioxide and zinc oxide nanoparticles in cosmetics. *J. Food Drug Anal.* **2015**, *23*, 587–594. [CrossRef]
52. Al-Abed, S.R.; Virkutyte, J.; Ortenzio, J.N.R.; McCarrick, R.M.; Degn, L.L.; Zucker, R.; Coates, N.H.; Childs, K.; Ma, H.; Diamond, S.; et al. Environmental aging alters $Al(OH)_3$ coating of TiO_2 nanoparticles enhancing their photocatalytic and phototoxic activities. *Environ. Sci. Nano* **2016**, *3*, 593–601. [CrossRef]
53. Camilleri, C.; Markich, S.J.; Noller, B.N.; Turley, C.J.; Parker, G.; Van Dam, R.A. Silica reduces the toxicity of aluminium to a tropical freshwater fish (*Mogurnda mogurnda*). *Chemosphere* **2003**, *50*, 355–364. [CrossRef]
54. Nnamdi, A.H.; Briggs, T.M.D.; Togunde, O.O.; Obanya, H.E. Antagonistic effects of sublethal concentrations of certain mixtures of metal oxide nanoparticles and the bulk (Al_2O_3, CuO, and SiO_2) on gill histology in clarias gariepinus. *J. Nanotechnol.* **2019**, *2019*, 7686597. [CrossRef]
55. Ríos, F.; Fernández-Arteaga, A.; Fernández-Serrano, M.; Jurado, E.; Lechuga, M. Silica micro- and nanoparticles reduce the toxicity of surfactant solutions. *J. Hazard. Mater.* **2018**, *353*, 436–443. [CrossRef]
56. Choudhary, R.; Khurana, D.; Kumar, A.; Subudhi, S. Stability analysis of Al_2O_3/water nanofluids. *J. Exp. Nanosci.* **2017**, *12*, 140–151. [CrossRef]
57. Negm, N.A.; Tawfik, S.M.; Abd-Elaal, A.A. Synthesis, characterization and biological activity of colloidal silver nanoparticles stabilized by gemini anionic surfactants. *J. Ind. Eng. Chem.* **2015**, *21*, 1051–1057. [CrossRef]
58. Shao, X.-R.; Wei, X.-Q.; Song, X.; Hao, L.-Y.; Cai, X.-X.; Zhang, Z.-R.; Peng, Q.; Lin, Y.-F. Independent effect of polymeric nanoparticle zeta potential/surface charge, on their cytotoxicity and affinity to cells. *Cell Prolif.* **2015**, *48*, 465–474. [CrossRef] [PubMed]
59. Zhu, Z.J.; Wang, H.; Yan, B.; Zheng, H.; Jiang, Y.; Miranda, O.R.; Rotello, V.M.; Xing, B.; Vachet, R.W. Effect of surface charge on the uptake and distribution of gold nanoparticles in four plant species. *Environ. Sci. Technol.* **2012**, *46*, 12391–12398. [CrossRef]
60. Jeon, S.; Kim, E.; Lee, J.; Lee, S. Potential risks of TiO_2 and ZnO nanoparticles released from sunscreens into outdoor swimming pools. *J. Hazard. Mater.* **2016**, *317*, 312–318. [CrossRef]
61. Levard, C.; Hotze, E.M.; Lowry, G.V.; Brown, G.E. Environmental transformations of silver nanoparticles: Impact on stability and toxicity. *Environ. Sci. Technol.* **2012**, *46*, 6900–6914. [CrossRef]
62. Lowry, G.V.; Gregory, K.B.; Apte, S.C.; Lead, J.R. Transformations of Nanomaterials in the Environment. *Environ. Sci. Technol.* **2012**, *46*, 6893–6899. [CrossRef]
63. Zhang, W.; Xiao, B.; Fang, T. Chemical transformation of silver nanoparticles in aquatic environments: Mechanism, morphology and toxicity. *Chemosphere* **2018**, *191*, 324–334. [CrossRef]

64. Peretyazhko, T.S.; Zhang, Q.; Colvin, V.L. Size-controlled dissolution of silver nanoparticles at neutral and acidic pH conditions: Kinetics and size changes. *Environ. Sci. Technol.* **2014**, *48*, 11954–11961. [CrossRef]
65. Yu, S.; Yin, Y.; Chao, J.; Shen, M.; Liu, J. Highly Dynamic PVP-Coated Silver Nanoparticles in Aquatic Environments: Chemical and Morphology Change Induced by Oxidation of Ag^0 and Reduction of Ag^+. *Environ. Sci. Technol.* **2014**, *48*, 403–411. [CrossRef]
66. Hou, W.C.; Stuart, B.; Howes, R.; Zepp, R.G. Sunlight-driven reduction of silver ions by natural organic matter: Formation and transformation of silver nanoparticles. *Environ. Sci. Technol.* **2013**, *47*, 7713–7721. [CrossRef]
67. Radwan, I.M.; Gitipour, A.; Potter, P.M.; Dionysiou, D.D.; Al-Abed, S.R. Dissolution of silver nanoparticles in colloidal consumer products: Effects of particle size and capping agent. *J. Nanopart. Res.* **2019**, *21*, 155. [CrossRef]
68. Wan, J.; Kim, Y.; Mulvihill, M.J.; Tokunaga, T.K. Dilution destabilizes engineered ligand-coated nanoparticles in aqueous suspensions. *Environ. Toxicol. Chem.* **2018**, *37*, 1301–1308. [CrossRef] [PubMed]
69. Hedberg, J.; Eriksson, M.; Kesraoui, A.; Norén, A.; Odnevall Wallinder, I. Transformation of silver nanoparticles released from skin cream and mouth spray in artificial sweat and saliva solutions: Particle size, dissolution, and surface area. *Environ. Sci. Pollut. Res.* **2020**, *28*, 12968–12979. [CrossRef] [PubMed]
70. Mitrano, D.M.; Rimmele, E.; Wichser, A.; Erni, R.; Height, M.; Nowack, B. Presence of nanoparticles in wash water from conventional silver and nano-silver textiles. *ACS Nano* **2014**, *8*, 7208–7219. [CrossRef] [PubMed]
71. Windler, L.; Lorenz, C.; Von Goetz, N.; Hungerbühler, K.; Amberg, M.; Heuberger, M.; Nowack, B. Release of titanium dioxide from textiles during washing. *Environ. Sci. Technol.* **2012**, *46*, 8181–8188. [CrossRef] [PubMed]
72. Kaegi, R.; Sinnet, B.; Zuleeg, S.; Hagendorfer, H.; Mueller, E.; Vonbank, R.; Boller, M.; Burkhardt, M. Release of silver nanoparticles from outdoor facades. *Environ. Pollut.* **2010**, *158*, 2900–2905. [CrossRef]
73. Labille, J.; Slomberg, D.; Catalano, R.; Robert, S.; Apers-Tremelo, M.L.; Boudenne, J.L.; Manasfi, T.; Radakovitch, O. Assessing UV filter inputs into beach waters during recreational activity: A field study of three French Mediterranean beaches from consumer survey to water analysis. *Sci. Total Environ.* **2020**, *706*, 136010. [CrossRef]
74. Gondikas, A.; Von Der Kammer, F.; Kaegi, R.; Borovinskaya, O.; Neubauer, E.; Navratilova, J.; Praetorius, A.; Cornelis, G.; Hofmann, T. Where is the nano? Analytical approaches for the detection and quantification of TiO_2 engineered nanoparticles in surface waters. *Environ. Sci. Nano* **2018**, *5*, 313–326. [CrossRef]
75. Sanchís, J.; Jiménez-Lamana, J.; Abad, E.; Szpunar, J.; Farré, M. Occurrence of Cerium-, Titanium-, and Silver-Bearing Nanoparticles in the Besòs and Ebro Rivers. *Environ. Sci. Technol.* **2020**, *54*, 3969–3978. [CrossRef]
76. Rouhnia, M.; Strom, K. Sedimentation from flocculated suspensions in the presence of settling-driven gravitational interface instabilities. *J. Geophys. Res. Ocean.* **2015**, *120*, 6384–6404. [CrossRef]
77. Liu, Y.; Nie, Y.; Wang, J.; Wang, J.; Wang, X.; Chen, S.; Zhao, G.; Wu, L.; Xu, A. Mechanisms involved in the impact of engineered nanomaterials on the joint toxicity with environmental pollutants. *Ecotoxicol. Environ. Saf.* **2018**, *162*, 92–102. [CrossRef]
78. Fang, J.; Shijirbaatar, A.; Lin, D.-h.; Wang, D.-j.; Shen, B.; Sun, P.-d.; Zhou, Z.-Q. Stability of co-existing ZnO and TiO_2 nanomaterials in natural water: Aggregation and sedimentation mechanisms. *Chemosphere* **2017**, *184*, 1125–1133. [CrossRef] [PubMed]
79. Foss Hansen, S.; Larsen, B.H.; Olsen, S.I.; Baun, A. Categorization framework to aid hazard identification of nanomaterials. *Nanotoxicology* **2007**, *1*, 243–250. [CrossRef]
80. Quadros, M.; Pierson, R.I.; Tulve, N.; Willis, R.; Rogers, K.; Thomas, T.; Marr, L.C. Release of Silver from Nanotechnology—Based Consumer Products for Children. *Environ. Sci. Technol.* **2013**, *47*, 8894–8901. [CrossRef] [PubMed]
81. Ale, A.; Liberatori, G.; Vannuccini, M.L.; Bergami, E.; Ancora, S.; Mariotti, G.; Bianchi, N.; Galdopórpora, J.M.; Desimone, M.F.; Cazenave, J.; et al. Exposure to a nanosilver-enabled consumer product results in similar accumulation and toxicity of silver nanoparticles in the marine mussel Mytilus galloprovincialis. *Aquat. Toxicol.* **2019**, *211*, 46–56. [CrossRef] [PubMed]
82. Hedberg, J.; Skoglund, S.; Karlsson, M.-E.E.; Wold, S.; Odnevall Wallinder, I.; Hedberg, Y. Sequential studies of silver released from silver nanoparticles in aqueous media simulating sweat, laundry detergent solutions and surface water. *Environ. Sci. Technol.* **2014**, *48*, 7314–7322. [CrossRef]
83. Mitrano, D.M.; Ranville, J.F.; Bednar, A.; Kazor, K.; Hering, A.S.; Higgins, C.P. Tracking dissolution of silver nanoparticles at environmentally relevant concentrations in laboratory, natural, and processed waters using single particle ICP-MS (spICP-MS). *Environ. Sci. Nano* **2014**, *1*, 248–259. [CrossRef]
84. Liu, C.; Leng, W.; Vikesland, P.J. Controlled Evaluation of the Impacts of Surface Coatings on Silver Nanoparticle Dissolution Rates. *Environ. Sci. Technol.* **2018**, *52*, 2726–2734. [CrossRef]
85. Jiménez-Lamana, J.; Slaveykova, V.I. Silver nanoparticle behaviour in lake water depends on their surface coating. *Sci. Total Environ.* **2016**, *573*, 946–953. [CrossRef] [PubMed]
86. Ma, R.; Levard, C.; Marinakos, S.M.; Cheng, Y.; Liu, J.; Michel, F.M.; Brown, G.E.; Lowry, G.V. Size-controlled dissolution of organic-coated silver nanoparticles. *Environ. Sci. Technol.* **2012**, *46*, 752–759. [CrossRef] [PubMed]
87. Borm, P.; Klaessig, F.C.; Landry, T.D.; Moudgil, B.; Pauluhn, J.; Thomas, K.; Trottier, R.; Wood, S. Research strategies for safety evaluation of nanomaterials, Part V: Role of dissolution in biological fate and effects of nanoscale particles. *Toxicol. Sci.* **2006**, *90*, 23–32. [CrossRef] [PubMed]
88. Hedberg, J.; Blomberg, E.; Odnevall Wallinder, I. In the Search for Nanospecific Effects of Dissolution of Metallic Nanoparticles at Freshwater-Like Conditions: A Critical Review. *Environ. Sci. Technol.* **2019**, *53*, 4030–4044. [CrossRef] [PubMed]

89. De Leersnyder, I.; Rijckaert, H.; De Gelder, L.; Van Driessche, I.; Vermeir, P. High Variability in Silver Particle Characteristics, Silver Concentrations, and Production Batches of Commercially Available Products Indicates the Need for a More Rigorous Approach. *Nanomaterials* **2020**, *10*, 1394. [CrossRef] [PubMed]
90. Geranio, L.; Heuberger, M.; Nowack, B. The Behavior of Silver Nanotextiles during Washing. *Environ. Sci. Technol.* **2009**, *43*, 8113–8118. [CrossRef] [PubMed]
91. Kim, J.B.; Kim, J.Y.; Yoon, T.H. Determination of silver nanoparticle species released from textiles into artificial sweat and laundry wash for a risk assessment. *Hum. Ecol. Risk Assess. Int. J.* **2017**, *23*, 741–750. [CrossRef]
92. Mackevica, A.; Olsson, M.E.; Hansen, S.F. Quantitative characterization of TiO_2 nanoparticle release from textiles by conventional and single particle ICP-MS. *J. Nanopart. Res.* **2018**, *20*, 6. [CrossRef]
93. Benn, T.M.; Westerhoff, P. Nanoparticle Silver Released into Water from Commercially Available Sock Fabrics. *Environ. Sci. Technol.* **2008**, *42*, 4133–4139. [CrossRef]
94. Baek, S.; Joo, S.H.; Blackwelder, P.; Toborek, M. Effects of coating materials on antibacterial properties of industrial and sunscreen-derived titanium-dioxide nanoparticles on Escherichia coli. *Chemosphere* **2018**, *208*, 196–206. [CrossRef]
95. Virkutyte, J.; Al-Abed, S.R.; Dionysiou, D.D.; Chen, Y.; Capco, D.D.G.D.; Capco, D.D.G.D. Depletion of the protective aluminum hydroxide coating in TiO_2-based sunscreens by swimming pool water ingredients. *Chem. Eng. J.* **2012**, *191*, 95–103. [CrossRef]
96. Labille, J.; Feng, J.; Botta, C.; Borschneck, D.; Sammut, M.; Cabie, M.; Auffan, M.; Rose, J.; Bottero, J.-Y. Aging of TiO_2 nanocomposites used in sunscreen. Dispersion and fate of the degradation products in aqueous environment. *Environ. Pollut.* **2010**, *158*, 3482–3489. [CrossRef]
97. Slomberg, D.L.; Catalano, R.; Ziarelli, F.; Viel, S.; Bartolomei, V.; Labille, J.; Masion, A. Aqueous aging of a silica coated TiO_2 UV filter used in sunscreens: Investigations at the molecular scale with dynamic nuclear polarization NMR. *RSC Adv.* **2020**, *10*, 8266–8274. [CrossRef]

MDPI

Article

Phototransformation of Graphene Oxide on the Removal of Sulfamethazine in a Water Environment

Fei-Fei Liu [1,*], Meng-Ru Li [1], Su-Chun Wang [1], Yu-Xue Zhang [1], Guang-Zhou Liu [1] and Jin-Lin Fan [2]

1 Institute of Marine Science and Technology, Shandong University, Qingdao 266237, China; limr666@163.com (M.-R.L.); suchunw@163.com (S.-C.W.); zhangyuxue@mail.sdu.edu.cn (Y.-X.Z.); liuguangzhou@sdu.edu.cn (G.-Z.L.)

2 Department of Science and Technology Management, Shandong University, Jinan 250100, China; fanjinlin@sdu.edu.cn

* Correspondence: liufeifei@sdu.edu.cn

Abstract: Graphene oxide (GO) is widely used in various fields and has raised concerns regarding its potential environmental fate and effect. However, there are few studies on its influence on coexisting pollutants. In this study, the phototransformation of GO and coexisting sulfamethazine (SMZ) under UV irradiation was investigated, with a focus on the role of reactive oxygen species. The results demonstrated that GO promoted the degradation of SMZ under UV irradiation. The higher the concentration of GO, the higher the degradation rate of SMZ, and the faster the first-order reaction rate. Two main radicals, ·OH and $^1O_2$, both contributed greatly in terms of regulating the removal of SMZ. Cl^-, $SO_4{}^{2-}$, and pH mainly promoted SMZ degradation by increasing the generation of ·OH, while humic acid inhibited SMZ degradation due to the reduction of ·OH. Moreover, after UV illumination, the GO suspension changed from light yellow to dark brown with increasing absorbance at a wavelength of 225 nm. Raman spectra revealed that the I_D/I_G ratio slightly decreased, indicating that some of the functional groups on the surface of GO were removed under low-intensity UV illumination. This study revealed that GO plays important roles in the photochemical transformation of environmental pollutants, which is helpful for understanding the environmental behaviors and risks of nanoparticles in aquatic environments.

Keywords: graphene oxide; sulfamethazine; phototransformation; free radicals

Citation: Liu, F.-F.; Li, M.-R.; Wang, S.-C.; Zhang, Y.-X.; Liu, G.-Z.; Fan, J.-L. Phototransformation of Graphene Oxide on the Removal of Sulfamethazine in a Water Environment. *Nanomaterials* **2021**, *11*, 2134. https://doi.org/10.3390/nano11082134

Academic Editor: Vivian Hsiu-Chuan Liao

Received: 15 July 2021
Accepted: 19 August 2021
Published: 22 August 2021

1. Introduction

As a kind of two-dimensional layered nanomaterial, graphene oxide (GO) possesses good mechanical, electrical, and thermal properties, and is widely applied in various fields, including biology, medicine, chemistry, and electronic engineering [1]. The global production of GO is expected to reach 3800 metric tons in 2027 [2]. Due to the presence of a large number of oxygen-containing functional groups, such as hydroxyl, carboxyl, and epoxy groups, GO has excellent hydrophilicity and a high probability of being present in natural aquatic environments, thus having uncertain environmental impacts and ecological risks. It has been reported that GO and its derivatives exhibited cytotoxicity to bacteria, biofilms, and algae [3,4]. Moreover, GO could cause developmental genotoxicity in aquatic animals such as zebrafish at trace concentrations [5], and could even accumulate in humans through the food chain [6]. Therefore, an increasing number of studies on the environmental behaviors of GO have received attention.

Once released into the environment, GO can interact with other pollutants mainly through π bonds, hydrophobic interactions, hydrogen bonds, and electrostatic interactions [7–10], thus affecting the transport and fate of coexisting compounds. For example, GO exhibited a high affinity for heavy-metal ions, which improved the transport ability of Pb^{2+} and Cd^{2+} in saturated porous media [11]. GO also facilitated the transport of antibiotics (levofloxacin, ciprofloxacin, and tetracycline) in saturated or unsaturated porous media because of the high sorption capacity of antibiotics by GO [12,13]. Furthermore, highly hydrophilic and mobile GO could serve as a carrier and promote the transport of nano-TiO_2 in porous media [14]. In addition, the interaction between GO and other pollutants would change their combined toxicity to organisms. GO enhanced Cd toxicity on photosynthesis, biomass, and cell membrane lipids in wheat seedlings [15]. GO also promoted lipotoxicity and hepatic function deficits caused by *cis*-bifenthrin exposure in tadpoles [16]. Cao et al. revealed that environmentally relevant concentrations of GO (1 mg/L) significantly increased the phytotoxicity of As (III) and As (V) in plants, which resulted in more severe oxidative stress and a significant reduction in nutrient content [17].

However, it should be noted that GO may be subjected to the phototransformation process in the environment because its special sp^2 domains can effectively adsorb sunlight, especially UV light [18–20]. GO was structurally degraded and chemically formed reduced GO under UV or sunlight irradiation [21]. After phototransformation, the toxicity of GO to bacteria (such as Gram-negative *Escherichia coli* and Gram-positive *Staphylococcus aureus*) and algal cells (*Chlorella pyrenoidosa*) was enhanced [22,23]. Meanwhile, GO can be regarded as a semiconductor with a zero energy gap to generate electrons, holes, and a series of reactive oxygen species (ROS) [24,25], which can mediate the transformation of the coexisting pollutants in the environment. For example, Cao et al. reported that silver nanoparticles could be formed from aqueous Ag^{2+} in the presence of GO under light [26]. Cu^{2+} on the surface of GO sheets could also trap e^- generated by GO and be reduced to Cu(I) and then form Cu_2O nanoparticles with the assistance of ROS, which suppressed the joint toxicity of GO and Cu^{2+} to freshwater algae after phototransformation [22]. In addition, GO could oxidize 42% of the adsorbed As (III) to As (V) under light irradiation, which was induced by electron-hole pairs on the surface of GO. However, coexposure to GO greatly enhanced the toxicity of As (III, V) to algae [27]. Therefore, it is of great significance to explore the photochemical transformation of GO on coexisting contaminants, especially when evaluating their environmental fate and possible toxicity and risks.

Antibiotics, as emerging contaminants, have gained increasing attention in recent years due to their widespread application and large production amounts [28,29]. As a result, antibiotics will inevitably find their way into the environment. Sulfamethazine (SMZ), one of the most common broad-spectrum antibiotics, is widely used in aquaculture, animal husbandry, hospitals, pharmaceutical factories, and other processes. Previous studies revealed that SMZ was frequently detected in wastewater, surface water, and even groundwater at concentrations ranging from ng L^{-1} to μg L^{-1} [30,31]. An increasing number of studies have focused on the environmental behaviors of SMZ, including its adsorption, migration, photooxidation, and so on [32–35].

Therefore, in this study, SMZ was selected as the model compound to reveal the effect of phototransformation of GO on coexisting contaminants. We systematically investigated the interaction between GO and SMZ under UV light, considering the influence of different environmental factors, including pH values, ionic strength and species, and natural organic matter (NOM). The phototransformation of GO together with the generation mechanisms of ROS were further explored to reveal the possible cotransformation pathways of antibiotics and GO.

2. Materials and Methods

2.1. Materials

GO was synthesized by an improved Hummers' method [36]. SMZ (≥99%) was purchased from Aladdin Biochemical Technology Co., Ltd. (Shanghai, China). The other reagents used in this study were obtained from Sinopharm Chemical Reagent Co., Ltd. (Shanghai, China). All aqueous samples were prepared with ultrawater.

2.2. Photochemistry Experiment

All experiments were conducted in a multichannel photocatalytic reaction system (PCX50C, Beijing Perfect Light Science and Technology Co., Ltd., Beijing, China). The system was operated at an average light intensity of 10.0 mW cm^{-2} with 5 W LED white lamps (365 nm). During the 6 h photochemical experiments, 50 mL of the reaction solutions was magnetically stirred at 300 rpm in quartz tubes that were maintained at constant temperature (22 ± 2 °C) with the circulating water bath of the reactor.

2.3. SMZ Degradation

One batch experiment was first conducted with a fixed amount of SMZ (5 μM) with GO ranging from 10 mg/L to 50 mg/L. Dark control experiments were also conducted under the same conditions. To investigate the effects of solution chemistry factors on the photochemical transformation, another three sets of experiments were also performed with 5 μM SMZ and 30 mg/L GO. The pH effect experiments were conducted with the solution pH ranging from 3.0 to 9.0, which was adjusted with 0.1 M HCl or NaOH. Ionic strength and species effect experiments were performed in the presence of 0–600 mM NaCl or 0–30 mM Na_2SO_4. In addition, the photochemical transformation of SMZ was tested in the presence of humic acid (HA) in the range of 0–10 mg/L. All the above experiments were performed in triplicate. During the experiments, 3 mL of solution was sampled at determined time intervals and filtered with 0.22 μm nylon membranes to remove GO. Then, SMZ was analyzed at a determination wavelength of 270 nm by high-performance liquid chromatography (HPLC, Shimadzu LC-20AT, Tokyo, Japan) with a UV detector using a C18 column (25 cm × 4.6 mm, 5 μm). The mobile phase was acetonitrile/0.05 M acetic acid (30:70, *v:v*) with a flow rate of 1 mL/min. The injection volume was 10 μL, and the column temperature was maintained at 40 °C.

2.4. ROS Generation

It should be noted that $O_2{\cdot}^-$ was not detected with the XTT sodium salt (probe for $O_2{\cdot}^-$) in this study; therefore, we only focused on the production of ·OH and $^1O_2$. Free radical quenching experiments were first carried out with L-histidine and potassium iodide (KI) as radical quenchers to identify the contribution of $^1O_2$ and ·OH, respectively [37,38]. The inhibition rate of SMZ degradation was determined after introducing free radical scavengers. In addition, 200 μM terephthalic acid (TPA) and 300 μM furfuryl alcohol (FFA) were used as indicators to quantify the amount of ROS [39,40]. TPA reacted with ·OH and produced 2-hydroxyterephthalic acid (HTPA), which could be measured by a fluorescence spectrophotometer (HITACHI, F-2500, Tokyo, Japan). The excitation and emission wavelengths were 315 nm and 425 nm, respectively [23,40]. FFA was analyzed by HPLC at 218 nm. The mobile phase was 30% acetonitrile and 70% phosphoric acid and run at 1.0 mL/min.

2.5. GO Characterization

To investigate the phototransformation of GO, the changes in GO in the photoreaction system were characterized by UV–vis spectrophotometry from 200 nm to 600 nm. Additionally, the Raman spectra were measured at 1000 cm^{-1} to 2000 cm^{-1} with 532 nm excitation with a Raman spectrometer (Renishaw inVia Reflex, New Mills, UK) before and after UV illumination.

3. Results and Discussion

3.1. SMZ Degradation

The effects of different concentrations of GO on SMZ degradation were first studied. GO did not adsorb SMZ much in the dark, and little degradation of SMZ occurred under UV illumination without GO (Figure S1). However, the degradation of SMZ was accelerated in the presence of GO. The degradation rates of SMZ were 33.32 ± 2.54%, 34.90 ± 2.69%, and 37.44 ± 2.12% in the presence of 10 mg/L, 30 mg/L, and 50 mg/L GO, respectively (Figure 1a). According to the first-order kinetic fitting of the reaction in the first two hours (Figure S2), the observed reaction rate constants (k_{obs}) were 0.0732 h^{-1}, 0.0964 h^{-1}, and 0.1129 h^{-1} (Figure 1b).

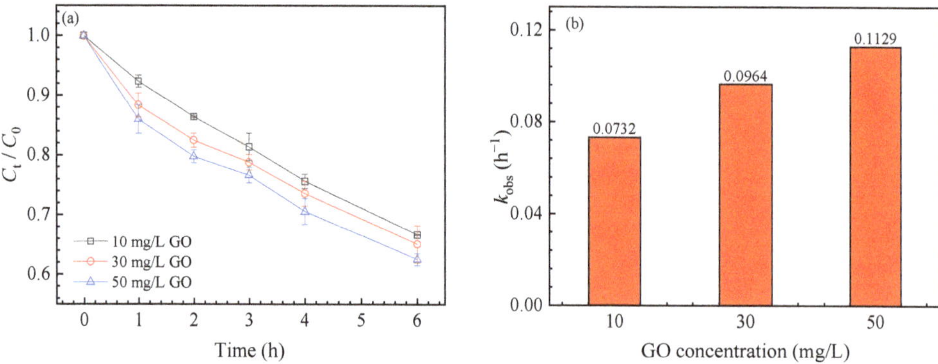

Figure 1. Effects of GO on SMZ degradation under UV light (**a**) and k_{obs} (**b**).

3.2. ROS Generation

Generally, nanoparticles can generate ROS under UV light irradiation, which can participate in the degradation of chemicals. ROS generation by GO is similar to that of semiconductors. A large number of oxygen-containing functional groups attached to the GO surface play an important role in electron transfer and promote ROS generation [21]. To further explore the mechanism of GO on SMZ transformation, free radical scavengers, including L-histidine and KI, were added to the reaction solution to identify the role of $^1O_2$ and ·OH. As shown in Figure 2a,b, 5 mM/10 mM L-histidine significantly inhibited SMZ degradation, reducing its degradation rate from 32.52 ± 4.34% to 6.57 ± 3.24% and 4.18 ± 1.63%, with k_{obs} decreasing from 0.1004 h^{-1} to 0.0080 h^{-1} and 0.0265 h^{-1} (Figure S3), respectively. Similar results were also observed in the presence of KI, where the decomposition of SMZ reduced to 26.36% for 10 mM KI and 18.59% for 50 mM KI. Compared with the initial k_{obs} of 0.1004 h^{-1}, k_{obs} decreased to only 0.0753 h^{-1} and 0.0457 h^{-1} (Figure S3), respectively. Thus, the above results showed that both $^1O_2$ and ·OH participated in SMZ degradation.

ROS quantification was performed during the photochemical experiments. Figure 2c,d shows that the free radical production of GO was proportional to the illumination time; 75.70 μM $^1O_2$ and 0.35 μM ·OH could be produced after 6 h of illumination in the presence of 30 mg/L GO. Based on the above experimental results, possible ROS generation pathways were further proposed, as shown in the following reaction formulas [23,41,42]:

$$GO + hv \rightarrow GO^* \left(e^-_{CB} - h^+_{VB}\right), \tag{1}$$

$$GO^* + O_2 \rightarrow O_2{}^1 + GO, \tag{2}$$

$$e^-_{CB} + O_2 \rightarrow O_2\cdot^-, \tag{3}$$

$$h^+_{VB} + H_2O \rightarrow \cdot OH + H^+, \tag{4}$$

$$h^+_{VB} + OH^- \rightarrow \cdot OH, \tag{5}$$

$$O_2\cdot^- + h^+_{VB} \rightarrow O_2{}^1, \tag{6}$$

$$O_2\cdot^- + e^-_{VB} + 2H^+ \rightarrow H_2O_2, \tag{7}$$

$$H_2O_2 + e^-_{CB} \rightarrow \cdot OH + OH^-, \tag{8}$$

3.3. Effects of Different Conditions on SMZ Degradation

3.3.1. Effect of pH

The pH value of the solution has a great influence on the photolysis of SMZ (Figure 3). The degradation rates were 26.02 ± 3.05%, 41.06 ± 4.23%, 49.33 ± 5.11%, and 51.14 ± 5.63% as the pH increased from 3.0 to 9.0, and k_{obs} were 0.0641 h^{-1}, 0.0814 h^{-1}, 0.1214 h^{-1}, and 0.1193 h^{-1} (Figure S4), respectively. The relative high degradation of SMZ at higher pH conditions was probably due to the following two reasons: firstly, SMZ (pK_{a1} = 2.6, pK_{a2} = 8.0) can be degraded more easily in its ionic forms compared with the neutral form [43]. Secondly, similar to semiconductors [44], GO produces holes after UV illumination, which can further react with OH^- to produce ·OH [45,46]. The generation of ·OH increased with increasing pH, resulting in the promotion of SMZ conversion. In addition, the dispersion of GO was higher at higher pH because of the deprotonation of oxygen-containing functional groups on the GO surface [47], which might result in an increase in the steady-state concentration of ROS. Therefore, SMZ degradation by GO was higher at high pH than at low pH.

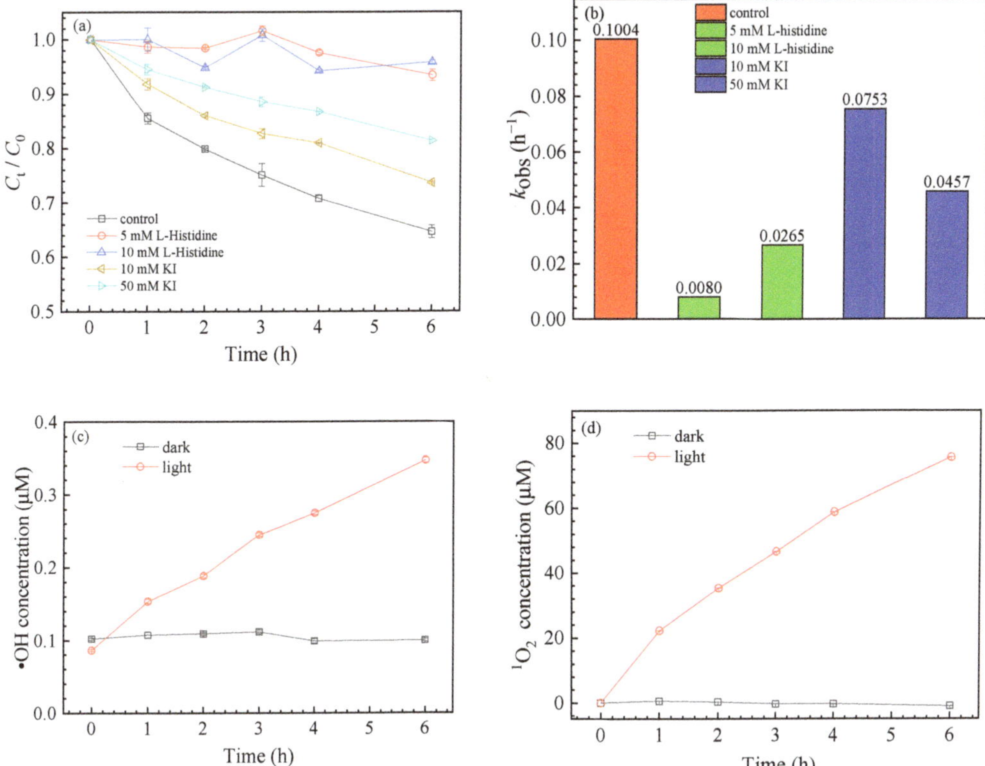

Figure 2. Degradation kinetics of SMZ (**a**) and k_{obs} (**b**) with free radical scavengers: L-histidine and KI; generation kinetics of ·OH (**c**) and $^1O_2$ (**d**).

3.3.2. Effect of Coexisting Anions

The effects of ionic strength and species on SMZ degradation are presented in Figure 4. NaCl improved the degradation of SMZ, with the degradation rate rising from 35.36 ± 1.69% to 43.83 ± 2.21%, 45.18 ± 2.88%, and 47.9 ± 2.79% in the presence of NaCl from 100 mM to 600 mM, respectively. Similarly, when 10 mM, 20 mM, and 30 mM Na_2SO_4 were added to the solution, the SMZ decomposition rate increased to 37.92 ± 2.38%, 41.94 ± 2.57%, and 46.52± 2.78%, respectively.

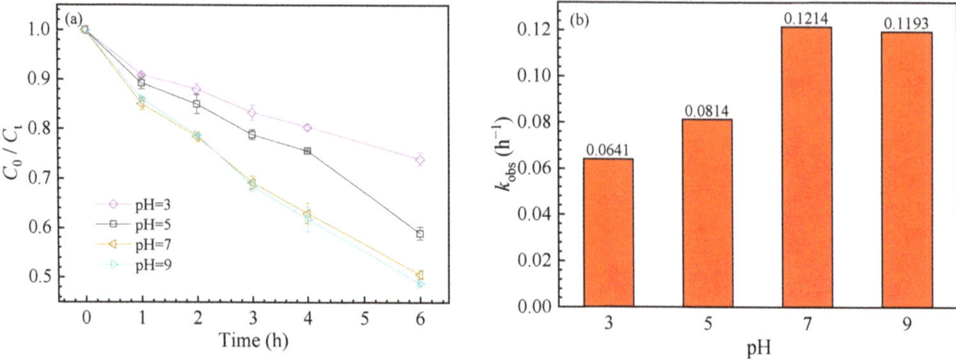

Figure 3. Effect of pH on SMZ degradation kinetics (**a**) and k_{obs} of SMZ degradation (**b**).

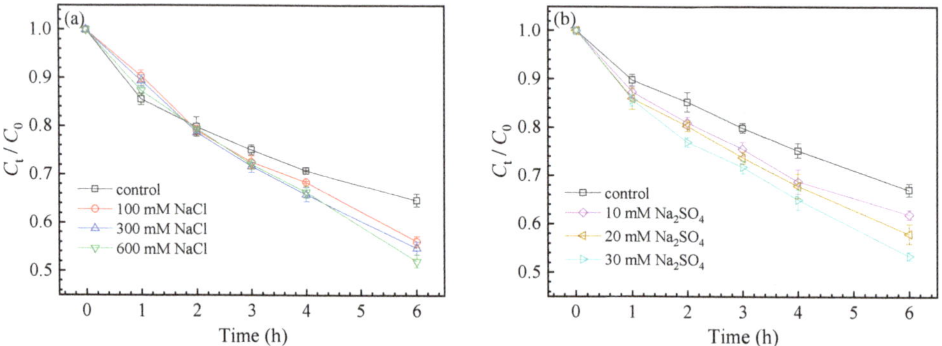

Figure 4. Effect of Cl^- (**a**) and SO_4^{2-} (**b**) on SMZ degradation kinetics.

To further explore the effect of Cl^- and SO_4^{2-} on the photolysis of SMZ, quantitative analysis of $^1O_2$ and ·OH was carried out (Figure 5). It was evident that Cl^- showed a negative influence on the production of $^1O_2$, whose level was reduced to 21.63 μM with increasing Cl^- concentration, compared with that of the control 75.70 μM. However, the presence of Cl^- accelerated the generation of ·OH, especially 100 mM NaCl, which increased the amount of ·OH by 1.6 times compared with the control. This could be explained by the fact that Cl^- generated hydrated electrons under UV irradiation, which were then transferred to nanomaterials to generate more ROS (Equation (9)) [48]. It should be noted that excessive Cl^- would agglomerate GO under high ionic strength [48,49], which would reduce the surface area of GO and the concentration of ROS. Thus, the steady-state concentration of ·OH first increased and then decreased with increasing NaCl concentration. The presence of Cl^- promoted the decomposition of SMZ, which was in accordance with the role of ·OH. Therefore, ·OH was expected to be the main ROS species that regulated SMZ degradation.

$$Cl^- + hv \rightarrow Cl + e_{aq}{}^-, \quad (9)$$

Similar to Cl^-, the presence of SO_4^{2-} also inhibited the generation of $^1O_2$ but prompted the production of ·OH. The concentration of $^1O_2$ decreased from 75.70 μM to 30.28 μM with increasing SO_4^{2-} from 0 to 30 mM, but the ·OH concentration gradually increased from 0.35 μM to 0.51 μM. Therefore, the introduction of SO_4^{2-} into the solution promoted SMZ degradation by increasing the steady-state concentration of ·OH. On the other hand, SO_4^{2-} existing on the GO surface would form reactive sulfate radicals by holes (Equation (10)) [46], which may also accelerate the transformation of SMZ [50].

$$SO_4^{2-} + h^+ \rightarrow SO_4^{\cdot -}, \quad (10)$$

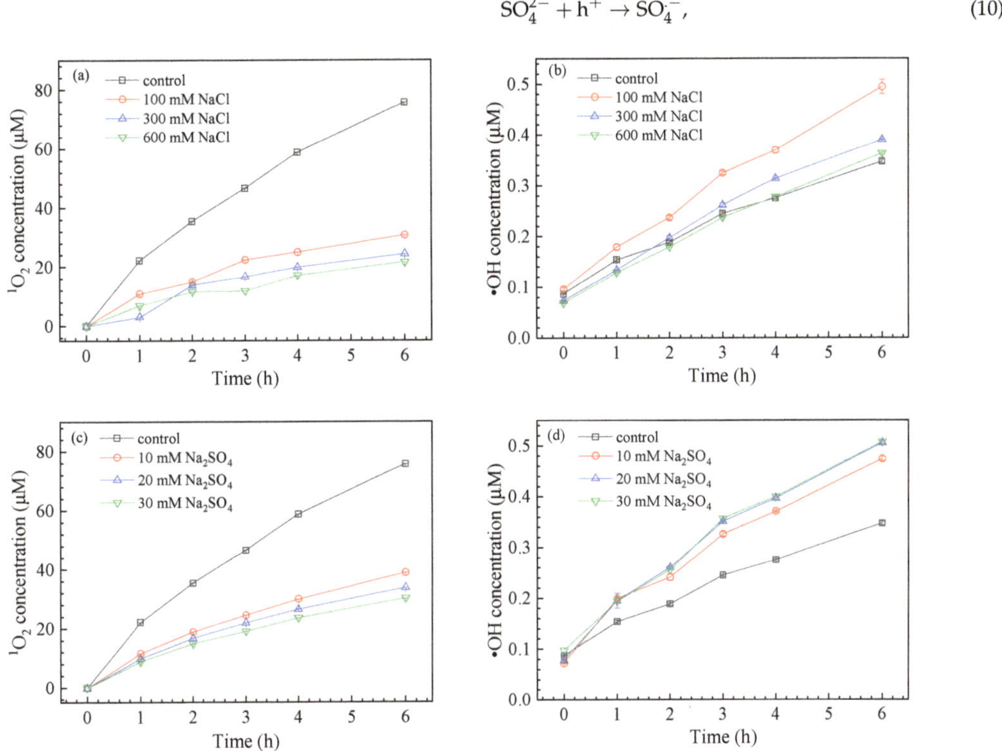

Figure 5. Effects of Cl^- and SO_4^{2-} on $^1O_2$ production (**a**,**c**) and ·OH production (**b**,**d**).

3.3.3. Effect of NOM

As the representative NOM, HA is a macromolecular polymer containing carboxyl, phenolic, and keto groups, which is widely distributed in natural waters. Previous studies showed that NOM might play different roles in the transformation of organic pollutants. For example, Chen et al. reported that HA could consume a large amount of ·OH under UV light [51], which decreased the degradation of diethyl phthalate. However, Niu et al. proposed that NOM could be transformed into excited-state substances or free radicals under UV irradiation, which enhanced the degradation of norfloxacin [38]. In the present study, as shown in Figure 6, the degradation of SMZ significantly decreased from 41.06 ± 2.34%, to 29.80 ± 2.64%, 24.72 ± 2.56%, and 23.95 ± 2.59% in the presence of 1 mg/L, 5 mg/L and 10 mg/L HA, respectively. HA had an inhibitory effect on the degradation of SMZ. As HA might react with ·OH, we only measured the production of $^1O_2$ in the presence of HA. As shown in Figure 6b, compared with the control, HA slightly influenced the generation of $^1O_2$, indicating that $^1O_2$ contributed little to SMZ degradation. Therefore, it could be speculated that HA mainly quenched ·OH to decrease the decomposition of SMZ.

3.4. *GO Transformation*

Under UV illumination, the color of the GO suspension changed from light yellow to dark brown (Figure S5), indicating that some oxygen-containing functional groups attached to the GO surface might be removed [19]. The variation of the solution absorbance with time was further determined by UV–vis spectrophotometry (Figure S6). The peak at 225 nm was attributed to the π -π^* transition of unsaturated C-C bonds of GO. After 6 h of UV irradiation, the absorbance at 225 nm increased, indicating that the sp^3 structure of GO was reduced and the sp^2 structure had been recovered [21]. It should be noted that our previous study demonstrated that UV light intensity greatly affected the absorbance of GO, and the absorption peak at 225 nm could be redshifted to 255 nm at a high light intensity of 54 mW cm^{-2} in 4 h [23]. In the present study, the light intensity was only 10 mW cm^{-2}; thus, the absorption peak did not shift significantly. Raman spectra were further used to analyze the GO samples before and after UV illumination. The D band at approximately 1350 cm^{-1} and the G band at approximately 1580 cm^{-1} are the two characteristic peaks of GO. Peak D represents the vibration of sp^3 carbon atoms, and peak G is the characteristic peak of carbon sp^2. The ratio of I_D/I_G is usually used as a qualitative measurement of the disorder degree caused by nonaromatic sp^3 carbon defects. After 6 h of illumination, I_D/I_G decreased only from 0.8481 to 0.8438 (Figure S7), indicating that the graphitization structure of GO was somewhat improved and that the sp^2 region was expanded. However, the insignificant decrease in I_D/I_G suggested that the UV light intensity was not high enough in the current study, which was in accordance with the changes in UV–vis absorbance.

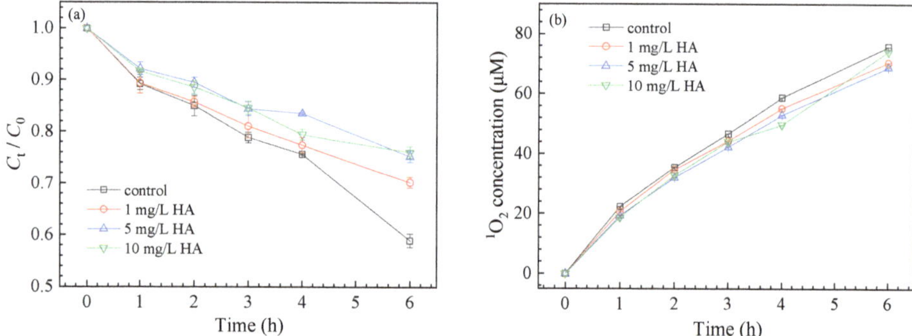

Figure 6. Effects of HA on SMZ degradation kinetics (**a**) and $^1O_2$ production (**b**).

Based on the above results, we proposed the possible cotransformation pathways of GO and SMZ (Figure 7). Similar to semiconductors, GO generated electrons and holes under illumination. Electrons could be captured by O_2 to generate $O_2\cdot^-$, which was further converted into $^1O_2$ and ·OH. Meanwhile, GO could form excited-state GO* under illumination, and then O_2 accepted excess energy and generated $^1O_2$. Therefore, ROS generated in the above ways promoted SMZ degradation. At the same time, GO could capture electrons to reduce its surface oxygen-containing functional groups.

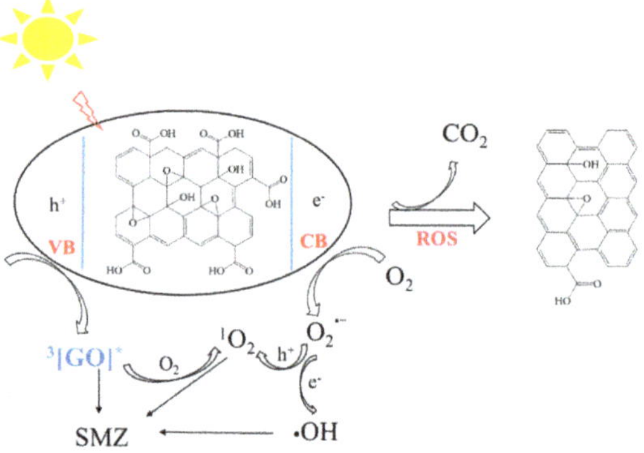

Figure 7. Proposed pathways for ROS generation and transformation of GO and SMZ under UV light.

4. Conclusions

In this study, the photochemical behaviors of GO and the degradation of SMZ were quantitatively analyzed. GO could promote the degradation of SMZ under UV light. ·OH and $^1O_2$ were the main free radicals participating in the cotransformation between GO and SMZ. High pH, Cl^-, and SO_4^{2-} improved the degradation of SMZ by affecting the formation of ·OH. However, the presence of HA consumed ·OH, leading to less degradation of SMZ. As for GO, its color changed from light yellow to dark brown under UV illumination. However, the absorption peak did not shift significantly, and the ratio of I_D/I_G was slightly smaller, which indicated that GO was somewhat reduced. The findings of this work may have significant implications for predicting the fate and assessing the potential risks of environmental pollutants and nanoparticles. However, to better understand the environmental behaviors of nanoparticles, long-term experiments under natural solar radiation are still needed.

Supplementary Materials: The following are available online at https://www.mdpi.com/2079-4991/11/8/2134/s1, Figure S1: Figure S1. Photolysis kinetics of SMZ (5 μM) under UV light without GO (a) and the adsorption of SMZ (5 μM) by GO (10 and 100 mg/L) in the dark within 6 h (b), Figure S2: Pseudo first-order fitting results for SMZ degradation kinetics under various GO concentrations (10-50 mg/L), Figure S3: Pseudo first-order kinetics fitting for kinetics of SMZ degradation with L-histidine and KI, Figure S4: Pseudo first-order kinetics fitting for kinetics of SMZ degradation at pH 3.0-9.0, Figure S5: Changes in the color of GO under UV light as a function of irradiation time, Figure S6: Variation of GO absorbance with time under light, Figure S7: Raman spectra of GO before and after UV illumination.

Author Contributions: Conceptualization, F.-F.L. and G.-Z.L.; methodology, M.-R.L.; software, M.-R.L. and S.-C.W.; validation, F.-F.L., M.-R.L., S.-C.W., Y.-X.Z. and G.-Z.L.; formal analysis, F.-F.L., Y.-X.Z. and J.-L.F.; investigation, M.-R.L.; resources, M.-R.L. and Y.-X.Z.; data curation, M.-R.L.; writing—original draft preparation, F.-F.L., M.-R.L. and Y.-X.Z.; writing—review and editing, F.-F.L., G.-Z.L. and J.-L.F.; visualization, F.-F.L. and M.-R.L.; supervision, F.-F.L. and G.-Z.L.; project administration, F.-F.L. and G.-Z.L.; funding acquisition, F.-F.L., G.-Z.L. and J.-L.F. All authors have read and agreed to the published version of the manuscript.

Funding: This research was funded by the National Natural Science Foundation of China-Shandong Joint Fund (U1906224), the Natural Science Foundation of Shandong Province (ZR2020MD114, ZR2019MEE026), the Key Research and Development Program of Shandong Province (2019JZZY010333), the Young Scholars Program of Shandong University, and Shandong University Interdisciplinary Research and Innovation Team of Young Scholars (2020QNQT20).

Institutional Review Board Statement: Not applicable.

Informed Consent Statement: Not applicable.

Data Availability Statement: The datasets generated during and/or analyzed during the current study are available from the corresponding author on reasonable request.

Acknowledgments: We thank the anonymous reviewers for their valuable comments.

Conflicts of Interest: The authors declare no conflict of interest.

References

1. Zhu, Y.; Murali, S.; Cai, W.; Li, X.; Suk, J.W.; Potts, J.R.; Ruoff, R.S. Graphene and graphene oxide: Synthesis, properties, and applications. *Adv. Mater.* **2010**, *22*, 3906–3924. [CrossRef]
2. Goodwin, D.G.; Adeleye, A.S.; Sung, L.; Ho, K.T.; Burgess, R.M.; Petersen, E.J. Detection and quantification of graphene-family nanomaterials in the environment. *Environ. Sci. Technol.* **2018**, *52*, 4491–4513. [CrossRef]
3. Guo, Z.L.; Xie, C.J.; Zhang, P.; Zhang, J.Z.; Wang, G.H.; He, X.; Ma, Y.H.; Zhao, B.; Zhang, Z.Y. Toxicity and transformation of graphene oxide and reduced graphene oxide in bacteria biofilm. *Sci. Total. Environ.* **2017**, *580*, 1300–1308. [CrossRef]
4. Zhao, J.; Cao, X.S.; Wang, Z.Y.; Dai, Y.H.; Xing, B.S. Mechanistic understanding toward the toxicity of graphene-family materials to freshwater algae. *Water Res.* **2017**, *111*, 18–27. [CrossRef]
5. Zhang, X.L.; Zhou, Q.X.; Zou, W.; Hu, X.G. Molecular mechanisms of developmental toxicity induced by graphene oxide at predicted environmental concentrations. *Environ. Sci. Technol.* **2017**, *51*, 7861–7871. [CrossRef] [PubMed]

6. Fadeel, B.; Bussy, C.; Merino, S.; Vázquez, E.; Flahaut, E.; Mouchet, F.; Evariste, L.; Gauthier, L.; Koivisto, A.J.; Vogel, U.; et al. Safety assessment of graphene-based materials: Focus on human health and the environment. *ACS Nano* **2018**, *12*, 10582–10620. [CrossRef] [PubMed]
7. Li, D.; Müller, M.B.; Gilje, S.; Kaner, R.B.; Wallace, G.G. Processable aqueous dispersions of graphene nanosheets. *Nat. Nanotechnol.* **2008**, *3*, 101–105. [CrossRef] [PubMed]
8. Pavagadhi, S.; Tang, A.L.L.; Sathishkumar, M.; Loh, K.P.; Balasubramanian, R. Removal of microcystin-LR and microcystin-RR by graphene oxide: Adsorption and kinetic experiments. *Water Res.* **2013**, *47*, 4621–4629. [CrossRef]
9. Wei, H.; Yang, W.S.; Xi, Q.; Chen, X. Preparation of Fe3O4@graphene oxide core-shell magnetic particles for use in protein adsorption. *Mater. Lett.* **2012**, *82*, 224–226. [CrossRef]
10. Yang, Z.; Yan, H.; Yang, H.; Li, H.B.; Li, A.M.; Cheng, R.S. Flocculation performance and mechanism of graphene oxide for removal of various contaminants from water. *Water Res.* **2013**, *47*, 3037–3046. [CrossRef]
11. Jiang, Y.; Zhang, X.; Yin, X.; Sun, H.; Wang, N. Graphene oxide-facilitated transport of Pb^{2+} and Cd^{2+} in saturated porous media. *Sci. Total Environ.* **2018**, *631–632*, 369–376. [CrossRef] [PubMed]
12. Sun, K.; Dong, S.; Sun, Y.; Gao, B.; Du, W.; Xu, H.; Wu, J. Graphene oxide-facilitated transport of levofloxacin and ciprofloxacin in saturated and unsaturated porous media. *J. Hazard. Mater.* **2018**, *348*, 92–99. [CrossRef] [PubMed]
13. Wang, M.; Song, Y.; Zhang, H.; Lu, T.; Chen, W.; Li, W.; Qi, W.; Qi, Z. Insights into the mutual promotion effect of graphene oxide nanoparticles and tetracycline on their transport in saturated porous media. *Environ. Pollut.* **2021**, *268*, 115730. [CrossRef] [PubMed]
14. Xia, T.; Lin, Y.; Guo, X.; Li, S.; Cui, J.; Ping, H.; Zhang, J.; Zhong, R.; Du, L.; Han, C.; et al. Co-transport of graphene oxide and titanium dioxide nanoparticles in saturated quartz sand: Influences of solution pH and metal ions. *Environ. Pollut.* **2019**, *251*, 723–730. [CrossRef]
15. Gao, M.; Yang, Y.; Song, Z. Effects of graphene oxide on cadmium uptake and photosynthesis performance in wheat seedlings. *Ecotoxicol. Environ. Saf.* **2019**, *173*, 165–173. [CrossRef]
16. Li, M.; Zhu, J.; Wu, Q.; Wang, Q. The combined adverse effects of *cis*-bifenthrin and graphene oxide on lipid homeostasis in *Xenopus laevis*. *J. Hazard. Mater.* **2021**, *407*, 124876. [CrossRef]
17. Cao, X.; Ma, C.; Chen, F.; Luo, X.; Musante, C.; White, J.C.; Zhao, X.; Wang, Z.; Xing, B. New insight into the mechanism of graphene oxide-enhanced phytotoxicity of arsenic species. *J. Hazard. Mater.* **2021**, *410*, 124959. [CrossRef]
18. Bai, H.; Jiang, W.; Kotchey, G.P.; Saidi, W.A.; Bythell, B.J.; Jarvis, J.M.; Marshall, A.G.; Robinson, R.A.; Star, A. Insight into the mechanism of graphene oxide degradation via the photo-Fenton reaction. *J. Phys. Chem. C* **2014**, *118*, 10519–10529. [CrossRef]
19. Gao, Y.; Ren, X.M.; Zhang, X.D.; Chen, C.L. Environmental fate and risk of ultraviolet- and visible-light-transformed graphene oxide: A comparative study. *Environ. Pollut.* **2019**, *251*, 821–829. [CrossRef]
20. Zhang, X.F.; Shao, X.; Liu, S. Dual fluorescence of graphene oxide: A time-resolved study. *J. Phys. Chem. A* **2012**, *116*, 7308–7313. [CrossRef]
21. Hou, W.C.; Chowdhury, I.; Goodwin, D.G.; Henderson, W.M.; Fairbrother, D.H.; Bouchard, D.; Zepp, R.G. Photochemical transformation of graphene oxide in sunlight. *Environ. Sci. Technol.* **2015**, *49*, 3435–3443. [CrossRef] [PubMed]
22. Zhao, J.; Ning, F.; Cao, X.; Yao, H.; Wang, Z.; Xing, B. Photo-transformation of graphene oxide in the presence of co-existing metal ions regulated its toxicity to freshwater algae. *Water Res.* **2020**, *176*, 115735. [CrossRef] [PubMed]
23. Zhao, F.F.; Wang, S.C.; Zhu, Z.L.; Wang, S.G.; Liu, F.F.; Liu, G.Z. Effects of oxidation degree on photo-transformation and the resulting toxicity of graphene oxide in aqueous environment. *Environ. Pollut.* **2019**, *249*, 1106–1114. [CrossRef] [PubMed]
24. Katsnelson, M.I. Graphene: Carbon in two dimensions. *Mater. Today* **2007**, *10*, 20–27. [CrossRef]
25. Meyer, J.C.; Geim, A.K.; Katsnelson, M.I.; Novoselov, K.S.; Booth, T.J.; Roth, S. The structure of suspended graphene sheets. *Nature* **2007**, *446*, 60–63. [CrossRef]
26. Cao, X.; Ma, C.; Zhao, J.; Guo, H.; Dai, Y.; Wang, Z.; Xing, B. Graphene oxide mediated reduction of silver ions to silver nanoparticles under environmentally relevant conditions: Kinetics and mechanisms. *Sci. Total Environ.* **2019**, *679*, 270–278. [CrossRef]
27. Cao, X.; Ma, C.; Zhao, J.; Musante, C.; White, J.C.; Wang, Z.; Xing, B. Interaction of graphene oxide with co-existing arsenite and arsenate: Adsorption, transformation and combined toxicity. *Environ. Int.* **2019**, *131*, 104992. [CrossRef]
28. Li, W.H.; Shi, Y.L.; Gao, L.H.; Liu, J.M.; Cai, Y.Q. Occurrence of antibiotics in water, sediments, aquatic plants, and animals from Baiyangdian Lake in North China. *Chemosphere* **2012**, *89*, 1307–1315. [CrossRef]
29. Zhang, Q.Q.; Ying, G.G.; Pan, C.G.; Liu, Y.S.; Zhao, J.L. Comprehensive evaluation of antibiotics emission and fate in the river basins of China: Source analysis, multimedia modeling, and linkage to bacterial resistance. *Environ. Sci. Technol.* **2015**, *49*, 6772–6782. [CrossRef]
30. Ma, Y.; Li, M.; Wu, M.; Li, Z.; Liu, X. Occurrences and regional distributions of 20 antibiotics in water bodies during groundwater recharge. *Sci. Total Environ.* **2015**, *518–519*, 498–506. [CrossRef]
31. Zhang, Q.; Jia, A.; Wan, Y.; Liu, H.; Wang, K.; Peng, H.; Dong, Z.; Hu, J. Occurrences of three classes of antibiotics in a natural river basin: Association with antibiotic-resistant *Escherichia coli*. *Environ. Sci. Technol.* **2014**, *48*, 14317–14325. [CrossRef]
32. Gao, Y.; Li, Y.; Zhang, L.; Huang, H.; Hu, J.J.; Shah, S.M.; Su, X.G. Adsorption and removal of tetracycline antibiotics from aqueous solution by graphene oxide. *J. Colloid. Interface Sci.* **2012**, *368*, 540–546. [CrossRef]

33. Latch, D.E.; Stender, B.L.; Packer, J.L.; Arnold, W.A.; McNeill, K. Photochemical fate of pharmaceuticals in the environment: Cimetidine and ranitidine. *Environ. Sci. Technol.* **2003**, *37*, 3342–3350. [CrossRef]
34. Li, Y.J.; Liu, X.L.; Zhang, B.J.; Zhao, Q.; Ning, P.; Tian, S.L. Aquatic photochemistry of sulfamethazine: Multivariate effects of main water constituents and mechanisms. *Environ. Sci. Process. Impacts* **2018**, *20*, 513–522. [CrossRef] [PubMed]
35. Liu, F.F.; Zhao, J.; Wang, S.G.; Xing, B.S. Adsorption of sulfonamides on reduced graphene oxides as affected by pH and dissolved organic matter. *Environ. Pollut.* **2016**, *210*, 85–93. [CrossRef] [PubMed]
36. Liu, F.F.; Zhao, J.; Wang, S.G.; Du, P.; Xing, B.S. Effects of solution chemistry on adsorption of selected pharmaceuticals and personal care products (PPCPs) by graphenes and carbon nanotubes. *Environ. Sci. Technol.* **2014**, *48*, 13197–13206. [CrossRef] [PubMed]
37. Liu, L.; Xu, X.; Li, Y.W.; Su, R.D.; Li, Q.; Zhou, W.Z.; Gao, B.Y.; Yue, Q.Y. One-step synthesis of "nuclear-shell" structure iron-carbon nanocomposite as a persulfate activator for bisphenol A degradation. *Chem. Eng. J.* **2020**, *382*, 12. [CrossRef]
38. Niu, X.Z.; Busetti, F.; Langsa, M.; Croue, J.P. Roles of singlet oxygen and dissolved organic matter in self-sensitized photo-oxidation of antibiotic norfloxacin under sunlight irradiation. *Water Res.* **2016**, *106*, 214–222. [CrossRef] [PubMed]
39. Li, Y.; Zhang, W.; Niu, J.F.; Chen, Y.S. Mechanism of photogenerated reactive oxygen species and correlation with the antibacterial properties of engineered metal-oxide nanoparticles. *ACS Nano* **2012**, *6*, 5164–5173. [CrossRef]
40. Qu, X.L.; Alvarez, P.J.J.; Li, Q.L. Photochemical transformation of carboxylated multiwalled carbon nanotubes: Role of reactive oxygen species. *Environ. Sci. Technol.* **2013**, *47*, 14080–14088. [CrossRef]
41. Chen, C.; Ma, W.; Zhao, J. Semiconductor-mediated photodegradation of pollutants under visible-light irradiation. *Chem. Soc. Rev.* **2010**, *39*, 4206–4219. [CrossRef]
42. Nosaka, Y.; Nosaka, A.Y. Generation and detection of reactive oxygen species in photocatalysis. *Chem. Rev.* **2017**, *117*, 11302–11336. [CrossRef]
43. Chen, W.; Chen, X.; Yu, H. Photochemical degradation of sulfamethazine in aqueous solution. *J. Agro-Environ. Sci.* **2016**, *35*, 346–352.
44. Zhao, Y.C.; Jafvert, C.T. Environmental photochemistry of single layered graphene oxide in water. *Environ. Sci. Nano* **2015**, *2*, 136–142. [CrossRef]
45. Khodja, A.A.; Sehili, T.; Pilichowski, J.F.; Boule, P. Photocatalytic degradation of 2-phenylphenol on TiO_2 and ZnO in aqueous suspensions. *J. Photochem. Photobiol. A* **2001**, *141*, 231–239. [CrossRef]
46. Li, M.R.; Liu, F.F.; Wang, S.C.; Cheng, X.; Zhang, H.; Huang, T.Y.; Liu, G.Z. Phototransformation of zinc oxide nanoparticles and coexisting pollutant: Role of reactive oxygen species. *Sci. Total Environ.* **2020**, *728*, 12. [CrossRef] [PubMed]
47. Duan, L.; Hao, R.; Xu, Z.; He, X.; Adeleye, A.S.; Li, Y. Removal of graphene oxide nanomaterials from aqueous media via coagulation: Effects of water chemistry and natural organic matter. *Chemosphere* **2017**, *168*, 1051–1057. [CrossRef] [PubMed]
48. Li, Y.; Zhao, J.; Shang, E.X.; Xia, X.H.; Niu, J.F.; Crittenden, J. Effects of chloride ions on dissolution, ROS generation, and toxicity of silver nanoparticles under UV irradiation. *Environ. Sci. Technol.* **2018**, *52*, 4842–4849. [CrossRef] [PubMed]
49. Chowdhury, I.; Duch, M.C.; Mansukhani, N.D.; Hersam, M.C.; Bouchard, D. Colloidal properties and stability of graphene oxide nanomaterials in the aquatic environment. *Environ. Sci. Technol.* **2013**, *47*, 6288–6296. [CrossRef]
50. Budarz, J.F.; Turolla, A.; Piasecki, A.F.; Bottero, J.Y.; Antonelli, M.; Wiesner, M.R. Influence of aqueous inorganic anions on the reactivity of nanoparticles in TiO_2 photocatalysis. *Langmuir* **2017**, *33*, 2770–2779. [CrossRef]
51. Chen, X.R.; Fang, G.D.; Liu, C.; Dionysiou, D.D.; Wang, X.L.; Zhu, C.Y.; Wang, Y.J.; Gao, J.; Zhou, D.M. Cotransformation of carbon dots and contaminant under light in aqueous solutions: A mechanistic study. *Environ. Sci. Technol.* **2019**, *53*, 6235–6244. [CrossRef] [PubMed]

Article

Relationship between Cytotoxicity and Surface Oxidation of Artificial Black Carbon

Yen Thi-Hoang Le [1,2], Jong-Sang Youn [3], Hi-Gyu Moon [4], Xin-Yu Chen [4,5], Dong-Im Kim [4], Hyun-Wook Cho [1], Kyu-Hong Lee [4,5,*] and Ki-Joon Jeon [1,2,*]

1 Department of Environmental Engineering, Inha University, Incheon 22212, Korea; lethihoangyen@inha.edu (Y.T.-H.L.); hyunwook@inha.ac.kr (H.-W.C.)
2 Program in Environmental and Polymer Engineering, Inha University, Incheon 22212, Korea
3 Department of Energy and Environmental Engineering, The Catholic University of Korea, 43 Jibong-ro, Bucheon-si 14662, Korea; jsyoun@catholic.ac.kr
4 Jeonbuk Department of Inhalation Research, Korea Institute of Toxicology, Jeongeup 53212, Korea; higyu.moon@kitox.re.kr (H.-G.M.); chen.xinyu@kitox.re.kr (X.-Y.C.); Dongim.kim@kitox.re.kr (D.-I.K.)
5 Department of Human and Environmental Toxicology, University of Science & Technology, Daejeon 34113, Korea
* Correspondence: khlee@kitox.re.kr (K.-H.L.); kjjeon@inha.ac.kr (K.-J.J.)

Abstract: The lacking of laboratory black carbon (BC) samples have long challenged the corresponding toxicological research; furthermore, the toxicity tests of engineered carbon nanoparticles were unable to reflect atmospheric BC. As a simplified approach, we have synthesized artificial BC (aBC) for the purpose of representing atmospheric BC. Surface chemical properties of aBC were controlled by thermal treatment, without transforming its physical characteristics; thus, we were able to examine the toxicological effects on A549 human lung cells arising from aBC with varying oxidation surface properties. X-ray photoelectron spectroscopy, as well as Raman and Fourier transform infrared spectroscopy, verified the presence of increased amounts of oxygenated functional groups on the surface of thermally-treated aBC, indicating aBC oxidization at elevated temperatures; aBC with increased oxygen functional group content displayed increased toxicity to A549 cells, specifically by decreasing cell viability to 45% and elevating reactive oxygen species levels up to 294% for samples treated at 800 °C.

Keywords: artificial black carbon (aBC); thermal treatment; cytotoxicity; reactive oxygen species (ROS)

Citation: Le, Y.T.-H.; Youn, J.-S.; Moon, H.-G.; Chen, X.-Y.; Kim, D.-I.; Cho, H.-W.; Lee, K.-H.; Jeon, K.-J. Relationship between Cytotoxicity and Surface Oxidation of Artificial Black Carbon. *Nanomaterials* **2021**, *11*, 1455. https://doi.org/10.3390/nano11061455

Academic Editors: Vivian Hsiu-Chuan Liao, Saura Sahu and Eleonore Fröhlich

Received: 20 April 2021
Accepted: 28 May 2021
Published: 31 May 2021

1. Introduction

Black carbon (BC) is an undesired byproduct from the incomplete combustion of fossil fuels and biomass [1–5]. BC, commonly referring to soot and found in fine particulate matter ($PM_{2.5}$), is the main component of atmospheric carbonaceous aerosols [1–3]. As an efficient light-absorbing carbonaceous material, BC has been mainly implicated as a short-lived climate forcer [4–6]. Besides its impact on climate change and the ecosystem, BC has also been associated with pulmonary, cardiovascular, and premature death [3,6,7].

BC is characterized as having fractal agglomerates, being insoluble in water or common organic solvents, and being refractory and potential toxic [2,8]. The factor of human health was less concerned with than climate change, however, a number of studies focus on health impact assessments, which have usually been conducted by estimating the related adverse health outcomes of the population based on the concentration of BC [3,9]. Due to the lack of laboratory BC samples and the inability to separate BC from atmospheric particulate matter, to date, no specific toxicity experiment induced by BC has been conducted [10].

On the other hand, engineered carbon nanoparticles with corresponding toxicity were widely carried out [11,12]. Carbon nanoparticles have been shown to induce inflammation, enhance oxidative stress, and transform cell signaling and gene expression in mammalian cells and organs, and toxic effects arising from their well-defined physical features have been described in numerous studies [13–16]. Cheng et al. tested ultrafine carbon nanopowder and pointed that carbon particles disrupted the keratinocyte differentiation and upregulated inflammation; carbon powder was claimed to mimic ambient ultrafine particles, however, the commercial carbon powder properties are far different from real-world atmospheric particles [17]. Engineered carbon black and BC share some similar features (black appearance, aggregate morphology, and elemental component) and even some common biological responses; therefore, some of the toxicology results have misunderstood carbon black interchangeably with BC [10,18]. Hong et al. [18], based on elemental analysis, claimed that engineered carbon black with a high, pure percentage of elemental carbon is different from BC, therefore, an experiment with BC that is regarded to have intensive toxicity is needed.

The diesel soot particulate matter is produced by the National Institute of Standards and Technology (NIST), which can mimic real-world diesel soot, however, the existence of polycyclic aromatic hydrocarbon (PAH) with other chemical compositions renders the identification of specific parameters responsible for inducing toxicity and the mechanisms by which they cause harm to the human body highly challenging [19]. Moreover, soot samples were also collected from several sources with a complicated mixture in which BC was barely separated [20–22]. With a mixture of inorganic compounds, trace metals, and PAH, it is challenging to determine the key factors of soot that play the main roles in inducing toxicity. Therefore, artificial BC (aBC) is essential for a toxicity test satisfying three principal requirements: (1) representing elementary atmospheric BC; (2) maintaining origin physical properties; and (3) having a surface that is chemically controllable.

Herein, we have synthesized aBC using an aerosol generator. We defined aBC in this study as the ultrafine particles generated from one origin source, graphite, to simplify the test of relation between toxicity and the chemical surface of aBC; aBC must possess a controllable chemical surface and common physical properties. Thermal treatment is a straightforward method for manipulating the chemical properties of aBC surfaces, especially to control oxygen functional group content while maintaining its physical characteristics. Because a dominant cytotoxicity mechanism of ultrafine particles is a cell–particle interaction accompanied by the overproduction of ROS [23–25], ROS generation and related cytotoxicity were the focus of this study. We demonstrated that an increase in oxygenated functional groups on the surface of BC triggered increased cytotoxicity and ROS levels in the human lung cancer cell line (A549).

2. Materials and Methods

2.1. Particle Generation and Thermal Treatment

Graphite was selected to produce aBC generated from a spark discharge soot generator (DNP Digital 3000, Palas GmbH, Karlsruhe, Germany). It operates on the principle of spark discharge to produce nanoscale particles of consistent concentration in high yields and negligible volatile content. The generator maintains a jump spark between two graphite electrodes at a high voltage of 2500 V. Graphite is vaporized by this spark, and then condenses to form particles with a size distribution and structure similar to those of diesel soot. The generator was operated under a constant flow of carrier inert Ar gas and purified dilution air at flow rates of 5 and 10 LPM, respectively (gas streams were controlled with a series of mass flow controllers).

The generated particles were produced in a tubular furnace at different ranges of temperatures. The first experiment was performed with the furnace temperature increasing continuously from room temperature (RT) to 800 °C to investigate particle properties as a function of the treatment temperature. Five treatment temperatures (RT, 200 °C, 400 °C, 600 °C and 800 °C) were applied separately in the second experiment to evaluate the effects of temperature on the physicochemical properties of aBC. A schematic of the experimental setup, including the sampling and measuring devices, is presented in Figure 1.

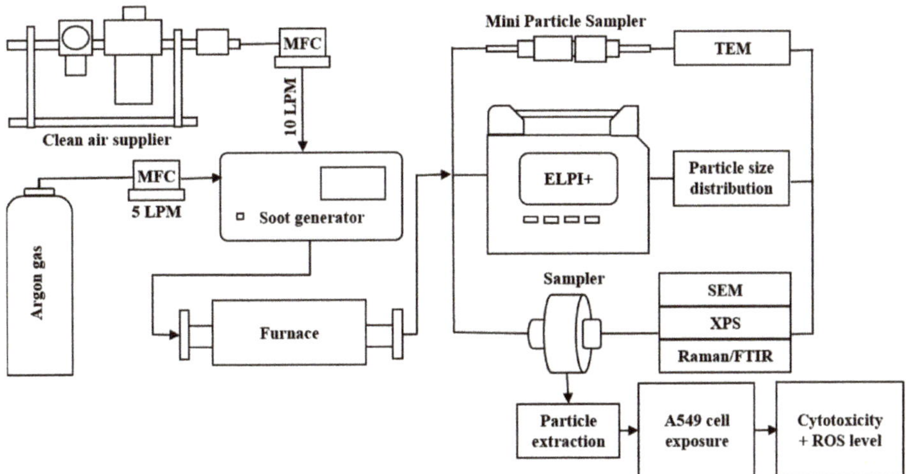

Figure 1. Schematic of the experimental setup.

2.2. Physicochemical Characterization

An electrical low-pressure impactor (ELPI+^{TM}, DEKATI, Finland) was employed to measure the concentration of generated particles in the range of 6 nm to 10 μm with a time resolution of 1 s. Furthermore, we estimated the surface area concentration of the deposited particles in the human respiratory tract based on ELPI+ stage-specific conversion factors. Teemu et al. estimated the ELPI+ response coefficient (β) by using an equivalent unit density (1 g/cm^3) for spherical particles, with a possible error in the mean diameter arising at each stage of each instrument 2% [26]. From the measured electrical current (I) carried by the aBC, the deposited area concentration of aBC (A_{dep}) was estimated for three regions, head airways, tracheobronchial, and alveolar, according to the following equation:

$$Adep = \beta \times I$$

The deposited area distribution of aBC was estimated to select suitable cell lines for cytotoxicity testing.

In the second experiment, samples were collected on pre-fired quartz fiber filters (six hours, 600 °C) with a diameter of 25 mm (FTQ25, Zefon, Ocala, FL, USA) and Teflon filters (R2PJ047, Pall Corporation, Port Washington, NY, USA) with a diameter of 47 mm using filter holders at five different temperatures (RT, 200 °C, 400 °C, 600 °C and 800 °C); these samples were utilized for chemical analysis and the in vitro test.

High-resolution transmission electron microscopy (HR-TEM; JEM-2100, JEOL, Akishima, Japan) was employed to observe the size and morphology of the aBC at an accelerating voltage of 200 kV. The aBC was collected on holey carbon TEM grids using a mini particle sampler (MPS, Ineris, Ecomesure, Saclay, France). For SEM analysis, aBC was dispersed on a heated (100 °C) silicon wafer by depositing a drop of particle suspension, created by sonicating aBC collected on a quartz filter in DI water. These samples were coated with a thin film of Pt and then observed by field emission-scanning electron microscopy (FE-SEM; S-4300, Hitachi, Tokyo, Japan).

The surface chemical compositions of the aBC samples were analyzed by X-ray photoelectron spectroscopy (XPS; Thermo Fisher Scientific Co., Walthem, MA, USA). Additionally, Raman spectra were acquired using an Xplora spectrometer (Horiba Jobin-Yvon, Palaiseau, France) equipped with a 532 nm solid state laser source. Fourier transform infrared vacuum spectroscopy (FTIR; Bruker VERTEX 80V, Coventry, UK) was employed to investigate structural changes occurring during the heating process.

2.3. Endocytosis, Cytotoxicity Assay, and Evaluation of Reactive Oxygen Species (ROS)

2.3.1. Cell Culture

A549 cells were purchased from the American Type Culture Collection (ATCC, Manassas, VA, USA). The cells were cultured in Roswell Park Memorial Institute (RABCI) 1640 (Gibco, Grand Island, NY, USA) containing 5% fetal bovine serum (FBS; Gibco) and 1% streptomycin and penicillin at 37 °C in an atmosphere containing 5% CO_2.

2.3.2. Cytotoxicity

For cytotoxicity experiments, A549 cells were seeded in 96-well plates at a density of 1×10^4 cells per well and left to attach overnight. The cells were then treated with various concentrations (50–2000 μg mL^{-1}) of aBC in the complete medium for 48 h. This was followed by the addition of 3-(4,5-dimethylthiazol-2-thiazyl)-2,5-diphenyl-tetrazolium bromide (MTT; Sigma Aldrich, St. Louis, MO, USA) solution to the final concentration of 1.0 mg mL^{-1}. After 3 h of incubation, the medium was removed, the formed blue formazan crystals were dissolved in 100 μL of dimethyl sulfoxide (Junsei Chemical Co., Tokyo, Japan) per well DMSO, and the absorbance at 570 nm (690 nm background control) was measured using a microplate reader (Biotek, Winooski, VT, USA). Low toxicity of aBC was observed in most treatments, with 75% of cells remaining viable after treatment with the highest dose (2.0 mg mL^{-1}) of aBC for 48 h. Therefore, these treatment conditions were used for all subsequent experiments.

2.3.3. Endocytosis of aBC in A549 Cells

The aBC-exposed cells were harvested using 0.25% trypsin-EDTA (Gibco, Thermo Fisher Scientific, Walthem, MA, USA) from culture plates and cyto-centrifuged (800 raBC; 10 min) onto glass microscope slides using Cytospin 4 (Thermo Fisher Scientific, Walthem, MA, USA). Cell smears were fixed in methanol for 1 min and stained using Diff-Quik solution (Dade Diagnostics, Aguada, Puerto Rico). Stained cells were analyzed under a light microscope (Leica, Wetzlar, Germany).

2.3.4. Measurement of Reactive Oxygen Species (ROS) Levels

A549 cells were incubated for 30 min at 37 °C in RABCI 1640 medium containing 3.3 μmol L^{-1} of 2,7-dichloro-fluorescein diacetate (DCF-DA) (Thermo Fisher Scientific, Walthem, MA, USA) for ROS detection. The cells were treated with aBC (2.0 mg mL^{-1} for 48 h), then washed with Dulbecco's phosphate-buffered saline (DPBS; Welgene Inc., Daegu, Korea). The DCF-DA intensity in the cells was immediately measured at 495 nm (excitation)/529 nm (emission) using a microplate reader. ROS production in the cells is represented as a percentage of DCF-DA intensity relative to cell viability in each well, which was defined as 100%.

3. Results

3.1. Emission Characteristics of Synthesized aBC

Figure 2 indicates that the generated particles experienced a change in the number concentration during thermal treatment. The left vertical axis represents the aerodynamic diameter of aBC, and the contour map reflects a reduction in the particle number concentration going from red to blue. This result demonstrated that the total particle number concentration was at a maximum of 1.6×10^8 cm^{-3}, decreasing with higher treatment temperatures and eventually reaching 2.2×10^7 cm^{-3} when the furnace temperature was 800 °C. The size distribution is also provided here, with the majority of the produced particles exhibiting aerodynamic diameters smaller than 100 nm. The aerodynamic diameter has been reported as the dictation of particle penetration into the lung, therefore, with those nano-scaled sizes, aBC has a higher chance to travel deep into the human lung, even meeting the bloodstream, cells, and tissues [27,28]. All size distributions were found to be unimodal, with a typical size of approximately 30 nm in aerodynamic diameter. The dominant size of the particles increased marginally at approximately 400 °C and 800 °C.

The slight increase in particle size and decrease in the number concentration at higher temperatures were likely caused by thermal coagulation. Higher temperatures enhance attractive forces and Brownian motion, resulting in an increased frequency of collisions [29]. At every collision, there is one less particle, and no new growth or nucleation occurs during aggregation, leading to a small increase in size, and the thermal degradation of particles chiefly decreases the number concentration [30].

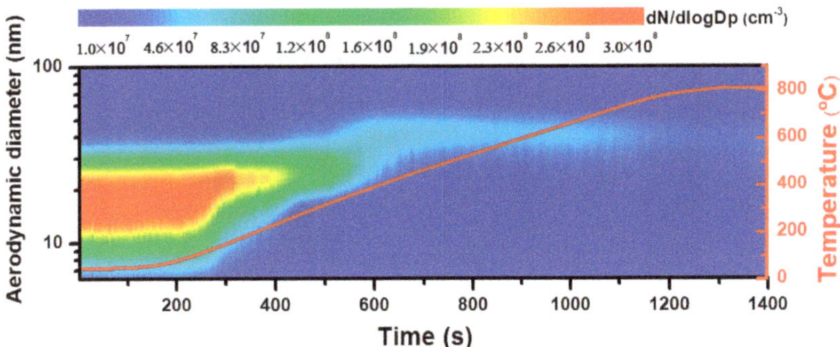

Figure 2. Size and time-resolved particle number concentration during heating from room temperature (RT) to 800 °C. The red line represents the furnace temperature.

3.2. TEM and SEM Image Analysis

TEM and SEM images of the particle samples were collected for the five treatment temperatures, as shown in Figure 3. SEM images indicated particle super-aggregation, with a tendency for extremely small particles to assemble into bigger particles. The majority of the aBC was approximately 27.7 ± 3.8 nm in diameter, with a near-spherical morphology (Figure 3a). In Figure 3b, the TEM results provide detailed images of the particles. Particles aggregated into branching structures and presented typical diesel soot morphologies, with nearly spherical and irregular carbonaceous particles. The primary particle size distribution is presented in Figure 3c, with a dominant size of approximately 30 nm, which is comparable to the real-time size distribution data shown in Figure 2. The general size and shape of the synthesized aBC revealed a uniform, near-spherical shape with a narrow size distribution. TEM and SEM images indicated that varying the treatment temperature did not result in significant differences in particle morphology.

3.3. Chemical Surface Properties

The XPS element and atomic concentration survey (Figure 4a) showed that the presence of Si2p is the highest from the blank sample due to utilization of a quartz filter. The concentration fraction of O1s slightly went up in 600 °C samples, and significantly increased in the 800 °C sample compared to others, indicating that the oxygen increased. The XPS carbon (C1s) spectra of the five samples are depicted in Figure 4b, and are de-convoluted into four peaks. The C–C bond, or graphitic carbon, was dominant, represented by the green peak at approximately 284.2 eV [31–33]. Other peaks at higher binding energies were related to carbon-oxygenated functional groups, including: C–O at 285.6 eV (red peak), –C–OH at 286.8 eV (pink peak), and –C=O at 288.7 eV (blue peak) [30–32]. The component fractions were assessed as functions of furnace temperature in Figure 4c. The C–C content decreased from 66% to 59%, while C–O increased from 16.4% to 19%, and C=O increased from 5.5% to 9.5% with increasing treatment temperature from RT to 800 °C. This change indicated that the percentage of oxygen functional groups had increased, resulting from higher amounts of oxygen present on the sample surfaces at higher temperatures. We assumed that the oxidation of graphitic carbon occurred above 600 °C.

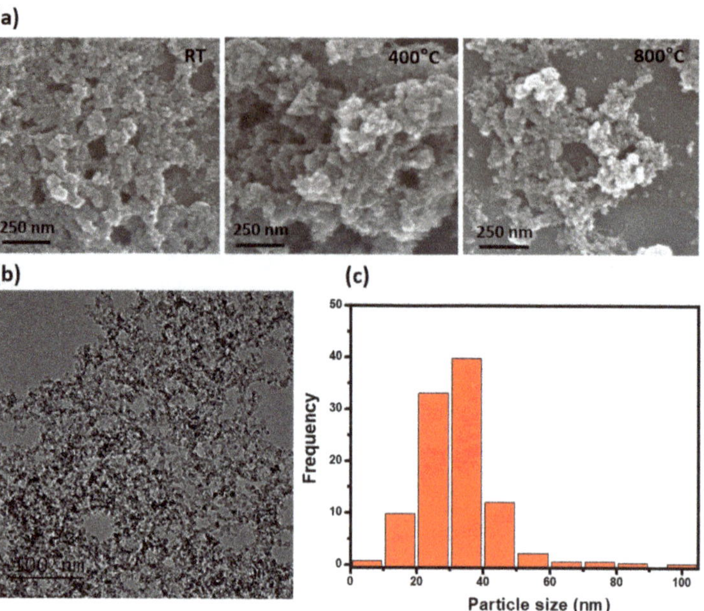

Figure 3. (**a**) SEM and (**b**) TEM images of aBC generated at various treatment temperatures, and (**c**) average particle size distribution for synthesized aBC.

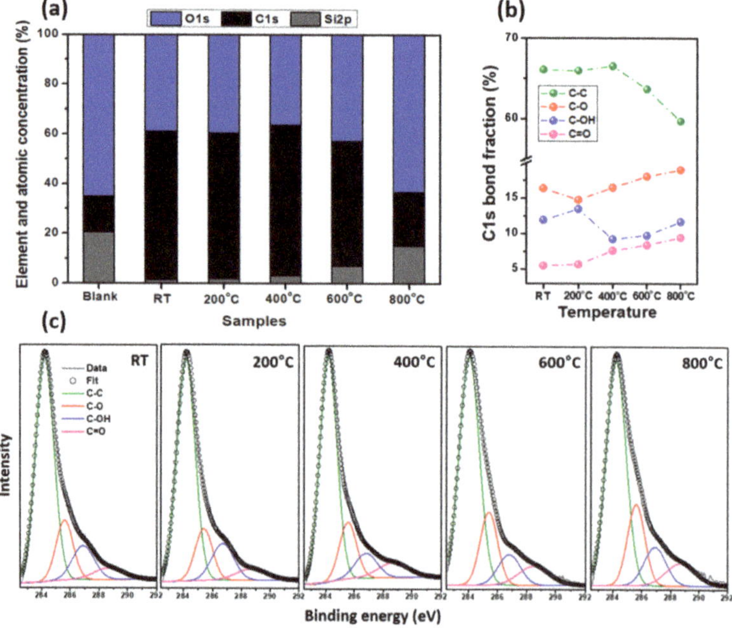

Figure 4. (**a**) Element and atomic concentration survey; (**b**) the C1s bond fraction indicates the variation in C–C, C–O, C–OH, and C=O content with increasing treatment; and (**c**) XPS spectra of synthesized aBC samples treated at five different temperatures.

To verify the oxidation of aBC, vibrational characterization was performed by Raman and FTIR analysis. Two typical overlapping peaks are visible in the Raman spectra displayed in Figure 5a, namely a D peak at 1340 cm^{-1} and a G peak at 1600 cm^{-1} [34–37]. The D band represents in-plane breathing vibrations of the aromatic ring structures (A_{1g} symmetry), and the G band is the in-plane stretching vibration of sp^2 carbons (E_{2g} symmetry) [38,39]. The intensity ratios between the D and G bands (I(D)/I(G)) were 0.80, 0.82, 0.82, 0.84, and 0.96 at RT, 200 °C, 400 °C, 600 °C, and 800 °C, respectively. The increase in the I(D)/I(G) ratio at 800 °C was ascribed to an increase in the in-plane breathing vibrations of the aromatic ring resulting from the appearance of a functionalized group on the aromatic ring. This result provides conclusive evidence of aBC oxidation by thermal processing, especially at 800 °C.

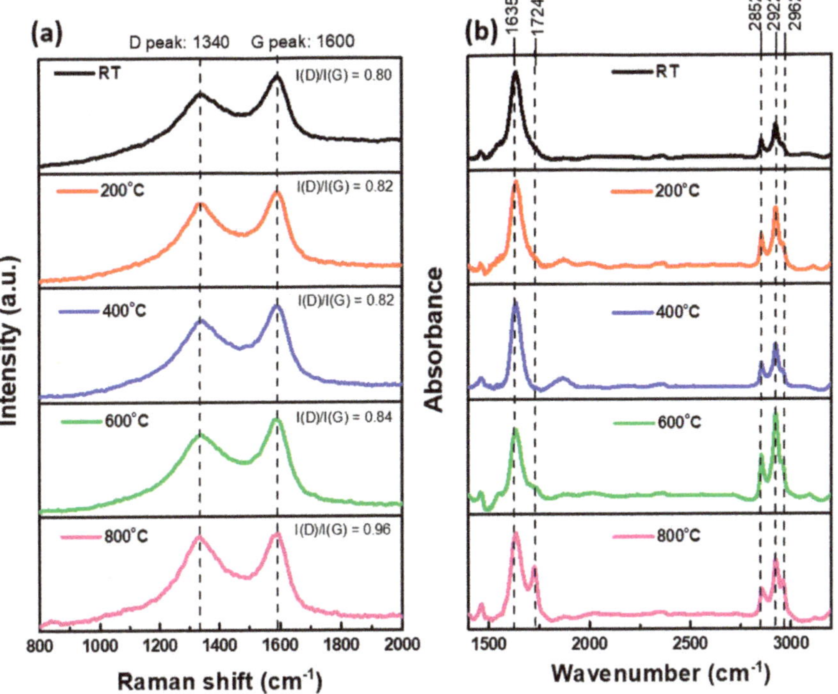

Figure 5. (**a**) Raman and (**b**) FTIR spectra obtained for five aBC samples generated at specific treatment temperatures.

Figure 5b shows the FTIR spectra of the five analyzed samples. No significant changes were observed among the spectra of the three samples treated in the range of RT to 400 °C. These spectra presented a sharp peak at approximately 1635 cm^{-1}, assigned to aromatic C=C. Three vibrations located at 2852, 2922, and 2962 cm^{-1} were assigned to asymmetric and symmetric C-H stretching of CH_3 and CH_2 aliphatic groups [40,41]. At 600 °C, the spectrum included an additional carbonyl C=O stretching (1720 cm^{-1}) shoulder, becoming a notable peak at 800 °C and indicating the presence of oxygen functionalities [41,42].

3.4. *In Vitro Toxicity of aBC*

Exposure to aBC results in harmful effects on human health, causing pulmonary and cardiovascular diseases [43,44]. The aBC is internalized by various immune and structural cell types, such as macrophages, lymphocytes, skin keratinocytes, and epithelial cells [45–47]. We deduced that aBC would be predominantly deposited in the alveolar region of the human lung by analyzing the electrical current carried by aBC (Figure 6). The estimation of the lung deposited surface area distribution based on the current charge carried by aBC implied that a higher fraction (50%) of aBC is possibly deposited in the alveolar region than on other regions of the human respiratory system. Alveolar macrophages and alveolar epithelial cells are highly likely to be the primary and secondary targets for aBC if inhaled by humans.

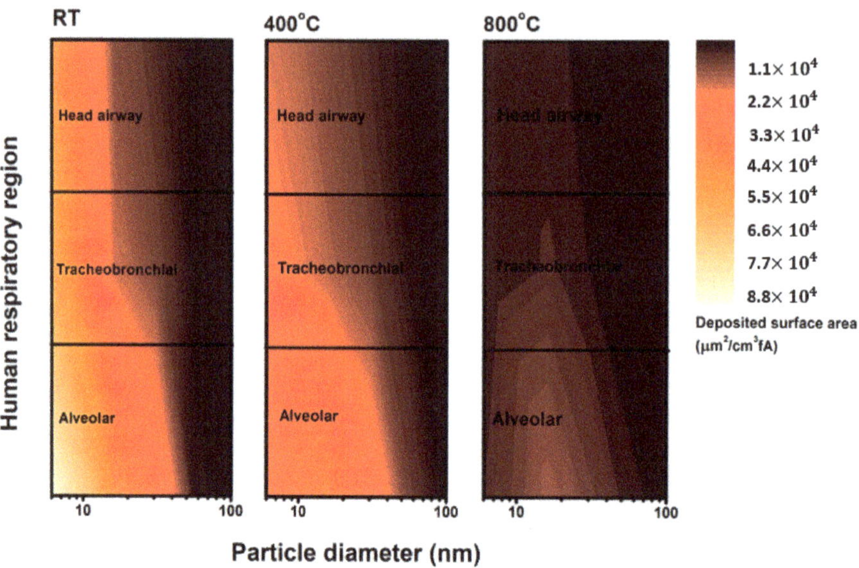

Figure 6. Deposited surface area distributions of aBC in three regions of the human respiratory tract, as predicted based on the electrical current recorded by ELPI+.

Herein, we selected the A549 human lung alveolar basal epithelial cell line for in vitro testing. To verify whether aBC directly affects epithelial cells, we confirmed endocytosis of aBC in A549 cells. Diff-Quik staining indicated the appearance of aBC in the cytoplasm and in the region near the nucleus (Figure 7), indicating the appearance of aBC in the cytomorphologic evaluation; however, to confirm the cellular uptake of aBC, further investigation is needed. The particle size is known to have significant effects on the interactions between nanoparticles and the cellular environment [13]. Here, nano-sized aBC has a greater opportunity for cellular uptake via endocytosis by membrane wrapping due to its effective binding to membrane receptors. When aBC interacts and penetrates through the membrane, the defense mechanism is activated, and cell damage occurs [48,49]. In addition, to determine the cytotoxicity of aBC, we evaluated cell viability using the MTT assay. Cytotoxic effects of the particles were clearly observed in epithelial cells 48 h after stimulation with aBC. The results revealed that aBC exhibited considerable cytotoxicity to A549 cells at a concentration of 2 mg/mL, reducing the survival rate of A549 cells to 75%, 68.2%, 68.5%, and 65% when synthesized at RT, 200 °C, 400 °C, and 600 °C, respectively. In particular, the 800 °C sample induced the strongest cytotoxic effect (cell viability was reduced up to 45%) compared to the naïve control (Figure 8a). Hence, aBC had adverse effects on the survival of A549 cells; aBC directly contacts the cell membrane, inducing membrane stress by disrupting and damaging it, resulting in cell death. Even exposed to a high dose of aBC, the decreased cell viability was still low compared to other literature due to the aggregation of aBC in the cell medium [11,50].

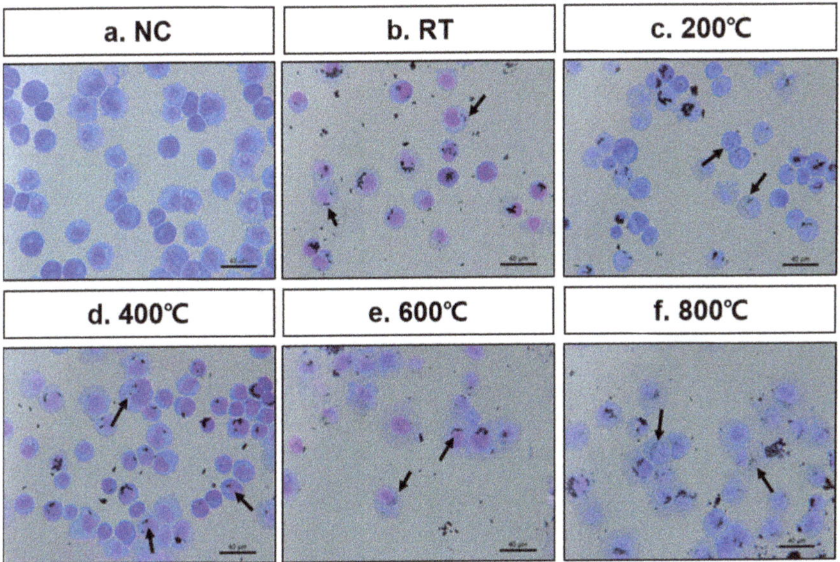

Figure 7. Diff-Quik staining images of (**a**) the naïve control (NC) and (**b–f**) A549 cells stimulated by aBC synthesized at five treatment temperatures. Arrows indicate the internalization of aBC in the A549 cell.

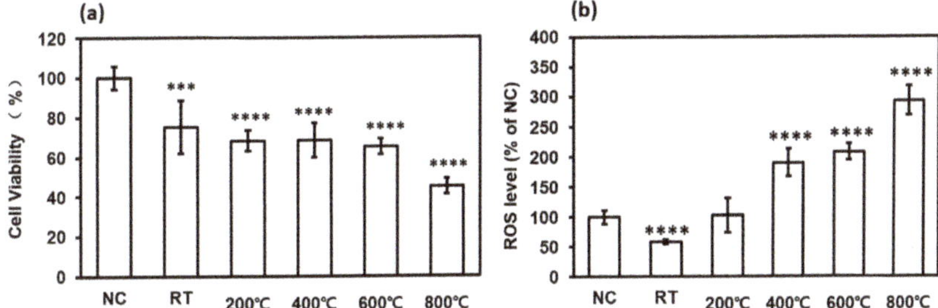

Figure 8. (**a**) Cell viability and (**b**) ROS production in the naïve control (NC) and synthesized aBC-stimulated A549 cells. Data are presented as the mean ± SD (n = 8). *** $p < 0.01$ and **** $p < 0.001$ compared to the NC.

Exposure to aBC induces cell damage via oxidative stress mediated by ROS [47,51]. Excess ROS levels trigger oxidative stress that disrupts cellular homeostasis and affects the oxidation of biomolecules, including DNA, lipids, and proteins [44,52]. ROS production caused by aBC in A549 cells was evaluated by measuring the fluorescence intensity of DCF-CytDA. The aBC generated at higher temperatures induced significant changes in ROS levels in A549 cells compared to the naïve controls (Figure 8b). As observed for cytotoxicity, the greatest oxidative stress was demonstrated in the 800 °C aBC samples for particle-stimulated A549 cells. These results indicated that synthesized aBC directly induced ROS-mediated human alveolar basal epithelial cell damage. The production of high levels of ROS causes significant damage to DNA, thereby affecting cell survival. The correlation of biomass burning aerosol chemical composition with ROS production was used for investigation, however, due to the complex effect of chemical compositions, soot-induced bio-toxicity with ROS production was barely interpreted. It is claimed that no significant correlation can be figured out in the ROS response with measured chemical compositions, and the aging process of biomass burning samples even drove the toxicity test to be more complicated [53]. Therefore, to minimize the factors on real-world soot, with the important roles played in the bio-toxicity results, we highlighted the ROS response with the change of the chemical surface on controlled aBC.

The physical characterization illustrated a uniform size and shape of aBC synthesized under varying treatment temperatures. On the other hand, Raman and FTIR analysis supported by XPS data led to the conclusion that a greater number of oxygenated functional groups were present on the surface of aBC treated at elevated temperatures. Cell viability and ROS level results indicated that cytotoxicity increased with increasing oxygenated functional group content on the surface of aBC. This agrees with the findings of Das et al., who investigated the correlation between cellular toxicity and oxygenated functional group density on graphene oxide (GO) [48]. They proved that the presence of organic functional groups on the surface of GO affected its interaction with mammalian cells at the "nano–bio" interface, and that an increase in oxygen functional groups rendered GO less biologically inert and resulted in elevated cytotoxicity. We can therefore infer from the present results that an increase in oxygen functional groups on the surface of aBC activates the "nano–bio" interface, thereby facilitating cell membrane disruption by aBC and resulting in higher cytotoxicity. Furthermore, oxidized flame soot increases ROS levels, as reported by A. Holder. This suggests that the increased cytotoxicity of oxidized soot is due to its ability to generate oxidants [50]. Consequently, studies centered on the evaluation of surface chemical properties associated with cytotoxicity suggest that oxygenated functional groups present on the surface of aBC and aBC oxidation are directly related to cell death and oxidative stress.

4. Conclusions

The aBC was successfully generated at various treatment temperatures without transforming the original physical features. The treatment temperature is known to significantly impact the surface chemical structure of aBC; hence, we were able to simplify the investigation into the effect of surface chemical properties of aBC on human epithelial cells; aBC with a nano size effectively penetrated the plasma membrane, leading to cell damage. It was found that an increase in the presence of oxygen functional groups on the surface of aBC directly affected cell viability and oxidative stress in A549 cells. The remaining limitation of this research is that aBC is still far from real-world BC, which has a complex mixture of chemical components, however, the results of the relationship between cytotoxicity and surface oxidation of aBC can be the foundation to conduct further research. Furthermore, aggregation and surface morphological changes of aBC in cell medium culture need further investigation, and aBC-induced bio-toxicity with proof of cell uptake is also needed to improve the quality of the cytotoxicity test. The semi-quantification of cell uptake should be investigated to support the mechanism of cell death. An in vivo test and in vitro test with an air–liquid interface approach would be needed for further toxicological models for our synthesized aBC, rather than a submerged cell culture to minimize the current limitations. Future studies will focus on aBC synthesized with sulfates, nitrates, metals, and organic compounds that are commonly emitted with BC to the atmosphere to better simulate real-world experimentation.

Author Contributions: Y.T.-H.L.: methodology, investigation, writing—origin draft; J.-S.Y.; supervision, writing—review & editing; X.-Y.C. and D.-I.K.; experimentation, formal analysis; H.-W.C.; experimentation, investigation; K.-H.L. and H.-G.M.; supervision, validation; K.-J.J.; conceptualization, supervision, validation, writing—review & editing. All authors have read and agreed to the published version of the manuscript.

Funding: This work was funded by the Korea Institute of Toxicology (KIT, KK−1905-02) and supported by the Korea Ministry of Environment (MOE) as part of the Environmental Health Action Program (2016001360005).

Data Availability Statement: Data are contained within the article.

Conflicts of Interest: The authors declare no conflict of interest.

References

1. Grange, S.; Lötscher, H.; Fischer, A.; Emmenegger, L.; Hueglin, C. Evaluation of equivalent black carbon (EBC) source appointment using observations from Switzerland between 2008 and 2018. *Atmos. Meas. Tech. Discuss.* **2019**, 1–27. [CrossRef]
2. Kang, S.; Zhang, Y.; Qian, Y.; Wang, H. A review of black carbon in snow and ice and its impact on the cryosphere. *Earth-Sci. Rev.* **2020**, *210*, 103346. [CrossRef]
3. Farzad, K.; Khorsandi, B.; Khorsandi, M.; Bouamra, O.; Maknoon, R. A study of cardiorespiratory related mortality as a result of exposure to black carbon. *Sci. Total Environ.* **2020**, *725*, 138422. [CrossRef]
4. Kahnert, M.; Kanngießer, F. Modelling optical properties of atmospheric black carbon aerosols. *J. Quant. Spectrosc. Radiat. Transf.* **2020**, *244*, 106849. [CrossRef]
5. Zhao, D.; Liu, D.; Yu, C.; Tian, P.; Hu, D.; Zhou, W.; Ding, S.; Hu, K.; Sun, Z.; Huang, M.; et al. Vertical evolution of black carbon characteristics and heating rate during a haze event in Beijing winter. *Sci. Total Environ.* **2020**, *709*, 136251. [CrossRef]
6. Lyu, Y.; Olofsson, U. On black carbon emission from automotive disc brakes. *J. Aerosol Sci.* **2020**, *148*, 105610. [CrossRef]
7. Moallemi, A.; Kazemimanesh, M.; Corbin, J.C.; Thomson, K.; Smallwood, G.; Olfert, J.S.; Lobo, P. Characterization of black carbon particles generated by a propane-fueled miniature inverted soot generator. *J. Aerosol Sci.* **2019**, *135*, 46–57. [CrossRef]
8. Wang, M.; Chen, Y.; Fu, H.; Qu, X.; Li, B.; Tao, S.; Zhu, D. An investigation on hygroscopic properties of 15 black carbon (BC)-containing particles from different carbon sources: Roles of organic and inorganic components. *Atmos. Chem. Phys.* **2020**, *20*, 7941–7954. [CrossRef]
9. Zhou, H.; Lin, J.; Shen, Y.; Deng, F.; Gao, Y.; Liu, Y.; Dong, H.; Zhang, Y.; Sun, Q.; Fang, J.; et al. Personal black carbon exposure and its determinants among elderly adults in urban China. *Environ. Int.* **2020**, *138*, 105607. [CrossRef]
10. Buseck, P.R.; Adachi, K.; Gelencsér, A.; Tompa, É.; Pósfai, M. Are black carbon and soot the same? *Atmos. Chem. Phys. Discuss.* **2012**, *12*, 24821–24846. [CrossRef]
11. Lindner, K.; Ströbele, M.; Schlick, S.; Webering, S.; Jenckel, A.; Kopf, J.; Danov, O.; Sewald, K.; Buj, C.; Creutzenberg, O.; et al. Biological effects of carbon black nanoparticles are changed by surface coating with polycyclic aromatic hydrocarbons. *Part. Fibre Toxicol.* **2017**, *14*, 1–17. [CrossRef]
12. Panomsuwan, G.; Chokradjaroen, C.; Rujiravanit, R.; Ueno, T.; Saito, N. In vitro cytotoxicity of carbon black nanoparticles synthesized from solution plasma on human lung fibroblast cells. *Jpn. J. Appl. Phys.* **2018**, *57*. [CrossRef]
13. Adeyemi, J.A.; Machado, A.R.T.; Ogunjimi, A.T.; Alberici, L.C.; Antunes, L.M.G.; Barbosa, F. Cytotoxicity, mutagenicity, oxidative stress and mitochondrial impairment in human hepatoma (HepG2) cells exposed to copper oxide, copper-iron oxide and carbon nanoparticles. *Ecotoxicol. Environ. Saf.* **2020**, *189*, 109982. [CrossRef]
14. Hou, L.; Guan, S.; Jin, Y.; Sun, W.; Wang, Q.; Du, Y.; Zhang, R. Cell metabolomics to study the cytotoxicity of carbon black nanoparticles on A549 cells using UHPLC-Q/TOF-MS and multivariate data analysis. *Sci. Total Environ.* **2020**, *698*, 1–9. [CrossRef]
15. Yuan, X.; Zhang, X.; Sun, L.; Wei, Y.; Wei, X. Cellular Toxicity and Immunological Effects of Carbon-based Nanomaterials. *Part. Fibre Toxicol.* **2019**, *16*. [CrossRef]
16. Fiorito, S.; Mastrofrancesco, A.; Cardinali, G.; Rosato, E.; Salsano, F.; Su, D.S.; Serafino, A.; Picardo, M. Effects of carbonaceous nanoparticles from low-emission and older diesel engines on human skin cells. *Carbon N. Y.* **2011**, *49*, 5038–5048. [CrossRef]
17. Cheng, Z.; Liang, X.; Liang, S.; Yin, N.; Faiola, F. A human embryonic stem cell-based in vitro model revealed that ultrafine carbon particles may cause skin inflammation and psoriasis. *J. Environ. Sci. (China)* **2020**, *87*, 194–204. [CrossRef] [PubMed]
18. Hong, J.; Moon, H.; Kim, J.; Lee, B.C.; Kim, G.B.; Lee, H.; Kim, Y. Differentiation of carbon black from black carbon using a ternary plot based on elemental analysis. *Chemosphere* **2021**, *264*, 128511. [CrossRef]
19. Li, J.; Yang, X.; Zhang, Z.; Xiao, H.; Sun, W.; Huang, W.; Li, Y.; Chen, C.; Sun, Y. Aggregation kinetics of diesel soot nanoparticles in artificial and human sweat solutions: Effects of sweat constituents, pH, and temperature. *J. Hazard. Mater.* **2021**, *403*, 123614. [CrossRef] [PubMed]
20. Jia, H.; Li, S.; Wu, L.; Li, S.; Sharma, V.K.; Yan, B. Cytotoxic Free Radicals on Air-Borne Soot Particles Generated by Burning Wood or Low-Maturity Coals. *Environ. Sci. Technol.* **2020**, *54*, 5608–5618. [CrossRef] [PubMed]
21. Rouse, R.L.; Murphy, G.; Boudreaux, M.J.; Paulsen, D.B.; Penn, A.L. Soot nanoparticles promote biotransformation, oxidative stress, and inflammation in murine lungs. *Am. J. Respir. Cell Mol. Biol.* **2008**, *39*, 198–207. [CrossRef] [PubMed]
22. Soto, K.F.; Garza, K.M.; Shi, Y.; Murr, L.E. Direct contact cytotoxicity assays for filter-collected, carbonaceous (soot) nanoparticulate material and observations of lung cell response. *Atmos. Environ.* **2008**, *42*, 1970–1982. [CrossRef]
23. Badran, G.; Verdin, A.; Grare, C.; Abbas, I.; Achour, D.; Ledoux, F.; Roumie, M.; Cazier, F.; Courcot, D.; Lo Guidice, J.M.; et al. Toxicological appraisal of the chemical fractions of ambient fine (PM2.5-0.3) and quasi-ultrafine (PM0.3) particles in human bronchial epithelial BEAS-2B cells. *Environ. Pollut.* **2020**, *263*, 1–12. [CrossRef]
24. Liu, L.; Zhou, Q.; Yang, X.; Li, G.; Zhang, J.; Zhou, X.; Jiang, W. Cytotoxicity of the soluble and insoluble fractions of atmospheric fine particulate matter. *J. Environ. Sci. (China)* **2020**, *91*, 105–116. [CrossRef]
25. Jan, R.; Roy, R.; Bhor, R.; Pai, K.; Satsangi, P.G. Toxicological screening of airborne particulate matter in atmosphere of Pune: Reactive oxygen species and cellular toxicity. *Environ. Pollut.* **2020**, *261*, 113724. [CrossRef]
26. Lepistö, T.; Kuuluvainen, H.; Juuti, P.; Järvinen, A.; Arffman, A.; Rönkkö, T. Measurement of the human respiratory tract deposited surface area of particles with an electrical low pressure impactor. *Aerosol Sci. Technol.* **2020**, *54*, 958–971. [CrossRef]
27. Donaldson, K.; Poland, C.A. Nanotoxicity: Challenging the myth of nano-specific toxicity. *Curr. Opin. Biotechnol.* **2013**, *24*, 724–734. [CrossRef] [PubMed]

28. Azarmi, S.; Tao, X.; Chen, H.; Wang, Z.; Finlay, W.H.; Löbenberg, R.; Roa, W.H. Formulation and cytotoxicity of doxorubicin nanoparticles carried by dry powder aerosol particles. *Int. J. Pharm.* **2006**, *319*, 155–161. [CrossRef]
29. Hinds, W.C. *Aerosol Technology: Properties, Behaviour, and Measurement of Airborne Particles*; Wiley: Hoboken, NJ, USA, 1982.
30. Zhang, R.; Khalizov, A.F.; Pagels, J.; Zhang, D.; Xue, H.; McMurry, P.H. Variability in morphology, hygroscopicity, and optical properties of soot aerosols during atmospheric processing. *Proc. Natl. Acad. Sci. USA* **2008**, *105*, 10291–10296. [CrossRef]
31. Mustafi, N.N.; Raine, R.R.; James, B. Characterization of exhaust particulates from a dual fuel engine by TGA, XPS, and Raman techniques. *Aerosol Sci. Technol.* **2010**, *44*, 954–963. [CrossRef]
32. Vander Wal, R.L.; Bryg, V.M.; Hays, M.D. XPS analysis of combustion aerosols for chemical composition, surface chemistry, and carbon chemical state. *Anal. Chem.* **2011**, *83*, 1924–1930. [CrossRef]
33. Gilham, R.J.J.; Spencer, S.J.; Butterfield, D.; Seah, M.P.; Quincey, P.G. On the applicability of XPS for quantitative total organic and elemental carbon analysis of airborne particulate matter. *Atmos. Environ.* **2008**, *42*, 3888–3891. [CrossRef]
34. Xu, C.; Shi, X.; Ji, A.; Shi, L.; Zhou, C.; Cui, Y. Fabrication and characteristics of reduced graphene oxide produced with different green reductants. *PLoS One* **2015**, *10*. [CrossRef] [PubMed]
35. Saffaripour, M.; Tay, L.L.; Thomson, K.A.; Smallwood, G.J.; Brem, B.T.; Durdina, L.; Johnson, M. Raman spectroscopy and TEM characterization of solid particulate matter emitted from soot generators and aircraft turbine engines. *Aerosol Sci. Technol.* **2017**, *51*, 518–531. [CrossRef]
36. Rosenburg, F.; Ionescu, E.; Nicoloso, N.; Riedel, R. High-temperature Raman spectroscopy of nano-crystalline carbon in silicon oxycarbide. *Materials (Basel)* **2018**, *11*, 93. [CrossRef] [PubMed]
37. Sadezky, A.; Muckenhuber, H.; Grothe, H.; Niessner, R.; Pöschl, U. Raman microspectroscopy of soot and related carbonaceous materials: Spectral analysis and structural information. *Carbon N. Y.* **2005**, *43*, 1731–1742. [CrossRef]
38. Ess, M.N.; Ferry, D.; Kireeva, E.D.; Niessner, R.; Ouf, F.X.; Ivleva, N.P. In situ Raman microspectroscopic analysis of soot samples with different organic carbon content: Structural changes during heating. *Carbon N. Y.* **2016**, *105*, 572–585. [CrossRef]
39. Gaddam, R.R.; Kantheti, S.; Narayan, R.; Raju, K.V.S.N. Graphitic nanoparticles from thermal dissociation of camphor as an effective filler in polymeric coatings. *RSC Adv.* **2014**, *4*, 23043–23049. [CrossRef]
40. Cain, J.P.; Gassman, P.L.; Wang, H.; Laskin, A. Micro-FTIR study of soot chemical composition - Evidence of aliphatic hydrocarbons on nascent soot surfaces. *Phys. Chem. Chem. Phys.* **2010**, *12*, 5206–5218. [CrossRef]
41. Santamaría, A.; Mondragón, F.; Molina, A.; Marsh, N.D.; Eddings, E.G.; Sarofim, A.F. FT-IR and 1H NMR characterization of the products of an ethylene inverse diffusion flame. *Combust. Flame* **2006**, *146*, 52–62. [CrossRef]
42. Paul, B.; Datta, A. FTIR characterization of soot precursor nano-particles from different heights of an isooctane/air premixed flame. *Int. J. Adv. Eng. Sci. Appl. Math.* **2014**, *6*, 97–105. [CrossRef]
43. Löndahl, J.; Swietlicki, E.; Rissler, J.; Bengtsson, A.; Boman, C.; Blomberg, A.; Sandström, T. Experimental determination of the respiratory tract deposition of diesel combustion particles in patients with chronic obstructive pulmonary disease. *Part. Fibre Toxicol.* **2012**, *9*, 30. [CrossRef] [PubMed]
44. Bates, J.T.; Weber, R.J.; Abrams, J.; Verma, V.; Fang, T.; Klein, M.; Strickland, M.J.; Sarnat, S.E.; Chang, H.H.; Mulholland, J.A.; et al. Reactive Oxygen Species Generation Linked to Sources of Atmospheric Particulate Matter and Cardiorespiratory Effects. *Environ. Sci. Technol.* **2015**, *49*, 13605–13612. [CrossRef]
45. Kim, D.I.; Song, M.K.; Kim, S.H.; Park, C.Y.; Lee, K. TF−343 Alleviates Diesel Exhaust Particulate-Induced Lung Inflammation via Modulation of Nuclear Factor- κ B Signaling. *J. Immunol. Res.* **2019**, *2019*. [CrossRef]
46. Colasanti, T.; Fiorito, S.; Alessandri, C.; Serafino, A.; Andreola, F.; Barbati, C.; Morello, F.; Alfè, M.; Di Blasio, G.; Gargiulo, V.; et al. Diesel exhaust particles induce autophagy and citrullination in Normal Human Bronchial Epithelial cells. *Cell Death Dis.* **2018**, *9*. [CrossRef] [PubMed]
47. Lodovici, M.; Bigagli, E. Oxidative stress and air pollution exposure. *J. Toxicol.* **2011**, *2011*. [CrossRef]
48. Wang, G.; Zhang, X.; Liu, X.; Zheng, J.; Chen, R.; Kan, H. Ambient fine particulate matter induce toxicity in lung epithelial-endothelial co-culture models. *Toxicol. Lett.* **2019**, *301*, 133–145. [CrossRef]
49. Santos, H.A.; Riikonen, J.; Salonen, J.; Mäkilä, E.; Heikkilä, T.; Laaksonen, T.; Peltonen, L.; Lehto, V.P.; Hirvonen, J. In vitro cytotoxicity of porous silicon microparticles: Effect of the particle concentration, surface chemistry and size. *Acta Biomater.* **2010**, *6*, 2721–2731. [CrossRef] [PubMed]
50. Holder, A.L.; Carter, B.J.; Goth-Goldstein, R.; Lucas, D.; Koshland, C.P. Increased cytotoxicity of oxidized flame soot. *Atmos. Pollut. Res.* **2012**, *3*, 25–31. [CrossRef]
51. Liu, S.; Zeng, T.H.; Hofmann, M.; Burcombe, E.; Wei, J.; Jiang, R.; Kong, J.; Chen, Y. Antibacterial activity of graphite, graphite oxide, graphene oxide, and reduced graphene oxide: Membrane and oxidative stress. *ACS Nano* **2011**, *5*, 6971–6980. [CrossRef]
52. Gurunathan, S.; Iqbal, M.A.; Qasim, M.; Park, C.H.; Yoo, H.; Hwang, J.H.; Uhm, S.J.; Song, H.; Park, C.; Do, J.T.; et al. Evaluation of graphene oxide induced cellular toxicity and transcriptome analysis in human embryonic kidney cells. *Nanomaterials* **2019**, *9*, 969. [CrossRef]
53. Li, J.; Li, J.; Wang, G.; Ho, K.F.; Dai, W.; Zhang, T.; Wang, Q.; Wu, C.; Li, L.; Li, L.; et al. Effects of atmospheric aging processes on in vitro induced oxidative stress and chemical composition of biomass burning aerosols. *J. Hazard. Mater.* **2021**, *401*. [CrossRef]

Article

Germination and Early Development of Three Spontaneous Plant Species Exposed to Nanoceria (nCeO$_2$) with Different Concentrations and Particle Sizes

Daniel Lizzi [1,2], Alessandro Mattiello [1], Barbara Piani [1], Guido Fellet [1], Alessio Adamiano [3] and Luca Marchiol [1,*]

1 DI4A—Department of Agriculture, Food, Environment and Animal Sciences, University of Udine, Via delle Scienze 206, 33100 Udine, Italy; lizzi.daniel.1@spes.uniud.it (D.L.); alessandro.mattiello@uniud.it (A.M.); barbara.piani@uniud.it (B.P.); guido.fellet@uniud.it (G.F.)
2 Department of Life Sciences, University of Trieste, Via Licio Giorgieri 10, 34127 Trieste, Italy
3 Institute of Science and Technology for Ceramics (ISTEC), National Research Council (CNR), Via Granarolo 64, 48018 Faenza, Italy; alessio.adamiano@istec.cnr.it
* Correspondence: marchiol@uniud.it

Received: 19 November 2020; Accepted: 15 December 2020; Published: 17 December 2020

Abstract: This study aimed to provide insight regarding the influence of Ce oxide nanoparticles (nCeO$_2$) with different concentrations and two different particle sizes on the germination and root elongation in seedlings of spontaneous terrestrial species. In a bench-scale experiment, seeds of the monocot, *Holcus lanatus* and dicots *Lychnis-flos-cuculi* and *Diplotaxis tenuifolia* were treated with solutions containing nCeO$_2$ 25 nm and 50 nm in the range 0–2000 mg Ce L^{-1}. The results show that nCeO$_2$ enters within the plant tissues. Even at high concentration, nCeO$_2$ have positive effects on seed germination and the development of the seedling roots. This study further demonstrated that the particle size had no influence on the germination of *L. flos-cuculi*, while in *H. lanatus* and *D. tenuifolia*, the germination percentage was slightly higher (+10%) for seeds treated with nCeO$_2$ 25 nm with respect to 50 nm. In summary, the results indicated that nCeO$_2$ was taken up by germinating seeds, but even at the highest concentrations, they did not have negative effects on plant seedlings. The influence of the different sizes of nCeO$_2$ on germination and root development was not very strong. It is likely that particle agglomeration and ion dissolution influenced the observed effects.

Keywords: nanomaterials; cerium oxide nanoparticles; wild herbs; seed germination; root length

1. Introduction

Nanoscience and nanotechnology are rapidly evolving in different applications having the potential to revolutionize human life. Considerable headways have been made for applications of engineered nanomaterials (ENMs) and nano-enabled products in medicine, energy, electronics, innovative materials and many others [1].

The flip side of nanotechnology is the release in the environment of tons of ENMs [2]. According to the ENMs flow models, soils and waters are the endpoints of such materials [3,4]. However, we still have patchy knowledge regarding the impacts of these materials on biota [5]. Since plant Kingdom represent about 82% of living organisms mass on Earth [6], and their ecological role is of paramount importance to understand the relationships between plants and ENMs. In particular, studying the behavior and fate of ENMs within plants is of great significance for exploring (i) ENMs uptake, translocation and storage in plant tissues, (ii) mechanisms of plant toxicity, and (iii) life cycle risk assessment of ENMs and risks of transfer to the trophic chain.

The early experimental demonstration regarding the negative influence of ENMs in higher plants was carried out not many years ago [7]. Subsequent studies reported physiological and morphological anomalies of plants exposed to nanomaterials [8,9]. Conversely, several studies of positive effects of ENMs applications to crops were reported. This is why applications of nano-enabled products in crop nutrition and protection are under investigation [10–12]. The first investigations revealed that the relationships between plants and ENMs are very complex. Up to now, the research has been paid almost exclusively to food crops, while the spontaneous plant species have been almost neglected. Although this was largely justified by the potential risks for ENMs human exposure, the potential negative impact of ENMs on primary producers could have very serious consequences on food webs and ecosystem services [13], and therefore, it should not be deemed less significant.

Experiments carried out on crops demonstrated that the chemical and physical properties of ENMs (e.g., size, shape, structure, composition, concentration, and others), the environmental conditions, the plant species and age contribute to determining the effects on plants [14,15]. It is not advisable to generalize the results on crops to other plants living on natural ecosystems, neither fertilized nor irrigated, and potentially more exposed to ENMs fluxes having a longer life-cycle than crops.

Studies have been conducted to investigate the flow of nanomaterials into aquatic and terrestrial environments. As regards plants, more aquatic [16–20] and wetland species [21–23] have been studied than terrestrial ones so far. To the best of our knowledge, *Pinus sylvestris* (L.) and *Quercus robur* (L.) are the only terrestrial wild plant species that have been investigated for exposure to silver nanoparticles (*n*Ag) and cerium oxide nanoparticles ($nCeO_2$) so far [24].

Investigations on the effects of metal nanoparticles (MeNPs) on plant physiology are based on the assumption that nanomaterials can be absorbed by plants and that the former can subsequently move within the plant tissues while maintaining the nanoform, or that they can release elements in ionic form. Hence, the experiments in this field must be designed in order to verify whether the nanomaterials are taken up by the plant roots or internalized through other pathways such as stomata, leaf cuticle/epidermis, and hydathodes [25].

Given the estimated global production of 100–1000 tons per year, $nCeO_2$ is among the most widely utilized metal oxide nanoparticle in Europe [26]. For this reason, the Organization for Economic Cooperation and Development (OECD) included $nCeO_2$ among the nanoparticles to be studied and analyzed for the risk assessment [27]. $nCeO_2$ could cause several effects on the plant system depending on $nCeO_2$ particle size, treatment concentration and plant species. Literature reports contradictory results. Positive effects in terms of germination, biomass yield, photosynthesis and nutritional status have been observed on several species [28]. Other papers report a reduction of germination rates, reduction or inhibition of root growth, restrictions of biomass growth, and crop yield [29–31].

In this study, we evaluated the influence of $nCeO_2$ having different concentrations and two particle sizes on the germination and root elongation in seedlings of the spontaneous monocot *Holcus lanatus* (L.), and the dicots *Lychnis flos-cuculi* (L.) and *Diplotaxis tenuifolia* (L.) DC. The plant species have been chosen since they are common and widespread in natural systems, highly competitive and easily adaptable to different ecological conditions. *Holcus lanatus* L. (common velvet grass) is a hairy, tufted, fibrous-rooted and meadow soft perennial grass, growing between 50 and 100 cm tall, belonging to the *Poaceae* family. It has a wide climatic range and occurs over a wide range of soil types and fertility conditions [32]. *Lychnis flos–cuculi* L. (ragged-robin) is an herbaceous perennial plant belonging to the *Caryophyllaceae* family, and it is native and distributed throughout Europe [33]. It is found in open habitats, along roads and in wet meadows and pastures. Finally, *Diplotaxis tenuifolia* L. DC. (perennial wall rocket) is a perennial flowering herbaceous Mediterranean species, but it is native to Europe and Western Asia [34]. It grows in temperate climates and could be found in different habitats, but in particular in ruderal plant associations.

2. Materials and Methods

2.1. Nanoparticles Characterization

The nCeO$_2$ with an average particle size of 25 nm and 50 nm, respectively, were purchased from Sigma-Aldrich (St. Louis, MO, USA). nCeO$_2$ has a density of 7.13 g mL^{-1} at 25 °C and 99.95% purity (81.25% of Ce).

The cerium oxide nanoparticles were suspended in deionized water and sonicated in a water bath for 60 min with a sonication power of 180 watts. The suspensions were characterized for Z–average size and hydrodynamic diameter (Hd), whose distributions were measured by dynamic light scattering (DLS) on a Zetasizer Nano ZS (Malvern Ltd., Worcestershire, UK) and relative polydispersity index (PDI). ζ—potentials at pH 7.0 were quantified by laser Doppler velocimetry as the electrophoretic mobility, using a disposable electrophoretic cell (DTS1061, Malvern Ltd., Worcestershire, UK). The size and average shape were measured with transmission electron microscopy (TEM, FEI Tecnai F20, FEI Company, Eindhoven, The Netherlands).

2.2. Experimental Setup

Seeds of *H. lanatus* and *L. flos–cuculi* were purchased from SemeNostrum (Udine, Italy), while seeds of *D. tenuifolia* were provided by Sementi Bruni (Corbetta, Milan, Italy). The experiment was carried out in controlled conditions. 30 seeds were placed into 15 mm Petri dishes containing filter paper soaked with 10 mL of deionized water (control) and 0.2, 2, 20, 200 and 2000 mg mL^{-1} of nCeO$_2$ 25 nm and nCeO$_2$ 50 nm suspensions. The suspensions of nanoceria were prepared and sonicated for 10 min to avoid aggregation. The Petri dishes were covered with aluminum paper to avoid light and set at room temperature (25 °C). The duration of the experiment was of two weeks. Each treatment was replicated three times. Germination was calculated as the ratio of germinated seeds out of the total seeds in each Petri dish. Seedlings were photographed, and Image J software [35] was used to measure roots length, which was calculated as the average of measures of all roots that emerged from seeds for each treatment.

2.3. Ce concentration in Plant Seedlings

To quantify the total content of Ce inside different plant species, seedlings were washed by agitation with HNO$_3$ 0.01 M for 15 min and rinsed with deionized water. The washed seedlings were oven-dried at 60 °C for three days, and 0.3 g of tissues were digested on a microwave oven (MARS Xpress, CEM, Matthews, NC, USA), using 9 mL of HNO$_3$ and 1 mL of hydrogen peroxide (H$_2$O$_2$) in Teflon cylinders at 180 °C. Plant extracts were diluted and filtered with Whatman 0.45 μm PTFE membrane filters. During the ICP–MS analysis, yttrium was the internal standard used for the analysis [36].

2.4. Internalization of nCeO$_2$ in Plant Tissues

At the end of the germination experiment, the uptake of nCeO$_2$ by plant seedlings was verified by enzymatic digestion. The digesting enzyme used was Macerozyme R−10 enzyme–pectinase from *Rhizopus* sp. (Sigma-Aldrich Co., St. Louis, MO, USA). The extraction of nCeO$_2$ from homogenized samples of these species was performed according to Jiménez-Lamana et al. (2016) [37]. In particular, 0.03 g of fresh plant samples were harvested, rinsed with deionized water and homogenized with 8 mL of 2 mM citrate buffer at pH 4.5, using an ultrasonic bath for 5 min. After the homogenization, 2 mL of the enzyme solution (0.05 g of enzyme powder for roots, shoots, leaves and seedlings, dissolved in 2 mL of MilliQ water) was added. The samples were shaken in a water bath at 37 °C for 24 h, and the obtained suspensions were filtered with a 0.45 μm cellulose filters to remove the solid parts of seedlings. The final supernatants were appropriately diluted and analyzed using the single particle inductively coupled plasma mass spectrometer (sp-ICP-MS) NexION 350 (PerkinElmer Waltham, MA, USA).

2.5. Data Analysis

Statistical analysis was carried out with three-, two- and one-way analysis of variance (ANOVA). When necessary, data were subjected to logarithmic transformation prior to analysis, which effectively homogenized the variances and produced approximately normal distributions. A posteriori comparison of individual means was performed using Tukey's test ($p < 0.05$). Differences between treatments for the different measured variables were tested using one-way ANOVA. Data are expressed as mean ± standard deviation (SD). Sp–ICP–MS data on nanoceria size distribution were processed by means of Syngistix Nano Application Module software (PerkinElmer Waltham, MA, USA) and interpolated with polynomial curves.

The $n\text{CeO}_2$ concentration range (0–2000 mg L^{-1}) was chosen considering that the large body of literature studies reporting the effects of $n\text{CeO}_2$ on plant physiology used Ce concentrations in the range 1–1000 mg L^{-1} [4,37], while the phytotoxicity test, as recommended by the USEPA approach, used the 2000 mg L^{-1} level [38].

3. Results

3.1. Characterization of nCeO$_2$

The results of the physicochemical characterization of $n\text{CeO}_2$ are reported in Table 1 and Figure 1A. In particular, the relative Z—averages are reported in Table 1, together with the relative polydispersity index (PDI) and the ζ—potentials of the particles. The Hd distribution of both the materials is in agreement with the value provided by the supplier. Both $n\text{CeO}_2$ 25 nm and 50 nm exhibit a monodisperse size particle distribution in the nanometric range with relatively low PDI, and the main size peak at 62.0 nm and 91.0 nm, respectively. The relative Z-averages were found to be much larger than these values; this was probably due to the presence of particle aggregates. Since a high net surface charge is typically associated with weak nanoparticle interactions and aggregation, these data are coherent with the higher Z-average detected for sample $n\text{CeO}_2$ 50 nm with respect to $n\text{CeO}_2$ 25 nm [39]. Figure 1B,C reports fields of TEM observation of $n\text{CeO}_2$ 25 and 50 nm suspensions.

Table 1. Z-average size, relative polydispersity index (PDI) and ζ—potentials of $n\text{CeO}_2$ 25 nm and 50 nm.

Material	Z-Average	PDI	ζ-Potential
	(nm)		(mV)
$n\text{CeO}_2$ 25 nm	126.7 ± 1.0	0.17 ± 0.01	39.2 ± 1.1
$n\text{CeO}_2$ 50 nm	205.7 ± 1.0	0.25 ± 0.02	24.1 ± 0.8

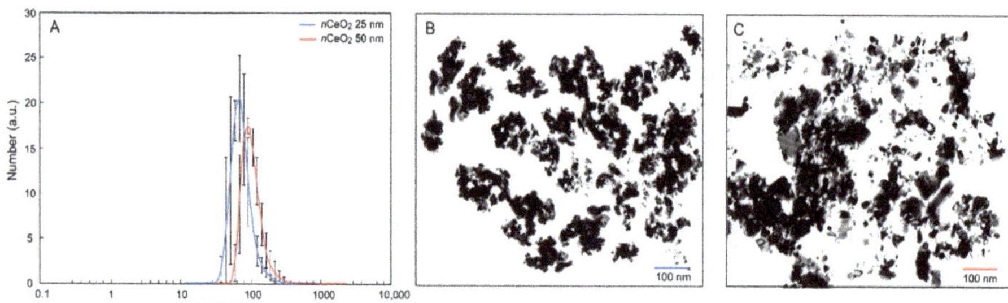

Figure 1. (**A**) Particle size distribution measured by dynamic light scattering (DLS) on suspensions of $n\text{CeO}_2$ 25 nm and 50 nm; (**B**) transmission electron microscopy (TEM) image of $n\text{CeO}_2$ 25 nm; (**C**) TEM $n\text{CeO}_2$ 50 nm.

3.2. nCeO$_2$ Plant Internalization

The early step of our study was devoted to verifying the entry of nCeO$_2$ into plant tissues. Control and treated seedlings of *H. lanatus*, *L. flos-cuculi* and *D. tenuifolia* were subjected to the extraction procedure and further analyzed by sp–ICP–MS. Size distributions of ceria nanoparticles in stock solutions and in the seedlings treated with 20 mg L^{-1} nCeO$_2$ 25 nm and 50 nm are reported in Figure 2.

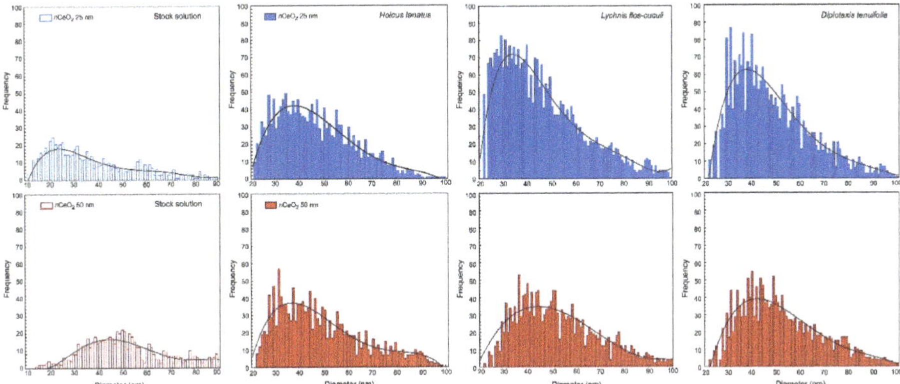

Figure 2. Particle size distributions of, respectively, nCeO$_2$ 25 nm and nCeO$_2$ 50 nm stock solutions (open bars) and after enzyme treatment of seedlings of *H. lanatus*, *L. flos-cuculi* and *D. tenuifolia* (closed bars) treated with 20 mg L^{-1} of nCeO$_2$.

As expected, in control seedlings, nCeO$_2$ was not detected, whereas in all treated species, (i) the presence of internalized nCeO$_2$ was verified, and (ii) the nCeO$_2$ have a different size distribution than stock solution suggesting aggregation phenomena between nanoparticles (Figure 2 and Table 2). Data from sp–ICP–MS analysis confirmed that nCeO$_2$ underwent agglomeration. The increase of the median diameter of nCeO$_2$ was evident for seedlings treated with nCeO$_2$ 25 nm, being 41.7 nm the average size of the nanoparticles extracted from seedlings (41 nm in *H. lanatus* and *L. flos-cuculi*, and 43 nm in *D. tenuifolia*). The mean size of particles extracted from seedlings treated with nCeO$_2$ 50 nm was 47 nm so in good agreement with the treatment (Figure 2 and Table 2).

Table 2. Most frequent size, mean size and number of peaks determined by sp–ICP–MS analysis after enzyme treatment of seedlings of *H. lanatus*, *L. flos-cuculi* and *D. tenuifolia* treated with 20 mg L^{-1} of nCeO$_2$ 25 nm and 50 nm.

Treatment	Species	Most Frequent Size (nm)	Mean Size (nm)	N. of Peaks (n)	Dissolved Ce (ppb)
	H. lanatus	40	41	1632	0.12
nCeO$_2$ 25 nm	*D. tenuifolia*	35	43	2630	0.08
	L. flos-cuculi	31	41	3079	0.03
	H. lanatus	41	44	1388	0.22
nCeO$_2$ 50 nm	*D. tenuifolia*	40	48	1871	0.20
	L. flos-cuculi	41	49	1842	0.21

The sp–ICP–MS results also show that the most frequent size of nanoparticles taken up by plants is similar for monocotyledons and dicotyledons for nCeO$_2$ 50 nm, whereas, for 25 nm treatments, the most frequent diameter is smaller in *L. flos-cuculi* and *D. tenuifolia* (respectively 31 and 35 nm) than *H. lanatus* (40 nm) (Figure 2 and Table 2). Regarding this aspect, some authors evidenced a size-dependent uptake and translocation of nCeO$_2$ in plants. In particular, nCeO$_2$ having a diameter smaller than 50 nm are present in all plant tissues and pass from roots to the aerial parts without dissolution and transformation.

3.3. Seed Germination and Root Length

A three-way ANOVA was run in order to have a general view regarding the effects of plant species, $nCeO_2$ size and Ce concentration on (i) percentage of germination, (ii) root length and (iii) Ce concentration in plant tissues. There were significant three-way interactions for percentage of germination ($p = 0.0463$ *) and root length ($p = 0.0000$ ***) (Table 3). Subsequently, the statistical analysis with two-way ANOVA within the species continued.

Table 3. Three-way ANOVA p values for the main effects of plant species, $nCeO_2$ size and Ce concentrations and their interactions on the percentage of germination, seedling root length and Ce concentration in seedling of *H. lanatus*, *L. flos-cuculi* and *D. tenuifolia*.

Effect	% Germination	Root Elongation	Ce Concentration in Seedlings
Species	0.0000 ***	0.0000 ***	0.0000 ***
$nCeO_2$ size	0.0000 ***	0.0000 ***	0.0000 ***
Ce concentration	0.0161 *	0.0000 ***	0.0000 ***
Species × $nCeO_2$ size	0.0000 ***	0.0000 ***	0.0000 ***
Species × Ce concentration	0.0115 *	0.0000 ***	0.0000 ***
$nCeO_2$ size × Ce concentration	0.7171 ns	0.0000 ***	0.0895 ns
Species × $nCeO_2$ size × Ce concentration	0.0463 *	0.0000 ***	0.0848 ns

ns: not significant at $p \leq 0.05$; *, ** and *** indicate significance at $p \leq 0.05$, $p \leq 0.01$ and $p \leq 0.001$.

As shown in Figure 3, treatments improve the germination percentage in all the three species if compared with controls. Indeed, germination increases more than 20% in several treatments in *H. lanatus*, 15% in *L. flos-cuculi* and 10% in *D. tenuifolia*, with respect to the control.

The evaluation of the effects induced by $nCeO_2$ of different sizes is the main objective of this study. Actually, we have not verified a clear trend related to the $nCeO_2$ size. In fact, the nanoparticle dimensions had no influence on the germination of *L. flos-cuculi*, while in both the other species, in some cases, they did. We observed that the germination percentage is higher for seeds treated with $nCeO_2$ 25 nm (about +10%) compared with 50 nm. (Figure 3). A statistically significant difference was found in *H. lanatus* at 2 and 200 mg L^{-1} (respectively, $p = 0.0282$ * and $p = 0.0132$ *) and *D. tenuifolia* at the at 0.2 and 2 mg L^{-1} (respectively, $p = 0.0072$ ** and $p = 0.0249$ *) (Figure 3). It is quite likely that in this species, the influence of the size of $nCeO_2$ on germination could have been observed even at the highest concentration, but the high variability of the data influenced the response of the statistics ($p = 0.0866$) (Figure 3).

Previously we demonstrated some relationships between the germination process and $nCeO_2$ size. Similar observations were carried out on the seedling root length (Figure 4).

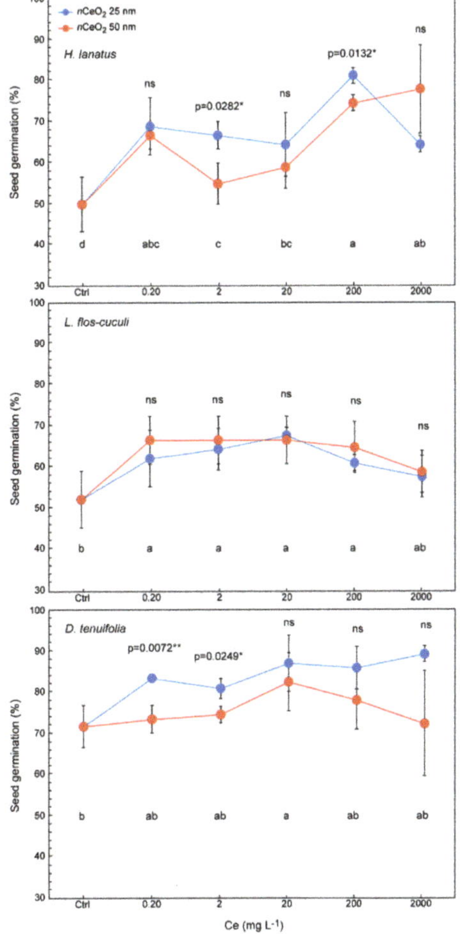

Figure 3. Percentage of seed germination in *H. lanatus*, *L. flos-cuculi* and *D. tenuifolia*, grown in Petri dishes and treated with solutions of nCeO$_2$ 25 nm and nCeO$_2$ 50 nm at 0, 2, 20, 200 and 2000 mg L^{-1}, respectively. The values are mean ± SD (standard deviation) of 3 replicates. Statistical significance of the treatments for each Ce concentration is reported: (i) figure upper part→comparison between nCeO$_2$ 25 nm and 50 nm; (ii) figure lower part→comparison between all treatments. Different letters indicate statistical differences. ns = not significant, * $p \leq 0.05$, ** $p \leq 0.01$.

In this case, the results demonstrate that, regardless of the Ce concentration, root length was not influenced by the nCeO$_2$ size. However, treatments stimulate the root growth in all three species, with a clear increase of length, in particular in *L. flos-cuculi* and *D. tenuifo*lia, if compared with control seedlings (Figure 4). Some differences in response to treatments are species-specific. The nCeO$_2$ of both sizes do not have any stimulating effect on the length of the roots of *H. lanatus*, which resulted in insensitivity to the treatments even at higher concentrations. On the other hand, we observed an increase in root length in treated seedlings of the other species. This effect was particularly intense in *L. flos-cuculi*, where the length of the roots of treated seedlings on average has almost doubled (+90.1%) compared to the control (14.7 mm and 7.73 mm, respectively). The stimulating effect of nCeO$_2$ demonstrated in *D. tenuifolia* is much less powerful but remarkable, where we found a 34% increase in root length compared to the control (22.8 mm and 16.9 mm, respectively).

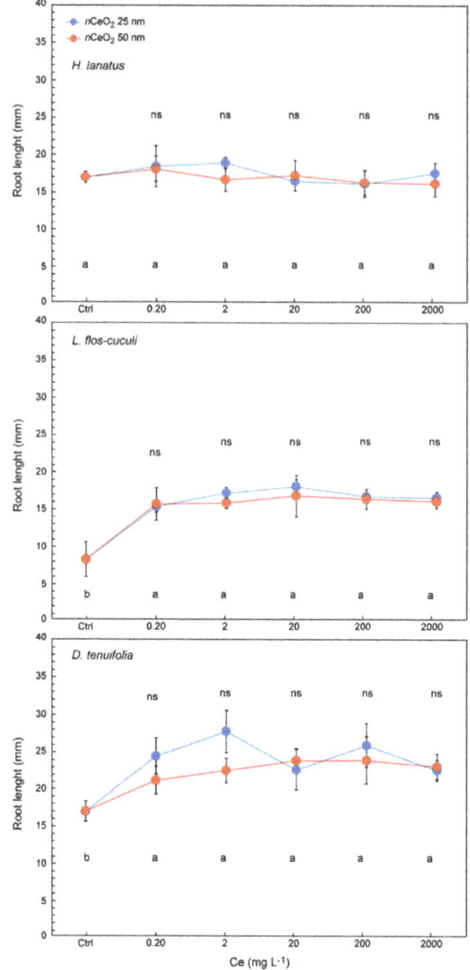

Figure 4. Root length in seedlings of *H. lanatus*, *L. flos-cuculi* and *D. tenuifolia*, grown in Petri dishes and treated with solutions of $n\text{CeO}_2$ 25 nm and $n\text{CeO}_2$ 50 nm at 0, 2, 20, 200 and 2000 mg L^{-1}, respectively. The values are mean ± SD (standard deviation) of 3 replicates. Statistical significance of the treatments for each Ce concentration is reported: (i) figure upper part→comparison between $n\text{CeO}_2$ 25 nm and 50 nm; (ii) figure lower part→comparison between all treatments. Different letters indicate statistical differences. ns = not significant, * $p \leq 0.05$, ** $p \leq 0.01$.

3.4. Ce Concentration in Plant Seedlings

To quantify the total content of Ce that was taken up by seedlings in the three plant species, we used an ICP–MS after the acid digestion of the samples. The elaborated data with the total concentration of Ce are presented in Figure 5.

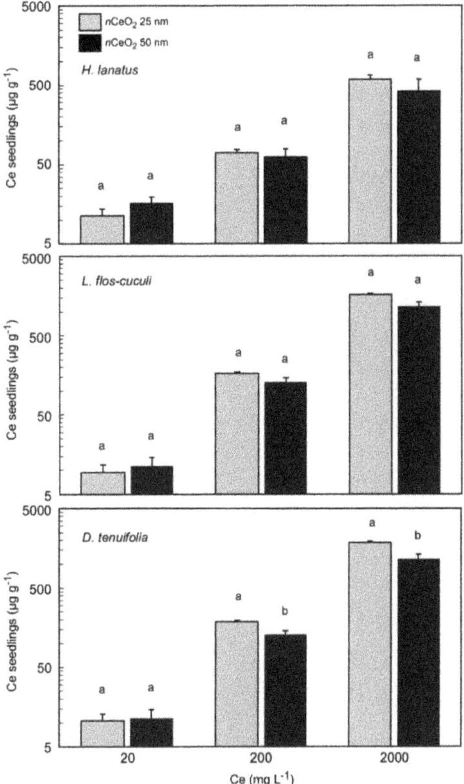

Figure 5. Ce concentration in seedlings of *H. lanatus*, *L. flos-cuculi* and *D. tenuifolia*, grown in Petri dishes and treated with solutions of $n\mathrm{CeO_2}$ 25 nm and $n\mathrm{CeO_2}$ 50 nm at 20, 200 and 2000 mg L^{-1}, respectively. The values are mean ± SD (standard deviation) of 3 replicates. Statistical significance of the treatments (ANOVA p value) for each Ce concentration is reported. Different letters indicate statistical differences ($p \leq 0.05$).

The concentration of total Ce in seedling tissues of *H. lanatus*, *L. flos-cuculi* and *D. tenuifolia* shows a different magnitude of accumulation according to the treatments. In fact, a statistically significant effect of treatments ($p < 0.05$) was verified for all the species. As already reported in Table 2, the interaction "species x Ce concentration" was highly statistically significant (p = 0.0000 ***). With regard to the $n\mathrm{CeO_2}$ size, we observed that the seedlings treated with the smaller $n\mathrm{CeO_2}$ reveal a higher concentration of Ce in their tissues than the ones treated with the 50 nm nanoparticles. This occurred in particular in *L. flos-cuculi* and *D. tenuifolia* and at the two highest concentrations of treatments, but not in *H. lanatus*. In *L. flos-cuculi*, the total content of Ce corresponds to 165 and 128 mg kg^{-1} DW at 200 mg L^{-1}; 1616 and 1151 mg kg^{-1} DW at 2000 mg L^{-1} (p = 0.0134 *), respectively at 25 nm and 50 nm. In *D. tenuifolia* we detected 189 and 114 mg kg^{-1} DW at 200 mg L^{-1}; 1841 and 1305 mg kg^{-1} DW at 2000 mg L^{-1} (p = 0.0465 *), respectively at 25 nm and 50 nm (Figure 5).

The size of the nanoparticles in our study was 25 nm and 50 nm. By looking at Figure 2, it turns out that the stock solutions of the two nominal sizes are actually a mixture of nanoparticles of different dimensions, with 25 nm and 50 nm being the dimensions among the ones having the highest frequencies. This makes one conclude that both the dispersions contain nanoparticles that can potentially enter the plant roots. There are two main factors, among others, that can influence the uptake and translocation of the nanoparticles: (i) the size of the pores in the cell membrane; (ii) the tendency of the nanoparticles to aggregate due to chemical interactions. Given the fact that the dimension of the nanoparticles at a given shape determines the surface to volume ratio, this can affect the entity of such aggregation (the smaller the dimension, the more likely the aggregation).

Tables S1 and S2 report the calculations of the $n\text{CeO}_2$ 25 nm and 50 nm ratios based on the assumptions that all the nanoparticles are spherical and equal in size (25 nm or 50 nm) in the dispersions as well as inside the seedlings and no aggregations occur. The theoretical ratio (Table S1) refers to an ideal scenario for which an equal mass of nanoparticles is taken up by both the plants exposed to the $n\text{CeO}_2$ 25 nm and the $n\text{CeO}_2$ 50 nm; the theoretical ratio is hence obtained by calculating the number of nanoparticles at a given size (25 nm or 50 nm) as the mathematical division of the total mass of Ce in the plant by the mass of a single nanoparticle. The observed ratio is calculated following the same procedure and assumptions, but considering the experimental mean Ce mass measured in planta for the different treatments (Table S2).

4. Discussion

The previously published studies carried out in controlled conditions, such as Petri dishes and hydroponics, reported that the toxicity of nanomaterials in the initial development stages of plant growth could be due to physicochemical properties, as well as particle size and shape [40,41]. In general, MeNPs show early negative consequences on the development stages of crops and this observation is confirmed in some publications [7,8,42,43].

Literature papers suggest that $n\text{CeO}_2$ generally enters plants through root uptake and may cause several effects on the early stages of plant development, such as reducing or increasing germination rates and improving, reducing or inhibiting radical growth [44,45]. When germinating seeds are exposed to nanoparticles, different effects could be verified, basically depending on the plant species and particle size and concentration [8].

It was demonstrated that $n\text{CeO}_2$ having a diameter comprised in the range 50–100 nm are taken up by roots, but they hardly move towards the aerial plant fractions, while $n\text{CeO}_2$ larger than 100 nm is not absorbed by roots [46,47]. We observed that the formation of particle agglomerates concerns, in particular $n\text{CeO}_2$ 25 nm. It is very likely that this was due to the higher specific surface than $n\text{CeO}_2$ 50 nm. At the same time, sp–ICP–MS analysis showed the largest number of peaks in all seedlings treated with $n\text{CeO}_2$ 25. Confirming previous literature findings [48], this suggests that the smaller particle size has the ability to enter into the roots more easily than 50 nm. Combining the previous evidence, we hypothesize that the $n\text{CeO}_2$ 25 nm agglomeration occurred inside the seedling tissues.

According to Layet et al. (2017) [49], we demonstrate that the two dicotyledons take up more $n\text{CeO}_2$ than *H. lanatus*. Since seedlings of the different species growing in the same conditions, it is likely that the uptake and translocation of $n\text{CeO}_2$ are influenced by species-specific physiological traits. The ability of $n\text{CeO}_2$ uptake by roots and subsequent translocation to the other parts of plants was already demonstrated in crop species [50–52]. We observed a similar particle size distribution for $n\text{CeO}_2$ 25 nm in *L. flos-cuculi* and *D. tenuifolia*. At the same time, no relevant changes were observed for $n\text{CeO}_2$ 50 nm.

We recorded a negligible dissolved concentration of Ce ions in all samples, indicating that $n\text{CeO}_2$ did not undergo dissolution after being absorbed by roots. On the other hand, we observed that small signals of dissolved forms of Ce correspond to the presence of bigger nanoparticles (50 nm), as previously reported [52,53]. Hence, it is likely that $n\text{CeO}_2$ 25 nm after being taken up by the seedling roots move through the vascular system forming aggregates. This process has been explained by the attraction between nanoparticles caused by van der Waals forces or chemical bonds [54,55]. However, since this also occurs within the plant tissues, it is still unclear whether and how species-specific factors can influence this process.

A large number of studies highlight that in some plant species, particle agglomeration happens before the passage from roots to the other parts of seedlings [42,56–58]. This statement could be justified by the plausible hypothesis that MeNPs pass through the apoplastic pathway [59] or cause the destruction of some cell walls, and in so doing, they pass through the enlarged pores [60]. However, nanoparticles could enter the vascular system where the Casparian strip is not formed [61,62] or through the lateral root junction [25,61,63,64].

Our data also indicated that the treatments with $n\mathrm{CeO_2}$ of different size influenced the size distribution of nanoparticles within the plant tissues. This could be due to the smaller size of the materials that lead to an exponential increase in surface area relative to volume in contact with roots. We can conclude that the $n\mathrm{CeO_2}$ entered inside the seedlings, and therefore, the results that will be described later are reasonably influenced by the experimental treatments. With regard to $n\mathrm{CeO_2}$ aggregation, in this study, we did not develop further observations.

With regard to the observed stimulating effect on germination of $n\mathrm{CeO_2}$, it must be said that on this point, the literature reports conflicting data. Low toxicity and reduction of seed germination were observed on *Lycopersicum esculentum* and *Zea mays* [44] and *Glycine max* [65], whereas germination of *Hordeum vulgare* was indifferent up to 2000 mg L^{-1} $n\mathrm{CeO_2}$ [66]. As a matter of principle, a direct comparison of data from different experiments is difficult. However, if we were looking to draw a conclusion from available literature data on this point, we can say that $n\mathrm{CeO_2}$ does not cause acute toxicity in the early stages of plant development.

Leaving the nanoscale, it can be confirmed that Ce influence positively seeds germination. As for other rare earth elements, it has been suggested that Ce may have a positive effect by enhancing the effects of phytohormones on germinating seeds [67,68]. In addition, it seems that eventually, monocots species are more tolerant of Ce than dicot ones [69].

Contradictory literature evidence regards the influence of $n\mathrm{CeO_2}$ on root development in different stages of plant growth. Positive effects on root growth were observed in *Zea mays*, *Cucumis sativus* and *Lactuca sativa* [44,59]. As occurred in our species, very high tolerance to $n\mathrm{CeO_2}$ was reported for *Cucumis sativus*, *Brassica oleracea*, *Brassica napus* and *Raphanus sativus*, whose root growth resulted not affected up to $n\mathrm{CeO_2}$ 2000 mg L^{-1}. Oppositely, a slowed root development in treated *Medicago sativa* and *Lycopersicum esculentum* was reported [44].

As regards the Ce concentration (or mass per volume) in the seedlings exposed to the two dispersions, it is possible to conclude that the amount of Ce is greater for $n\mathrm{CeO_2}$ 25 nm. Actually, this is not very informative in terms of the number of nanoparticles inside the plants if we consider the fact that the two dispersions had the same quantity of Ce in mass but different amounts in terms of nanoparticles. If we assume that the nanoparticles of the two dispersions are of the same shape and that the frequency of the dimension is 100% for both the dimensional class (25 and 50 nm), then we can conclude that the seedlings exposed to the $n\mathrm{CeO_2}$ 25 nm were in fact exposed to 8 times the number of nanoparticles when compared to the 50 nm dispersion, since the volume goes with the cubic pattern.

Following this reasoning and assuming that the cell membrane pore size is not limiting the entrance of the nanoparticles because the mean size is greater than the sizes of the nanoparticles [70,71], it can be shown that the ratio between the number of nanoparticles taken up by the plant, calculated from the mass, is not in a ratio 1 to 8 as it would be if the two types of nanoparticles were taken up at the same level, but it is 1 to a greater number. This can lead us to speculate that the number of 50 nm nanoparticles inside the seedlings is lower than expected, and this could be due to the aggregation of the nanoparticles outside the plants that limited the entrance of the 50 nm nanoparticles more than the 25 nm nanoparticles despite their higher tendency to form clusters. On the other side, we may assume that the pores of the cell membrane act as filters and so explain why the $n\mathrm{CeO_2}$ 50 nm reached a lower concentration in the seedlings. Of course, this is based on some assumptions that are not likely to occur in real conditions (the shape and size of the nanoparticle are far from being homogeneous and aggregation occurs). In addition, in more complex conditions, such as at field condition or in experiments that use real soil, the strategies that plants use to absorb nutrients from the substrate can influence, for instance, the solubility of the nanoparticles (e.g., by acidification of the rhizosphere which may change the solubility of the different chemical forms of Ce).

5. Conclusions

Under our experimental conditions, the presence of $n\text{CeO}_2$—even at high concentrations—did not cause negative effects on *H. lanatus*, *L. flos-cuculi* and *D. tenuifolia*. On the contrary, $n\text{CeO}_2$ had a stimulating effect in the early stages of development of the plants. The plants' response with respect to the different $n\text{CeO}_2$ sizes has not been very evident. This aspect requires important insights that must take into account the aggregation/dissolution dynamics of $n\text{CeO}_2$ and the forms of Ce taken-up by plants, as well as the fate of the $n\text{CeO}_2$ assimilated by plants. Our results are quite in accordance with the literature, in which—it must be remembered—there are still rather conflicting results, obtained to a very large extent by observing crop species.

Finally, the knowledge of the effects of the exposure of plants to ENM is limited. It is acknowledged that the flux of ENM in the ecosystem involves the primary producers. That is the reason why it is important to focus the research on this field as well as to develop new methods of investigation suitable to non-food species.

As previously reported, the number of works dedicated to the study of the effects resulting from the exposure of spontaneous terrestrial species to ENMs is very low. This constitutes "per se" the major novelty element of this paper. Our observations were made during the early stages of vegetative development. It will be necessary to extend the study to evaluate the effects of the treatments over the whole plant life-cycle. It will be equally important to compare the responses of plants with respect to single and repeated treatments over time.

Supplementary Materials: The following are available online at http://www.mdpi.com/2079-4991/10/12/2534/s1, Table S1: Theoretical ratio calculated at a hypothetical equal mass of nanoparticles uptaken by plants exposed to nCeO_2 25 nm and 50 nm. The ratio is calculated by dividing the number of $n\text{CeO}_2$ 25 nm by the number of $n\text{CeO}_2$ 50 nm; both numbers are calculated dividing the mass by the estimated the mass of a single nanoparticle (g/g). Table S2: The observed ratio calculated at the measured mean Ce uptake by the plants exposed to the two $n\text{CeO}_2$ 25 nm and 50 nm for the treatments 200 and 2000 ppm and the two species *L. flos-cuculi* and *D. tenuifolia*. The ratio is calculated by dividing the number of $n\text{CeO}_2$ 25 nm by the number of $n\text{CeO}_2$ 50 nm; both numbers are calculated dividing the mass of Ce derived from the measured mean Ce by the estimated the mass of a single nanoparticle (g/g).

Author Contributions: Conceptualization, D.L. and L.M.; formal analysis, D.L. and L.M.; funding acquisition, L.M.; investigation, D.L.; methodology, A.M., B.P. and A.A.; supervision, L.M.; writing—original draft, L.M.; writing—review and editing, G.F. All authors have read and agreed to the published version of the manuscript.

Funding: This research was funded by Regione Friuli Venezia Giulia, General Directorate of Environment, grant "Nanomateriali 2017–20".

Conflicts of Interest: The authors declare no conflict of interest.

References

1. Bainbridge, W.S.; Roco, M.C. Science and technology convergence: With emphasis for nanotechnology-inspired convergence. *J. Nanopart. Res.* **2016**, *18*, 211. [CrossRef]
2. Mortimer, M.; Holden, P.A. Fate of engineered nanomaterials in natural environments and impacts on ecosystems. In *Exposure to Engineered Nanomaterials in the Environment*, 1st ed.; Marmiroli, N., White, J., Song, J., Eds.; Elsevier: Amsterdam, The Netherlands, 2019; pp. 61–103. [CrossRef]
3. Keller, A.A.; McFerran, S.; Lazareva, A.; Suh, S. Global life cycle releases of engineered nanomaterials. *J. Nanopart. Res.* **2013**, *15*, 1692. [CrossRef]
4. Holden, P.A.; Gardea-Torresdey, J.L.; Klaessig, F.; Turco, R.F.; Mortimer, M.; Hund-Rinke, K.; Cohen Hubal, E.A.; Avery, D.; Barceló, D.; Behra, R.; et al. Considerations of environmentally relevant test conditions for improved evaluation of ecological hazards of engineered nanomaterials. *Environ. Sci. Technol.* **2016**, *50*, 6124–6145. [CrossRef] [PubMed]
5. Reddy Pullagurala, V.L.; Adisa, I.O.; Rawat, S.; White, J.C.; Zuverza-Mena, N.; Hernandez-Viezcas, J.A.; Peralta-Videa, J.R.; Gardea-Torresdey, J.L. Fate of engineered nanomaterials in agroenvironments and impacts on agroecosystems. In *Exposure to Engineered Nanomaterials in the Environment*, 1st ed.; Marmiroli, N., White, J., Song, J., Eds.; Elsevier: Amsterdam, The Netherlands, 2019; pp. 105–142. [CrossRef]

6. Bar-On, Y.M.; Phillips, R.; Milo, R. The biomass distribution on Earth. *Proc. Natl. Acad. Sci. USA* **2018**, *5*, 16506–16511. [CrossRef]
7. Priester, J.H.; Ge, Y.; Mielke, R.E.; Horst, A.M.; Moritz, S.C.; Espinosa, K.; Gelb, J.; Walker, S.L.; Nisbet, R.M.; An, Y.J.; et al. Soybean susceptibility to manufactured nanomaterials with evidence for food quality and soil fertility interruption. *Proc. Natl. Acad. Sci. USA* **2012**, *109*, 2451–2456. [CrossRef]
8. Miralles, P.; Church, T.L.; Harris, A.T. Toxicity, uptake, and translocation of engineered nanomaterials in vascular plants. *Environ. Sci. Technol.* **2012**, *46*, 9224–9239. [CrossRef]
9. Zuverza-Mena, N.; Martínez-Fernández, D.; Du, W.; Hernàndez-Viezcas, J.A.; Bonilla-Bird, N.; López-Moreno, M.L.; Komárek, M.; Peralta-Videa, J.R.; Gardea-Torresdey, J.L. Exposure of engineered nanomaterials to plants: Insights into the physiological and biochemical responses. A review. *Plant Physiol. Biochem.* **2017**, *110*, 236–264. [CrossRef]
10. Lowry, G.V.; Avellan, A.; Gilbertson, L.M. Opportunities and challenges for nanotechnology in the agri-tech revolution. *Nat. Nanotechnol.* **2019**, *14*, 517–522. [CrossRef]
11. Kah, M.; Tufenkji, N.; White, J.C. Nano-enabled strategies to enhance crop nutrition and protection. *Nat. Nanotechnol.* **2019**, *14*, 532–540. [CrossRef]
12. Marchiol, L.; Iafisco, M.; Fellet, G.; Adamiano, A. Nanotechnology support the next agricultural revolution: Perspectives to enhancement of nutrient use efficiency. *Adv. Agron.* **2020**, *161*, 27–116. [CrossRef]
13. Hawthorne, J.; De la Torre Roche, R.; Xing, B.; Newman, L.A.; Ma, X.; Majumdar, S.; Gardea-Torresdey, J.G.; White, J.C. Particle-size dependent accumulation and trophic transfer of cerium oxide through a terrestrial food chain. *Environ. Sci. Technol.* **2014**, *48*, 13102–13109. [CrossRef] [PubMed]
14. Du, W.; Xu, Y.; Yin, Y.; Ji, R.; Guo, H. Risk assessment of engineered nanoparticles and other contaminants in terrestrial plants. *Curr. Opin. Environ. Sci. Health* **2018**, *6*, 21–28. [CrossRef]
15. Spielman-Sun, E.; Avellan, A.; Bland, G.; Tappero, R.V.; Acerbo, A.S.; Unrine, J.M.; Giraldo, J.P.; Lowry, G.V. Nanoparticle surface charge influences translocation and leaf distribution in vascular plants with contrasting anatomy. *Environ. Sci. Nano* **2019**, *6*, 2508–2519. [CrossRef]
16. Asztemborska, M.; Bembenek, M.; Jakubiak, M.; Stęborowski, R.; Bystrzejewska-Piotrowska, G. The effect of nanoparticles with sorption capacity on the bioaccumulation of divalent ions by aquatic plants. *Int. J. Environ. Res.* **2018**, *12*, 245–253. [CrossRef]
17. Ding, Y.; Bai, X.; Ye, Z.; Ma, L.; Liang, L. Toxicological responses of Fe_3O_4 nanoparticles on *Eichhornia crassipes* and associated plant transportation. *Sci. Total Environ.* **2019**, *671*, 558–567. [CrossRef]
18. Movafeghi, A.; Khataee, A.; Abedi, M.; Tarrahi, R.; Dadpour, M.; Vafaei, F. Effects of TiO_2 nanoparticles on the aquatic plant *Spirodela polyrrhiza*: Evaluation of growth parameters, pigment contents and antioxidant enzyme activities. *J. Environ. Sci.* **2018**, *64*, 130–138. [CrossRef]
19. Ekperusia, A.O.; Sikokic, F.D.; Nwachukwud, E.O. Application of common duckweed (*Lemna minor*) in phytoremediation of chemicals in the environment: State and future perspective. *Chemosphere* **2019**, *223*, 285–309. [CrossRef]
20. Geitner, N.K.; Cooper, J.L.; Avellan, A.; Castellon, B.T.; Perrotta, B.G.; Bossa, N.; Simonin, M.; Anderson, S.M.; Inoue, S.; Hochella, M.F.; et al. Size-based differential transport, uptake, and mass distribution of ceria (CeO_2) nanoparticles in wetland mesocosms. *Environ. Sci. Technol.* **2018**, *52*, 9768–9776. [CrossRef]
21. Yin, L.; Colman, B.P.; McGill, B.M.; Wright, J.P.; Bernhardt, E.S. Effects of silver nanoparticle exposure on germination and early growth of eleven wetland plants. *PLoS ONE* **2012**, *7*, e47674. [CrossRef]
22. Jacob, D.L.; Borchardt, J.D.; Navaratnam, L.; Otte, M.L.; Bezbaruah, A.N. Uptake and translocation of Ti nanoparticles in crops and wetland plants. *Int. J. Phytoremediation* **2013**, *15*, 142–153. [CrossRef]
23. Song, U.; Lee, S. Phytotoxicity and accumulation of zinc oxide nanoparticles on the aquatic plants *Hydrilla verticillata* and *Phragmites australis*: Leaf-type-dependent responses. *Environ. Sci. Pollut. Res.* **2016**, *23*, 8539–8545. [CrossRef] [PubMed]
24. Aleksandrowicz-Trzcińska, M.; Bederska-Błaszczyk, M.; Szaniawski, A.; Olchowik, J.; Studnicki, M. The effects of copper and silver nanoparticles on container-grown Scots pine (*Pinus sylvestris* L.) and Pedunculate oak (*Quercus robur* L.) seedlings. *Forests* **2019**, *10*, 269. [CrossRef]
25. Dietz, K.J.; Herth, S. Plant nanotoxicology. *Trends Plant Sci.* **2011**, *16*, 582–589. [CrossRef] [PubMed]
26. Sun, T.Y.; Gottschalk, F.; Hungerbuhler, K.; Nowack, B. Comprehensive probabilistic modelling of environmental emissions of engineered nanomaterials. *Environ. Pollut.* **2014**, *185*, 69–76. [CrossRef]

27. OECD. List of manufactured nanomaterials and list of endpoints for phase one of the OECD testing programme. In *Safety of Manufactured Nanomaterials No. 6;* ENV/JM/MONO(2008)13/REV; Organisation for Economic Co-operation and Development: Paris, France, 2008.
28. Ramirez-Olvera, S.M.; Trejo-Téllez, L.I.; García-Morales, S.; Pérez-Sato, J.R.; Gomez-Merino, F.C. Cerium enhances germination and shoot growth, and alters mineral nutrient concentration in rice. *PLoS ONE* **2018**, *13*, e0194691. [CrossRef] [PubMed]
29. Lizzi, D.; Mattiello, A.; Marchiol, L. Impacts of Cerium Oxide Nanoparticles ($n CeO_2$) on Crop Plants: A Concentric Overview. In *Nanomaterials in Plants, Algae and Micro-Organisms. Concepts and Controversies;* Tripathi, D.K., Ahmad, P., Sharma, S., Chauhan, D., Eds.; Elsevier: Amsterdam, The Netherlands, 2017; Volume 1, pp. 311–322. [CrossRef]
30. Skiba, E.; Wolf, W.M. Cerium oxide nanoparticles affect heavy metals uptake by Pea in a divergent way than their ionic and bulk counterparts. *Water Air Soil Pollut.* **2019**, *230*, 248. [CrossRef]
31. Adisa, I.O.; Rawat, S.; Reddy Pullagurala, V.L.; Dimkpa, C.O.; Elmer, W.E.; White, J.C.; Hernandez-Viezcas, J.A.; Peralta-Videa, J.R.; Gardea-Torresdey, J.L. Nutritional status of tomato (*Solanum lycopersicum*) fruit grown in Fusarium-infested soil: Impact of cerium oxide nanoparticles. *J. Agric. Food Chem.* **2020**, *68*, 1986–1997. [CrossRef]
32. Thompson, J.D.; Turkington, R. The biology of Canadian weeds—*Holcus lanatus* L. *Can. J. Plant Sci.* **1988**, *68*, 131–147. [CrossRef]
33. Jalas, J.; Suominen, J. *Atlas Florae Europaeae (AFE)—Distribution of vascular plants in Europe 7 (Caryophyllaceae (Silenioideae));* The Committee for Mapping the Flora of Europe and Societas Biologica Fennica Vanamo: Helsinki, Finland, 1998.
34. Hall, M.K.D.; Jobling, J.J.; Rogers, G.S. Some perspectives on rocket as a vegetable crop: A review. *Veg. Crops Res. Bull.* **2012**, *76*, 21–41. [CrossRef]
35. Schneider, C.; Rasband, W.; Eliceiri, K. NIH Image to ImageJ: 25 years of image analysis. *Nat. Methods* **2012**, *9*, 671–675. [CrossRef]
36. Packer, A.P.; Lariviere, D.; Li, C.; Chen, M.; Fawcett, A.; Nielsen, K.; Mattson, K.; Chatt, A.; Scriver, C.; Erhardt, L.S. Validation of an inductively coupled plasma mass spectrometry (ICP-MS) method for the determination of cerium, strontium and titanium in ceramic materials used in radiological dispersal devices (RDDs). *Anal. Chim. Acta* **2007**, *588*, 166–172. [CrossRef] [PubMed]
37. Jiménez-Lamana, J.; Wojcieszek, J.; Jakubiak, M.; Asztemborska, M.; Szpunar, J. Single particle ICP-MS characterization of platinum nanoparticles uptake and bioaccumulation by *Lepidium sativum* and *Sinapis alba* plants. *J. Anal. At. Spectrom.* **2016**, *31*, 2321. [CrossRef]
38. Zhang, W.; Yu, Z.; Rao, P.; Lo, I.M.C. Uptake and toxicity studies of magnetic TiO_2-based nanophotocatalyst in *Arabidopsis thaliana*. *Chemosphere* **2019**, *224*, 658–667. [CrossRef] [PubMed]
39. Mukherjee, B.; Weaver, J.W. Aggregation and charge behavior of metallic and nonmetallic nanoparticles in the presence of competing similarly-charged inorganic ions. *Environ. Sci. Technol.* **2010**, *44*, 3332–3338. [CrossRef]
40. Yang, L.; Watts, D.J. Particle surface characteristics may play an important role in phytotoxicity of alumina nanoparticles. *Toxicol. Lett.* **2005**, *158*, 122–132. [CrossRef]
41. Lin, D.H.; Xing, B.S. Phytotoxicity of nanoparticles: Inhibition of seed germination and root growth. *Environ. Pollut.* **2007**, *150*, 243–250. [CrossRef]
42. Lin, D.H.; Xing, B.S. Root uptake and phytotoxicity of ZnO nanoparticles. *Environ. Sci. Technol.* **2008**, *42*, 5580–5585. [CrossRef]
43. Gardea-Torresdey, J.L.; Rico, C.M.; White, J.C. Trophic transfer, transformation, and impact of engineered nanomaterials in terrestrial environments. *Environ. Sci. Technol.* **2014**, *48*, 2526–2540. [CrossRef]
44. López-Moreno, M.L.; De La Rosa, G.; Hernàndez-Viezcas, J.A.; Peralta-Videa, J.R.; Gardea-Torresdey, J.L. X-ray absorption spectroscopy (XAS) corroboration of the uptake and storage of CeO_2 nanoparticles and assessment of their differential toxicity in four edible plant species. *J. Agric. Food Chem.* **2010**, *58*, 3689–3693. [CrossRef]
45. Andersen, C.P.; King, G.; Plocher, M.; Storm, M.; Pokhrel, L.R.; Johnson, M.G.; Rygiewicz, P.T. Germination and early plant development of ten plant species exposed to titanium dioxide and cerium oxide nanoparticles. *Environ. Toxicol. Chem.* **2016**, *35*, 2223–2229. [CrossRef]

46. Slomberg, D.L.; Schoenfisch, M.H. Silica nanoparticle phytotoxicity to *Arabidopsis thaliana*. *Environ. Sci. Technol.* **2012**, *46*, 10247–10254. [CrossRef] [PubMed]
47. Zhang, P.; Ma, Y.; Zhang, Z.; He, X.; Zhang, J.; Guo, Z.; Tai, R.; Zhao, Y.; Chai, Z. Biotransformation of ceria nanoparticles in cucumber plants. *ACS Nano* **2012**, *6*, 9943–9950. [CrossRef] [PubMed]
48. Zhang, Z.; He, X.; Zhang, H.; Ma, Y.; Zhang, P.; Ding, Y.; Zhao, Y. Uptake and distribution of ceria nanoparticles in cucumber plants. *Metallomics* **2011**, *3*, 816–822. [CrossRef]
49. Layet, C.; Auffan, M.; Santaella, C.; Chevassus-Rosset, C.; Montes, M.; Ortet, P.; Barakat, M.; Collin, B.; Legros, S.; Bravin, M.N.; et al. Evidence that soil properties and organic coating drive the phytoavailability of cerium oxide nanoparticles. *Environ. Sci. Technol.* **2017**, *51*, 9756–9764. [CrossRef] [PubMed]
50. Dan, Y.; Ma, X.; Zhang, W.; Liu, K. Single particle ICP-MS method development for the determination of plant uptake and accumulation of CeO_2 nanoparticles. *Anal. Bioanal. Chem.* **2016**, *408*, 5157–5167. [CrossRef] [PubMed]
51. Zhang, P.; Ma, Y.; Liu, S.; Wang, G.; Zhang, J.; He, X.; Zhang, J.; Rui, Y.; Zhang, Z. Phytotoxicity, uptake and transformation of nano-CeO_2 in sand cultured romaine lettuce. *Environ. Pollut.* **2017**, *220*, 1400–1408. [CrossRef]
52. Zhang, W.; Dan, Y.; Shi, H.; Ma, X. Elucidating the mechanisms for plant uptake and in-planta speciation of cerium in radish (*Raphanus sativus* L.) treated with cerium oxide nanoparticles. *J. Environ. Chem. Eng.* **2017**, *5*, 572–577. [CrossRef]
53. Ma, Y.; Zhang, P.; Zhang, Z.; He, X.; Zhang, J.; Ding, Y.; Zhang, J.; Zheng, L.; Guo, Z.; Zhang, L.; et al. Where does the transformation of precipitated ceria nanoparticles in hydroponic plants take place? *Environ. Sci. Technol.* **2015**, *9*, 10667–10674. [CrossRef]
54. Bao, D.; Oh, Z.G.; Chen, Z. Characterization of silver nanoparticles internalized by Arabidopsis plants using single particle ICP-MS analysis. *Front. Plant Sci.* **2016**, *7*, 1–8. [CrossRef]
55. Corredor, E.; Testillano, P.S.; Coronado, M.J.; Gonzalez-Melendi, P.; Fernandez-Pacheco, R.; Marquina, C.; Ibarra, M.R.; De La Fuente, J.M.; Rubiales, D.; Perez De Luque, A.; et al. Nanoparticle penetration and transport in living pumpkin plants: In situ subcellular identification. *BMC Plant Biol.* **2009**, *9*, 45–56. [CrossRef]
56. Avellan, A.; Schwab, F.; Masion, A.; Chaurand, P.; Borschneck, D.; Vidal, V.; Rose, J.; Santaella, C.; Levard, C. Nanoparticle uptake in plants: Gold nanomaterial localized in roots of Arabidopsis thaliana by X-ray computed nanotomography and hyperspectral imaging. *Environ. Sci. Technol.* **2017**, *51*, 8682–8691. [CrossRef] [PubMed]
57. Geisler–Lee, J.; Wang, Q.; Yao, Y.; Zhang, W.; Geisler, M.; Li, K.; Huang, Y.; Chen, Y.; Kolmakov, A.; Ma, X. Phytotoxicity, accumulation and transport of silver nanoparticles by *Arabidopsis thaliana*. *Nanotoxicology* **2013**, *7*, 323–337. [CrossRef]
58. Ma, Y.H.; He, X.; Zhang, P.; Zhang, Z.Y.; Guo, Z.; Tai, R.Z.; Xu, Z.J.; Zhang, L.J.; Ding, Y.Y.; Zhao, Y.L.; et al. Phytotoxicity and biotransformation of La_2O_3 nanoparticles in a terrestrial plant cucumber (*Cucumis sativus*). *Nanotoxicology* **2011**, *5*, 743–753. [CrossRef] [PubMed]
59. Ma, X.; Geiser-Lee, M.J.; Deng, Y.; Kolmakov, A. Interactions between engineered nanoparticles (ENPs) and plants: Phytotoxicity, uptake and accumulation. *Sci. Total Environ.* **2010**, *408*, 3053–3061. [CrossRef] [PubMed]
60. Nair, R.; Varghese, S.H.; Nair, B.G.; Maekawa, T.; Yoshida, Y.; Kumar, D.S. Nanoparticulate material delivery to plants. *Plant Sci.* **2010**, *179*, 154–163. [CrossRef]
61. Lv, J.T.; Zhang, S.Z.; Luo, L.; Zhang, J.; Yang, K.; Christie, P. Accumulation, speciation and uptake pathway of ZnO nanoparticles in maize. *Environ. Sci. Nano* **2015**, *2*, 68–77. [CrossRef]
62. Schymura, S.; Fricke, T.; Hildebrand, H.; Franke, K. Elucidating the role of dissolution in CeO_2 nanoparticle plant uptake by smart radiolabeling. *Angew. Chem. Int. Ed.* **2017**, *56*, 7411–7414. [CrossRef]
63. McCully, M. How do real roots work? (Some new views of root structure). *Plant Physiol.* **1995**, *109*, 1–6. [CrossRef]
64. Dan, Y.; Zhang, W.; Xue, R.; Ma, X.; Stephan, C.; Shi, H. Characterization of gold nanoparticle uptake by tomato plants using enzymatic extraction followed by single-particle inductively coupled plasma-mass spectrometry analysis. *Environ. Sci. Technol.* **2015**, *49*, 3007–3014. [CrossRef]

65. López-Moreno, M.L.; de la Rosa, G.; Hernàndez-Viezcas, J.A.; Castillo Michel, H.; Botez, C.E.; Peralta-Videa, J.R.; Gardea-Torresdey, J.L. Evidence of the differential biotransformation and genotoxicity of ZnO and CeO_2 nanoparticles on soybean (*Glycine max*) plants. *Environ. Sci. Technol.* **2010**, *44*, 7315–7320. [CrossRef]
66. Mattiello, A.; Pošćić, F.; Musetti, R.; Giordano, C.; Vischi, M.; Filippi, A.; Bertolini, A.; Marchiol, L. Evidences of genotoxicity and phytotoxicity in Hordeum vulgare exposed to CeO_2 and TiO_2 nanoparticles. *Front. Plant Sci.* **2015**, *6*, 1043. [CrossRef] [PubMed]
67. Shtangeeva, I. Europium and cerium accumulation in wheat and rye seedlings. *Water Air Soil Pollut.* **2014**, *225*, 1964–1973. [CrossRef]
68. Ramos, S.J.; Dinali, G.S.; Oliveira, C.; Martins, G.C.; Moreira, C.G.; Siqueira, J.O.; Guilherme, L.R.G. Rare earth elements in the soil environment. *Curr. Pollut. Rep.* **2016**, *2*, 28–50. [CrossRef]
69. Pošćić, F.; Schat, H.; Marchiol, L. Cerium negatively impacts the nutritional status in rapeseed. *Sci. Total Environ.* **2017**, *593–594*, 735–744. [CrossRef]
70. Zhao, J.; Stenzel, M.H. Entry of nanoparticles into cells: The importance of nanoparticle properties. *Polym. Chem.* **2018**, *9*, 259–272. [CrossRef]
71. Banerjee, K.; Pramanik, P.; Maity, A.; Joshi, D.C.; Wani, S.H.; Krishnan, P. Methods of using nanomaterials to plant systems and their delivery to plants (Mode of entry, uptake, translocation, accumulation, biotransformation and barriers). In *Advances in Phytonanotechnology: From Synthesis to Application*; Ghorbanpour, M., Wani, S.H., Eds.; Elsevier: Amsterdam, The Netherlands, 2019; pp. 123–152. [CrossRef]

Publisher's Note: MDPI stays neutral with regard to jurisdictional claims in published maps and institutional affiliations.

MDPI

Review

Characteristics, Toxic Effects, and Analytical Methods of Microplastics in the Atmosphere

Huirong Yang [1,2,3,†], Yinglin He [1,2,†], Yumeng Yan [1,2], Muhammad Junaid [1] and Jun Wang [1,2,4,*]

1 College of Marine Sciences, South China Agricultural University, Guangzhou 510642, China; hry@scau.edu.cn (H.Y.); 20202150006@stu.scau.edu.cn (Y.H.); ymyan@stu.scau.edu.cn (Y.Y.); junaid@scau.edu.cn (M.J.)
2 Guangdong Laboratory for Lingnan Modern Agriculture, South China Agricultural University, Guangzhou 510642, China
3 Zhongshan Innovation Center, South China Agricultural University, Zhongshan 528400, China
4 Guangxi Key Laboratory of Marine Natural Products and Combinatorial Biosynthesis Chemistry, Biophysical and Environmental Science Research Center, Institute of Eco-Environmental Research, Guangxi Academy of Sciences, Nanning 530007, China
* Correspondence: wangjun2016@scau.edu.cn; Tel./Fax: +86-20-87571321
† Co-first author: Huirong Yang and Yinglin He.

Abstract: Microplastics (MPs) (including nanoplastics (NPs)) are pieces of plastic smaller than 5 mm in size. They are produced by the crushing and decomposition of large waste plastics and widely distributed in all kinds of ecological environments and even in organisms, so they have been paid much attention by the public and scientific community. Previously, several studies have reviewed the sources, occurrence, distribution, and toxicity of MPs in water and soil. By comparison, the review of atmospheric MPs is inadequate. In particular, there are still significant gaps in the quantitative analysis of MPs and the mechanisms associated with the toxic effects of inhaled MPs. Thus, this review summarizes and analyzes the distribution, source, and fate of atmospheric MPs and related influencing factors. The potential toxic effects of atmospheric MPs on animals and humans are also reviewed in depth. In addition, the common sampling and analysis methods used in existing studies are introduced. The aim of this paper is to put forward some feasible suggestions on the research direction of atmospheric MPs in the future.

Keywords: microplastics; atmosphere; distribution; characteristics; toxicity; quantitative analysis

Citation: Yang, H.; He, Y.; Yan, Y.; Junaid, M.; Wang, J. Characteristics, Toxic Effects, and Analytical Methods of Microplastics in the Atmosphere. *Nanomaterials* **2021**, *11*, 2747. https://doi.org/10.3390/nano11102747

Academic Editor: Vivian Hsiu-Chuan Liao

Received: 24 September 2021
Accepted: 14 October 2021
Published: 17 October 2021

1. Introduction

A significant number of previous studies have emphasized the ubiquitous presence of microplastics (MPs) in the oceans [1], freshwater bodies [2,3], and soil [4], along with food items, drinks [5], seasonings [6], and aquatic organisms (Figure 1). Microplastic pollution in the environment can be caused by several factors, including landfills [7]; dumping and application of sewage sludge [8]; fiber shedding of synthetic textiles; transportation (wear of tires, brakes, road signs, etc.) [9]; and other human activities, including industrial plastic pellet preproduction [10], plastic mulching and grinding in agricultural [11], fisheries, and tourism [12]. After being released into the environment through different pathways, these microplastics experience the process of degradation (physicochemical fragmentation, chemical aging, biological degradation, etc.) [11] and translocation [13] under different environmental conditions. Finally, they enter animals and humans through skin contact, oral ingestion, inhalation, and other ways and continue to get enriched [14]. After entering the body, microplastics may produce a variety of negative effects, such as decreased growth rate [15], inflammatory response [16] and oxidative stress [17], and metabolic disorders [18]. In severe cases, they can penetrate organs [19,20], tissues [21], and even cells [22], causing toxic effects. Because microplastics have a large specific surface area and strong adsorption capacity, they easily adsorb various inorganic pollutants, such as Cu, Pb, and Cd, [23] and organic pollutants, e.g., PCBs [24], PAHs [18], and polybrominated diphenyl ethers [25]. Even some microplastics themselves contain additives with a certain toxicity, and some studies have shown that microplastics combined with these pollutants pose a serious threat to organisms.

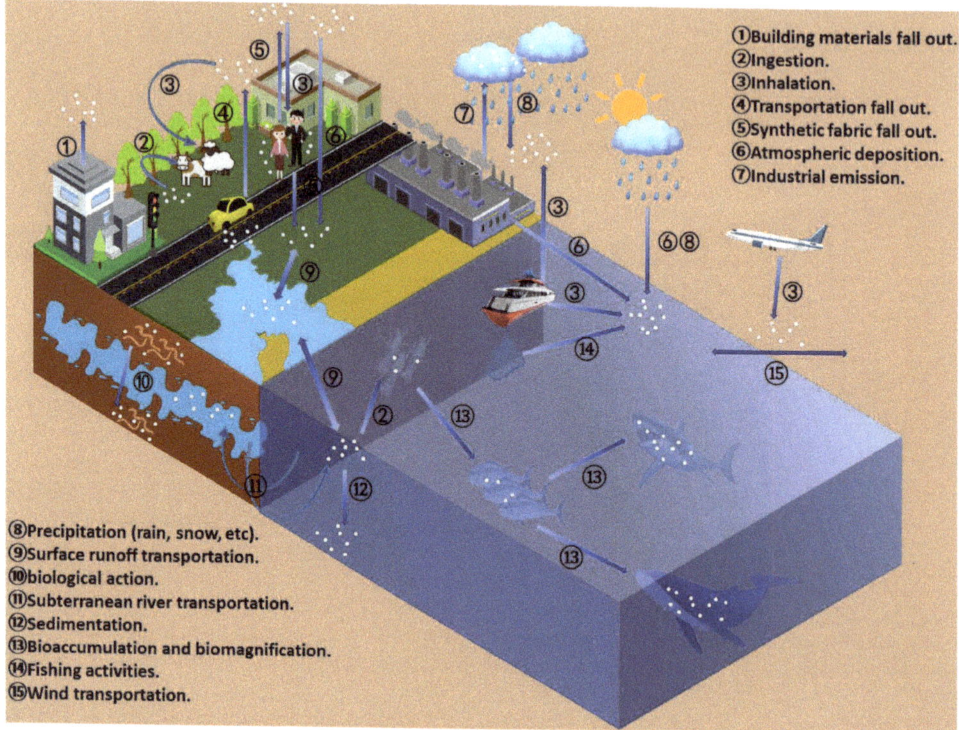

Figure 1. Cycle of microplastic pollution in ecosystems.

MPs in the air have been identified as particulate air pollutants and paid great attention recently [26]. However, at present, the research on environmental MPs is mainly focused on the aquatic ecosystem, and the number of studies on atmospheric MPs is limited, which is a limitation to further understanding the environmental characteristics and negative effects of atmospheric MPs. The small size of MPs, especially nanoplastics (NPs), facilitates their emissions into the air [27] and long-distance transportation [28] and can cause adverse effects on animals and humans through respiratory inhalation [29]. Critical analysis is urgently needed to open new ways of thinking about atmospheric MPs in the future.

In this review, the research progress on atmospheric MPs in recent years is summarized, including the following: (1) the global distribution of atmospheric MPs and their influencing factors, (2) the origin and fate of atmospheric MPs, (3) advances in sampling and analysis of atmospheric MPs, (4) toxicological impacts of MPs in the atmosphere on animals and humans, and last but not the least (5) the existing gaps in each part and the corresponding future research directions.

2. The Global Distribution of Atmospheric MPs and Associated Influencing Factors

2.1. Distribution Profile

At present, the relevant studies on the distribution of MPs mainly pay attention to the water and soil environment and the number of studies about atmospheric MPs is limited (Table 1). The earliest research on the distribution of atmospheric MPs can be traced back to Dris et al. collected and analyzed samples of outdoor air in Paris's urban areas, where the concentration of MPs ranged between 29 and 280 items/m^2/day. The size of MPs ranged from 0.1 to 5 mm, and the shapes mainly included fibers and fragments [30]. In another study, Dris et al. measured that the number of MPs in indoor air in Paris reached 5.4 items/m^3, while that in outdoor air in the same area was only 0.9 item/m^3, indicating indoor human activities among the major sources of MPs in indoor settings [31].

Table 1. Distribution and abundance of plastics in the atmosphere.

Location	Year	Sample Type	MP Type	Shape	Concentration (Item/Particle Number)	Size	Reference
Paris	2014	Urban outdoor air deposition	NA	Fiber, fragment	29–280/m^2/day	0.1–5 mm	[30]
Paris	2014–2015	Urban outdoor air deposition	NA	Fiber	110 ± 96/m^2/day	0.05–5 mm	[32]
Paris	2014–2015	Suburban outdoor air deposition	NA	Fiber	53 ± 38/m^2/day	0.05–5 mm	[32]
Paris	2015	Urban indoor air	PA, PP, PE	Fiber	0.4–59.4 (5.4)/m^3	0.05–3.25 mm	[31]
Paris	2015	Urban outdoor air	PA, PP, PE	Fiber	0.3–1.5 (0.9)/m^3	0.05–1.65 mm	[31]
Dongguan	2016	Urban outdoor air deposition	PE, PP, PS	Fiber, foam, fragment, film	175–313/m^2/day	Minimum: <0.2 mmMaximum: >4.2 mm	[33]
Yantai	2016	Urban outdoor air deposition	PET, PVC, PE, PS	Fiber, fragment, film, foam	2.33×10^{13}/160 km^2/year	0.05–1 mm	[34]
Sakarya	2016–2017	Crowded area outdoor air	PA, PUR, PE, PP, PES	Fiber, fragment	9067–30,793/L	0.05–0.5 mm	[35]
Edinburg	2017	Indoor air of houses	NA	Fiber	5 ± 33/sample	NA	[36]
Trent catchment	2017–2018	River catchment air deposition	NA	Fiber	2.9–128.42/m^2/day	NA	[37]
Shanghai	2018	Municipal outdoor air	PET, PE, PES, PAN, PAA, RY, EVA, EP, ALK	Fiber, fragment, granule	0–4.18 (1.42 ± 1.42)/m^3	23.07–9554.88 µm	[38]
Shanghai	2019	Urban outdoor air	PET, EP, PE, ALK, RY, PP, PA, PS	Fiber, fragment, microbead	0–2 (0.41)/m^3	12.35–2191.32 µm	[39]
Asaluyeh	2017	Urban and industrial outdoor air	NA	Fiber, fragment, film	0.3–1.1/m^3	2–100 µm	[40]
West Pacific Ocean	2018–2019	Ocean air	PET, PE, PE-PP, PES, ALK, EP, PA, PAN, PR, PMA, PP, PS, PVA, PVC	Fiber, fragment, granule, microbead	0–1.37 (0.06 ± 0.16)/m^3	16.14–2086.69 µm	[41]

Table 1. *Cont.*

Location	Year	Sample Type	MP Type	Shape	Concentration (Item/Particle Number)	Size	Reference
Pyrenees	2017–2018	Remote air deposition	PS, PE, PP, PVC, PET	Fiber, fragment, film	$365 \pm 69/m^2/day$	Minimum: <0.025 mmMaximum: >2.6 mm	[42]
Hamburg	2017–2018	Urban and rural outdoor air deposition	PE, EVA, PTFE, PVA, PET	Fragment, fiber	136.5–512/m^2/day	Minimum: <0.063 mmMaximum: >0.3 mm	[43]
Aarhus	2017	Indoor air of apartments	PES, PA, PS, PE, PUR	Fragment, fiber	1.7–16.2 (9.3 ± 5.8)/m^3	4–398 μm	[44]
London	2018	Urban outdoor air deposition	PAN, PES, PA, PP, PVC, PE, PET, PS, PUR, petroleum, resin, acrylic	Fragment, film, granule, foam	$771 \pm 167/m^2/day$	75–1080 μm	[45]
Karimata Strait	2019	Strait air	PET	Fiber	0–0.8/100 m^3	382.15	[46]
Pearl River Estuary	2019	River estuary air	PA, PEP, PET, PP	Fiber	3–7.7/100 m^3	288.2–1117.62 μm	[46]
South China Sea	2019	Ocean air	PET, PEVA, PP	Fiber, fragment	0–3.1/100 m^3	286.1–1861.78 μm	[46]
East Indian Ocean	2019	Ocean air	PAN-AA, PET, PP, PR	Fiber, fragment	0–0.8/100 m^3	58.591–988.37 μm	[46]
Beijing	NA	Urban outdoor air deposition	NA	Fiber	Surface layer: 5.7×10^{-3}/mLRoof: 5.6×10^{-3}/mL	5–200 μm	[47]
Alcalá de Henares-Guadalajara, Valladolid	2020	Rural and sub-rural PBL air	PET, PA, acrylic	Fiber	Rural area: 1/sampleSub-rural area: 3/sample	0–9.8 μm	[27]
Guadalajara	2020	Urban PBL air	PU, PS, PA, acrylic	Fragment, fiber	6/sample	NA	[27]
Madrid	2020	Urban PBL air	PA, PU, PET, PB, PE, PP	Fragment, fiber	12/sample	NA	[27]

PA: polyamide; PP: polypropylene; PE: polyethylene; PS: polystyrene; PET: polyethylene terephthalate; PVC: polyvinyl chloride; PUR: polyurethane; PES: polyester; PAN: polyacrylonitrile; PAA: poly(N-methyl acrylamide); RY: rayon; EVA: ethylene vinyl acetate; EP: epoxy resin; ALK: alkyd resin; PR: phenoxy resin; PTFE: Teflon; PVA: polyvinyl acetate; PEVA: poly(ethylene-co-vinyl acetate); PEP: poly(ethylene-co-propylene); PAN-AA: poly(acrylonitrile-coacrylic acid); PB: polybutadiene; PU: polyurethane; PBL: planetary boundary layer; NA: not available.

Subsequently, studies on atmospheric MPs have been carried out around the world, including a dozen countries and regions in Asia [33,34,38,39,47], Europe [36,43,45], and the Arctic [48]. The survey areas include urban sites, such as municipal areas [38], apartments [31], offices [31], industrial areas [40], terminals [35], and universities [47], as well as suburbs [32], rural areas [27], mountains [42], straits [46], estuaries [46], oceans [41,46], glaciers [28], and even the planetary boundary layer (PBL) [27]. These studies suggest that MPs exist in the atmosphere worldwide, from near the ground level up to 1.5 km high.

2.2. *Influencing Factors*

2.2.1. Vertical Concentration Gradient

Similar to other air particle pollutants, the concentration of atmospheric MPs near the ground is much higher than that at high altitudes due to the influence of gravity [49]. Li et al. also demonstrated this phenomenon by collecting atmospheric plastic particle deposition on the ground and on the roof of buildings in the urban areas of Beijing, where the concentration of particles in the former was higher than that in the latter [47]. However, this does not mean that MPs in the atmosphere are not worth taking seriously. Gonzalez-Pleiter et al. found MPs in the atmospheric particulate matter samples collected by the spacecraft in the planetary boundary layer (PBL) [27]. These studies suggest that MPs are ubiquitous in atmospheric environments, ranging from near the ground to high altitudes.

2.2.2. Meteorological Conditions

Dris et al. found that rainfall affects the sedimentation rate of microplastic fibers in the atmosphere of Paris. When rainfall is 0–0.2 mm/day, 2–34 items/day are recorded. When the rainfall reaches 2–5 mm/day, the amount of sediment increases to 11–355 items/day [32]. This indicates that rainfall has a significant effect on the precipitation behavior of atmospheric MPs. Boucher et al. pointed out that 7% of atmospheric MPs are transported into the ocean by wind [50], indicating that low-density atmospheric plastic particles can pollute other ecosystems through the wind as a medium. However, Prata et al. believe that atmospheric microplastic particles have properties similar to other particulate pollutants and meteorological factors such as wind, precipitation, and temperature will have an impact on their concentration changes [49]. These studies indicate that meteorological conditions are a significant factor influencing the distribution characteristics of atmospheric MPs.

2.2.3. Indoor and Outdoor Atmospheric Settings

Numerous studies have shown that the concentration of MPs in the indoor air environment is much higher than that in the outdoor within the same area. Dris et al. conducted a study on the outskirts of Paris in 2015, where they analyzed microplastic fibers in the air outside and inside apartments and found that these man-made fibers are mostly polypropylene (PP). Moreover, the fiber concentration in indoor air (0.3–1.5 (0.9) items/m^3) was significantly higher than that in outdoor air (0.4–59.4 (5.4) items/m^3) [31]. When the outdoor environment is crowded enough, the concentration of MPs in the atmosphere can reach very high levels. Kaya et al. analyzed the concentrations of atmospheric MPs in universities and terminals with a large population in Sakarya Province, Turkey, in 2016–2017, and found concentrations of particles as high as 10,495–30,822 particles/L [35]. The high concentration of airborne MPs in indoor and crowded outdoor environments may be attributed to similar conditions of high population density and poor particle dispersion capacity.

2.2.4. Regional Environmental Conditions

Varying distribution of atmospheric MPs was observed in different regions. The concentrations and types of atmospheric MPs in different urban areas have similarities and differences. Cai et al. analyzed the concentration of MPs (175–313 items/m^2/day) in the urban air of Dongguan in 2016 and found that the content of fibers in the air was the highest (90.1%), followed by fragments (6.8%), film (2.9%), and foam (0.2%)) [33]. In the same year, Zhou et al. carried out a similar investigation in the urban area of Yantai and found that the concentration of atmospheric MPs was 130–640 particles/m^2/day, with a higher occurrence of fiber (95.05%), followed by fragments (4.04%), film (0.73%), and foam (0.18%) [34]. The concentration of atmospheric MPs varies in different regions, which may be influenced by local meteorological conditions, the topography, the urban heat island effect, and other factors [51]. However, the types of atmospheric MPs in different regions and urban areas show high similarity. Fibers are observed as the absolute dominant shape for atmospheric MPs, while fragment, film, and foam MPs appear in significantly low quantities. It is known that microplastic fibers mainly come from synthetic textiles. Fragments may come from disposable plastic bags, film may be obtained by breaking thick plastic products, and foam may come from foamed plastics [33]. It is suggested that the atmospheric MPs in central urban areas mainly come from the shedding of synthetic textiles.

Atmospheric MP concentrations often differ between urban and suburban areas. Dris et al. collected and analyzed the deposition from the air in the urban and suburban areas of Paris during 2014–2015, and the concentration of MPs in the former (110 $\pm$ 96/m^2/day) was much higher than that in the latter (53 $\pm$ 38/m^2/day) [32]. The higher intensity of human activity in urban areas may have contributed to higher concentrations of atmospheric MPs. The similarity of occurrence between the two (mainly fiber) may be due to atmospheric migration that leads to regional homogenization.

However, it appears that not all suburban and remote areas have a low atmospheric microplastic distribution. Allen et al. collected atmospheric sediments from the Pyrenees Mountains in France in 2017–2018 and reported an MP concentration of 365 $\pm$ 69 items/m^2/day. It was dominated by polystyrene and polyethylene. The occurrence was fragments (68.0%), film (20.0%), and fiber (12.0%) [42]. Ambrosini et al. collected and analyzed atmospheric sediment samples from the Forni Glacier in the Alps in 2018, and the concentration of MPs in the samples was 74.4 $\pm$ 28.3 items/kg of sediment. MP polymer types included polyester (39%), polyethylene (9%), polyamide (9%), and polypropylene (4%). Fibers accounted for 65.2%, and fragments accounted for 34.8% [52]. It can be seen that remote areas may also have high concentrations of atmospheric MPs and there are great differences in the concentrations and types among different regions.

At present, due to large differences, it is still difficult to draw a clear picture of the regional distribution of atmospheric MPs, to identify the most polluted areas, so it is necessary to master more methods to study atmospheric MPs.

2.3. Gaps in and Prospective Research on Distribution Characteristics of Atmospheric MPs

From the above-reviewed studies, we found that the current research on the distribution of atmospheric MPs is relatively limited and there is a lack of clear and systematic studies. To this end, we propose the following for future research:

(1) It is difficult to confirm the extent of MP pollution in the atmosphere around the world. It is suggested that systematic spatial and temporal studies be conducted on the distribution of MPs in the atmosphere, to further clarify the concentrations, types, and occurrence of atmospheric MP pollution in different regions and determine the sources, distribution, and fate of atmospheric MPs in different regions.

(2) We found that the experimental methods for studying atmospheric MPs in the past papers were different and no standard methods for collection and characterization of MPs were validated, which greatly reduced the experimental efficiency. In addition, the measurement criteria and units used were so varied that it is hard to intuitively make a comparison with the experimental findings of researchers using different standards (e.g., there is no way to compare the concentrations of MPs in units of m^2/day and m^3/day). It is suggested that the use of more efficient sampling and analysis methods be unified and the industry standards for measuring MP concentration, type, and occurrence be standardized.

3. Sources of Atmospheric MPs

3.1. Sources

3.1.1. Synthetic Textiles

Synthetic textiles are a major source of atmospheric MPs. The output of synthetic textiles in the world has been growing gradually year by year. Since the annual output exceeded 60 million tons in 2016, it has been growing steadily by about 6%/year [53]. The commonly used plastic raw materials in synthetic textiles include fiber with polymers of polyamide (PA), polypropylene (PP), polyacrylonitrile (PAN), polyester (PET), polyvinyl formaldehyde (PVDF), polytetrafluoroethylene (PTFE), etc. In the use of synthetic textiles, fine fibers fall off from the fabric and are released into the air due to grinding and cutting in the textile industry, as well as wearing, washing, and drying clothes in daily life. It has been reported that thousands of fibers can be shed from a single gram of PAN fabric [54]. According to a study of De Falco et al., there is a significant correlation between the shedding of microplastic fibers (MFS) during the wearing of synthetic clothing and the type of fabric. Taking PES as an example, short silk fabrics release more MFS than filament fabrics, which may be because short fibers are easier to shed during movement, friction, and other behaviors. However, knitted garments are more likely to shed MFS than woven garments, which may be due to the looser arrangement of fabric fibers in knitted garments [55]. COVID-19 is wreaking havoc around the world, leading to a global surge in the production and use of masks and protective clothing. Mask and protective clothing materials are known to include PP, polyethylene (PE), polyurethane (PU), PTFE, PET, and ethylene side-by-side (ES) polymer plastics [56]. These anti-epidemic fabrics may become a major source of atmospheric MPs in coming years. In addition, because the inside of a mask is close to your mouth and nose, the MPs they shed can be easily inhaled.

3.1.2. Transportation

Transportation also contributes a lot to atmospheric MPs, for example, in the form of wear particles from the tires and brakes of cars as well as from road surfaces and aircraft tires [9,57]. Existing research suggests that the composition of tire and road wear particles (TRWPs) is about 50% natural or synthetic polymers, which include a large number of plastic components, such as styrene–butadiene rubber (SBR). It shows that a great number of plastic particles enter the surroundings each year because of TRWP emission [58], accompanied by a certain amount of preservatives, antioxidants, desiccants, plasticizers, and other additives. According to Wagner et al., road traffic alone produces 1.327 million tons of TRWPs per year in Europe. These TRWPs can easily pollute the air environment through direct discharge or resuspension of road dust [59]. The world produces 0.2–5.5 kg of TRWPs per person per year, of which the contribution to PM10 emissions accounts for 11%. Moreover, TRWPs make up more than 50% of MP emissions in Denmark and Norway, as well as about 30% in Germany [60]. In addition, TRWPs are usually emitted in heterogeneous aggregates with other wear particles present in traffic (brake wear, road wear, etc.) [61].

3.1.3. Dust

The concentration of MPs in ambient dust is very high, both in deposited dust and in suspended dust. In terms of deposited MP particles, the distribution of MPs of different shapes and sizes is very uneven under the influence of external forces (natural or human activities), and they are easily suspended in the atmospheric environment due to external forces. However, the form of MPs in floating dust is mainly fine fibers. Compared with other types of plastics, the lower-density characteristics of MPs facilitate their suspension in air [40]. Liu et al. studied and analyzed indoor and outdoor dust samples from 39 cities in China and found that PET majorly contributes to the content of MPs in indoor dust. This may be because indoor MPs are mainly derived from synthetic fabrics and PET is the main component of polyester in commonly used synthetic fabrics, which are easy to use, wash, and dry. Secondly, polycarbonate (PC), which is widely used in electronic equipment, hardware, and food packaging, can also easily fall off and enter the atmosphere [62,63].

3.1.4. Other Small Sources

Atmospheric MPs are also likely to come from the degradation of large plastics, such as building materials and synthetic furniture; landfills; synthetic particles from gardening; and industrial and other emissions-related activities. However, compared with the major sources of MPs such as synthetic textiles, transportation, and dust, the actual contribution of these sources to atmospheric MPs may be very small and remains in the stage of idea and speculation, without the availability of data [32,33,38].

3.1.5. Gaps in and Prospective Research on the Sources of Atmospheric MPs

From the above series of articles, it is found that the current articles on the sources of atmospheric MPs are mostly focused on synthetic textiles, transportation, dust deposition, and resuscitation, while other sources are often ignored, and the contribution of these sources to atmospheric MPs cannot be quantified. Detailed data on their pollution concentrations, types, and occurrence are scarce. In the analysis of the sources of atmospheric MPs in many pieces of literature, the proportion of unknown sources in the experimental results is relatively high, indicating that there is still a considerable part of atmospheric MP sources that is unclear. In addition, many research methods to identify the origin remain in the characterization and chemical composition analysis. To this end, we propose the following for future research:

(1) Continue to optimize the methods and tools for atmospheric MP characterization and component analysis and identification to more clearly identify the sources of MPs and avoid unclear and inaccurate source identification caused by rough differentiation.
(2) Establish a pollution source localization method suitable for atmospheric MPs, which can trace the source more accurately than characterization or component analysis.

3.2. *Transportation and Fate of Atmospheric MPs*

3.2.1. Migration

Under certain conditions, MPs in the air can migrate to other ecological environments (soil, water, etc.). Their behavior, transportation, concentration, and deposition are affected by various factors, such as vertical pollution concentration gradient (VPCG) [64,65], meteorological conditions (rainfall, precipitation, temperature, humidity, wind (occurrence, velocity, duration, intensity, and direction), etc.) [39,42,43,66], population density, human activities [32,39], urban topography, thermal cycling [51,64], local elevation, and geographical environment [39]. Some researchers have found that MPs travel long distances in the air. For example, MPs are found in extremely remote areas and even in the snow and ice of high-altitude glaciers. It is speculated that MPs cause cross-border and global pollution through air and wind currents [27,28].

3.2.2. Inhalation

Since the MP concentration indoors is much higher than that outdoors, atmospheric MPs enter the body through human inhalation [31]. Indoor air is an essential source of human exposure to airborne MPs because people stay indoors for longer periods and dispersing machines are less capable of removing plastic particles [49]. Factors affecting MP behavior and transmission in indoor air are ventilation, airflow, and room spacing [67]. The concentration of MP particles indoors is higher than that outdoors, which may be influenced by textiles, furniture, building materials, and human activities [31]. Compared with MPs in other environments, MPs indoors are more easily inhaled directly and continuously and cause health risks [49,53].

One of the earliest criteria for determining airborne MPs is 0.3–1.5 particle/m^3 outdoors and 0.4–56.5 particles per cubic meter indoors (33% polymer) [31]. According to statistics, each person inhales between 26 and 130 MP particles from the air per day [49].

Based upon air samples taken from mannequins, men who exercise lightly can expect to inhale 272 particles/day [44]. Estimates vary depending on sampling methods and space use factors.

3.2.3. Gaps in and Prospective Research on the Destinations of Atmospheric MPs

Based on the above discussion, the studies on the fate of MPs in the atmosphere are limited. So far, only limited studies have traced the atmospheric migration paths of MPs; therefore, it is nearly impossible to quantify the various environmental factors and human activities affecting the behavior and transmission of atmospheric MPs. Further, the quantification and characterization of atmospheric MPs in different parts of the human body and associated health impacts is challenging. To this end, we propose the following for future research:

(1) Further explore the factors affecting the fate of atmospheric MPs and understand the different destinations of MPs in the atmospheric environment under different conditions.
(2) Establish a spatial model and related software suitable for integrating the diffusion and migration trajectories of atmospheric MPs and the pollutants adsorbed by them.
(3) Investigate the difference in the quantities and proportions of MPs absorbed by people in different areas and under different conditions in the same area as well as body burden and associated risks of atmospheric MPs.

4. Toxic Effects

After the body inhales MPs, they enter the lungs along the trachea, enter the blood vessels through migration, and then spread through the circulatory system throughout the body, causing various degrees of toxic effects on the cells, tissues, organs, and systems of the body (Figure 2) [19,68–70].

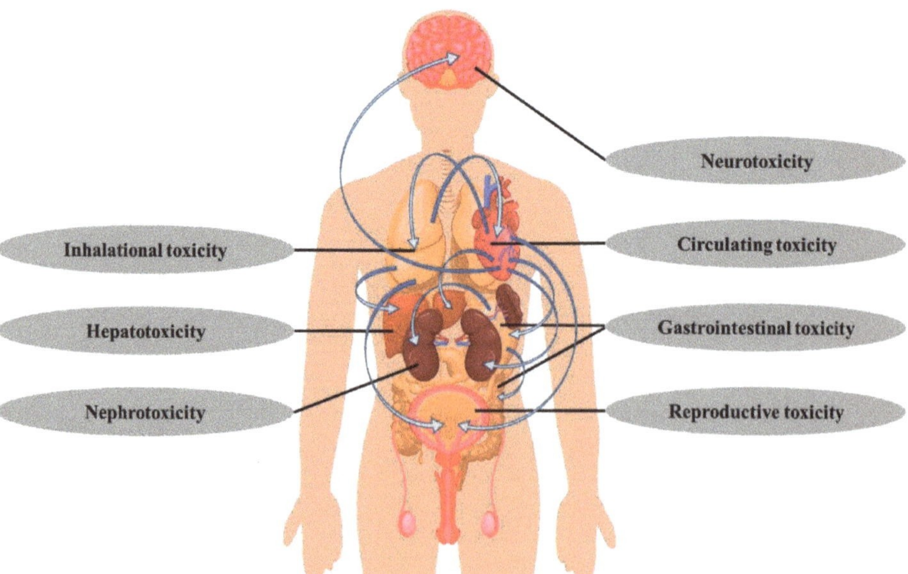

Figure 2. Toxic effects of atmospheric microplastics on different organs and systems.

4.1. Inhalational-Based Toxicity

MPs and NPs in the air mainly enter animals and humans through inhalation and first contact with the respiratory system. Due to the clearance mechanisms of the respiratory barriers (such as the nasal cavity, trachea, the bronchus, and alveolar macrophages), a considerable proportion of these plastic particles cannot enter the body and form deposition, and the deposition coefficient is mainly affected by the particle size, density, etc. [49]. Smaller nanoparticles (<2.5 μm) can avoid the clearance mechanism and penetrate the lung and respiratory barrier [19,68–70]. Macrophages can deposit plastic particles in the respiratory system, remove these from the respiratory system, or help them migrate in the respiratory system and allow them to enter the circulatory system, leading to dust metastasis. Further, the presence of MPs and NPs in the respiratory system and their large surface area can also lead to the release of chemokines that affect the migration of macrophages, resulting in particle overload [68]. The influx and inflammation of neutrophils associated with polystyrene granules (64 nm) in rat lungs, as well as the expression of pro-inflammatory genes in epithelial cells, are caused by granule-induced oxidation [71]. Due to the large specific surface area of MPs, a large number of oxidizing substances (such as metals) will be adsorbed on the surface of MPs, which may produce excessive reactive oxygen species and lead to excessive antioxidant reaction in vivo, namely oxidative stress [72,73].

There are only a few pieces of research on the toxic effects of atmospheric MPs on animals and humans. Inhalation of MPs has the most significant toxic effects on the respiratory system. Studies have shown that PVC prepared by emulsion polymerization (2 mm) has significant cytotoxic and hemolytic effects on rat and human lung cells in vitro [74]. Xu et al. 2002 [74] evaluated the toxic effects of PS-NPs on human alveolar epithelial cells and found that PS-NPs can rapidly internalize into cells; significantly reduce cell viability; affect cell cycle and apoptosis, as well as related gene transcription and protein expression; and promote inflammatory response. The smaller the diameter, the faster the corresponding velocity. Lim et al., using metabonomics to investigate the toxic effects of PS-NPs on bronchial epithelial cells, found that nanoparticles interfere with cell energy metabolism, accompanied by oxidative stress, and mediate the increase of intermediate metabolites to reduce cell resistance to toxicity [75]. Dong et al. found that PS-MPs can induce cytotoxicity and inflammation in human lung epithelium by inducing reactive oxygen species formation. Low concentrations of PS-MPs disrupt the lung barrier, while high concentrations of PS-MPs also induce decreased levels of α 1-antitrypsin, which increases the risk of chronic obstructive pulmonary disease [76]. Paget et al. explored particle internalization and cell damage after aminated PS-NPs acted on human lung epithelial cells and macrophages. Cells in the experimental group showed high glutathione depletion, excessive reactive oxygen species, and significant DNA damage. It is suggested that micro- and nanoplastics may exhibit strong genotoxicity when absorbed and internalized by the respiratory system [77]. In in vitro experiments, seven functional groups with different charges were added to an aminated PS-NPs and eight PS-NPs, including the original particles, were sent into different rats by pharyngeal aspiration. Most of the particles were phagocytized by alveolar macrophages and showed different acute lung inflammation. The particle potential showed excellent correlation with pneumonia-related parameters, indicating that surface charge is a key factor affecting lung inflammation induced by micro- and nanoplastics [78].

Respiratory lesions have been found in workers exposed to synthetic textiles, fusing, vinyl chloride, or PVC, which is quite different from the general population [79–83]. Inhaled MPs and NPs are not easily cleared by the human lungs, and they may stay in the lungs for a considerable period, which can cause an inflammatory response in the lungs [53]. It is reported that artificial glass fiber can cause DNA damage, which can induce cancer [84]. In short, Inhaling MPs can have toxic effects on animals and humans.

4.2. Other Toxic Effects

The inhaled MPs can also have negative effects on other systems of the body, such as diffusion or translocation. In vivo experiments have confirmed that microplastic particles can enter the circulatory system through the migration of macrophages after inhalation [68]. MPs in the circulation system may cause inflammation, vascular occlusion [85], and other blood toxicity [86]. In vitro experiments showed that polystyrene nanoparticles cause red blood cell (RBC) aggregation, while polypropylene particles increase hemolysis [87]. Inhaled MPs may enter the gastrointestinal tract by clearing upper respiratory tract cilia [53]. MPs entering the digestive system may alter the permeability of intestinal epithelial cells and cause changes in microbial composition [88]. Recent reports have found that after maternal lung exposure to NPs, the plastics can enter the placenta and the fetus through translocation and get deposited in the fetal liver, kidney, nervous system, and circulatory system. [21]. There is little research on the toxic effects of atmospheric MPs on systems other than the respiratory system, and more research must be done to explore their effects on animals and humans.

4.3. Joint Toxic Effects

MPs usually contain various additives, e.g., catalysts (organotin), flame retardants, polybrominated diphenyl ethers [89], antioxidants (nonylphenol), antibacterial agents (triclosan), and plasticizers (phthalate PAEs) [90]). All of them are harmful to animals and humans. In addition, MPs adsorb many inorganic pollutants in the environment, e.g., Au [16] and Cu [91], and organic pollutants, e.g., polycyclic aromatic hydrocarbons [18] and polychlorinated biphenyls [92]. These pollutants are mutagenic and carcinogenic substances that are widely present in the environment. In addition, many plastic monomers themselves (polystyrene (PS), polyvinyl chloride (PVC), etc.) have mutagenic and carcinogenic toxic effects on animals and humans [93]. Exposure of animals and humans to additives or adsorbents in these plastic particles results in combined toxicity (Table 2). MPs in the atmosphere are most likely to combine with other pollutants in atmospheric environments (such as POPs and Cu) and serve as carriers for the long-distance transport of these pollutants in the atmosphere [53]. Some articles have studied the negative effects of MPs combined with these pollutants on animals and the human body, such as a series of oxidative stress [16], inflammatory reaction [91], and metabolic disorders [18]. There is a risk of mutation and carcinogenesis [94]. Most of these studies have involved oral ingestion or in vitro studies. However, there are few studies on the toxicity mechanism of these binding substances after they enter animals and humans through inhalation, and more experimental data are needed.

Table 2. Toxic effects of chemicals in microplastics or nanoplastics on animals.

Classification	Chemicals	Affected Species	Resulting Toxicity	Reference
Ingredient	C_6H_6	*Human*	Mutagenic risk	[94]
	C_6H_5OH	*Human*	Mutagenic risk	[94]
	BD	*Human*	Cancer risk	[94]
	VCM	*Human*	Cancer risk	[94]
Adsorption	Au	*Danio rerio*	Embryo: ① Oxidative stress ② Inflammation	[16]
	CBz	*Mytilus galloprovincialis*	Larva: Excessive oxidation of digestive glands	[95]
	Cu	*Danio rerio*	Inflammation	[91]
	PAHs	*Danio rerio*	Metabolic disorders	[18]
	PCBs	*Human*	Neurotoxicity	[92]
Dyestuff	Pyrene	*Mytilus galloprovincialis*	① Immune responses ② Lysosomal compartment dysfunction ③ Peroxisome dysfunction ④ Antioxidant system disruption ⑤ Neurotoxic effects	[96]

Table 2. *Cont.*

Classification	Chemicals	Affected Species	Resulting Toxicity	Reference
Flame retardants	PBDEs	*Human*	① Thyroid homeostasis disruption ② Neurotoxicity ③ Reproductive changes ④ Cancer risk	[89]
Paint coat	TiO_2	*Caenorhabditis elegan*	Oxidative stress	[97]
Plasticizer	BPA	*Danio rerio*	Neurotoxicity	[98]
		Rat	Estrogen disorder	[99]
		Human	① Enzyme abnormality and damage of the liver ② Pancreatic cell dysfunction ③ Thyroid hormone disorder ④ Promotion of obesity ⑤ Cardiovascular disease ⑥ Low insulin levels	[99]
	DEHP, MEHP	*Rat*	Inhibition of estrogen levels	[90]
	PAEs	*Human*	① Increased risk of cardiovascular disease ② Reproductive system disruption	[90]

BD: butadiene; VCM: vinyl chloride monomer; CBz: carbamazepine; PAHs: polycyclic aromatic hydrocarbons; PCBs: polychlorinated biphenyls; PBDEs: polybrominated diphenyl ethers; BPA: bisphenol A; DEHP: dioctyl phthalate; MEHP: mono(2-ethylhexyl) phthalate; PAEs: phthalic acid esters.

4.4. Gaps in and Prospective Research on the Toxic Effects of Atmospheric MPs on Animals and Humans

In conclusion, studies on the quantitative analysis and toxicity mechanism of atmospheric MPs inhaled by animals or humans are scarce and these aspects need to be further studied. To this end, we propose the following for future research:

(1) In vivo experiments should be conducted to explore the different negative effects of MPs or NPs with different physical or chemical properties (such as different types, sizes, occurrences, crystallinity, and surface charge) on animal and human health after inhalation. An animal model should be established to study the movement trajectory and deposition proportion of and harmful substances released by atmospheric MPs in the body.

(2) To better understand the harmful additive impact of atmospheric MPs adsorbed with other pollutants as a pollutant internalization carrier, more research into the toxic additive effect of the two is required.

5. Existing Analytical Methods and Gaps in Measuring Atmospheric MPs

There are two main sampling methods for atmospheric MPs commonly used. One way is to collect the passive fallout from the atmosphere and filter it. The depositions in the atmosphere are collected through a non-plastic funnel (such as stainless steel or glass) the pipe of which drops into a glass collection bottle below. MPs can then be easily filtered out of the sediment [32,33,42,43]. Another method is active pump sampling and filtration, mainly through a set of pumping and filtration system, in which air is collected through the pump unit and then filtered through the filter to retain the plastic particles [31,38,40]. After the sample is collected, different efficient quantitative analysis methods can be used to analyze the types and sizes of particles (Table 3).

Table 3. Outstanding methods for the analysis of microplastics and nanoplastics in the atmospheric environment.

Methods	Tools	Medium	Plastic Components	Optimal Size	Advantages	Disadvantages	References
Spectral analysis	FT-IR	Water, oil, air	RY, PE, PET, PAA	>20 μm	① It does not destroy the sample. ② Pretreatment is simple. ③ The type of plastic particles can be determined.	It is difficult to identify the types of plastic particles that are aged or have contaminated surfaces.	[100–103]
	RM	Water, air	PA, PC, PE, PP, PS, PET, PVC, PMP, PCL, PMMA	0.5–20 μm	① It does not destroy the sample. ② It supports nano-sample imaging. ③ It supports low sample amount identification. ④ It is environmentally friendly.	① The measurement time is long. ② Fluorescence interference is easy to produce. ③ The signal-to-noise ratio is low. ④ The use of laser as the light source leads to background emission and sample degradation.	[104–108]
Thermal analysis	TGA-DSC	NA	PE, PP, etc.	NA	① The operation is simple. ② Less sample is required (1–20 mg). ③ Accuracy is high.	① It is difficult to distinguish the polymers with similar transition temperatures. ② It is difficult to identify copolymers. ③ The samples are destroyed. ④ It cannot identify the morphology, size, and quantity of the plastic particles.	[108–112]
	Py-GC-MS	NA	PA, PC, PE, PS, PP, rubber, PET, PVC, PMMA	NA	① Less sample is required (5–200 μg). ② The microplastic type and weight and additives can be identified simultaneously without pretreatment. ③ The accuracy is high. ④ It recognizes copolymers.	① The samples are destroyed. ② It cannot identify the morphology, size, and quantity of the plastic particles.	[110–115]
	TED-GC-MS	NA	PA, PE, PP, PS, PET	NA	It involves simple pretreatment and operation.		[108,110,112,116]
Other analytical methods	SEM-EDS	Majority	Majority	≥1 nm	① Imaging is at the nanoscale. ② Elements can be identified	① It is expensive. ② Work efficiency is low.	[108,110,112,117,118]
	MS	Majority	Majority	≥1 nm	It can identify the structure, molecular weight, degree of polymerization, functional group, and end group structure of the plastic particles.	Different samples require different ionizing reagents (poor applicability).	[108,119,120]
	XPS	Majority	Majority	>10 nm	It can identify elemental composition and content, chemical state, molecular structure, and chemical bonds.	It cannot identify the nanoplastic types definitely.	[121,122]
	RMR	Water, oil	Majority	>50 μm	① The cost is low. ② It is convenient for real-time field detection.	① It is only used to detect the concentration. ② It requires specific calibration samples.	[123]

FT-IR: Fourier-transform infrared spectroscopy; RM: Raman spectroscopy; TGA-DSC: thermogravimetric analysis-differential scanning calorimeter; Py-GC-MS: pyrolysis gas chromatography-mass spectrum; TED-GC-MS: thermal desorption gas chromatography-mass spectrum; SEM-EDS: scanning electron microscope-energy dispersive spectrometer; MS: mass spectrometry; XPS: X-ray photoelectron spectroscopy; RMR: resonance microwave reflectometry. RY: rayon; PE: polyethylene; PET: polyethylene terephthalate; PAA: polyacrylic acid; PA: polyamide; PC: polycarbonate; PP: polypropylene; PS: polystyrene; PVC: polyvinyl chloride; PMP: polymethylpentene; PCL: polycaprolactone; PMMA: polymethylmethacrylate; NA: not available.

Although a set of primary collection and analysis methods has been established, there are still many limitations and deficiencies. The above methods can only identify some common types of plastic particles at the ground level or near the ground level [27]. Moreover, it is difficult to accurately identify the size of the nanoplastics or the types of particles adsorbed by organic matter on their surfaces [101,103,104]. To this end, we propose the following for future research:

(1) To further develop and research some efficient methods and instruments. On the one hand, a great number of MPs should be sampled and accurately identified in a short period. On the other hand, we should be able to further identify more types of plastics and plastics of smaller sizes through microfiltration and various pollutants adsorbed on them.
(2) To develop a set of uniform standard methods for sampling and identification. For this, scientific data generated in different regions for atmospheric MPs should be compared.

6. Conclusions

MPs have been well studied in marine and freshwater environments, but MPs in the atmosphere have received little attention from researchers and society. Microplastics in the atmosphere enter the body mainly through inhalation and further systemic exposure, causing toxic reactions and disorders in various organs and systems and even posing a potential risk of cancer to animals and humans.

Due to the lack of practical methods for detection and analysis, we are still unable to gain a more detailed understanding of the global distribution, sources, and fate of atmospheric MPs, let alone further elucidate the mechanisms of toxic action of atmospheric MPs on animals and humans. We suggest that the scientific community conduct in-depth research on atmospheric MPs in the future, especially to explore relevant sampling and detection methods and establish a common industry standard. Further quantitative analysis of atmospheric MPs of different types and properties will be conducted to explore the toxicity mechanism and additive effect of their combination with other pollutants through in vivo experiments.

Author Contributions: Conceptualization, H.Y. and J.W.; methodology, Y.Y.; software, Y.H.; validation, H.Y. and J.W.; formal analysis, Y.H. and Y.Y.; investigation, M.J.; resources, H.Y. and Y.H.; writing—original draft preparation, H.Y. and Y.H.; writing—review and editing, H.Y. and Y.H.; visualization, Y.H.; supervision, H.Y. and J.W.; project administration, H.Y. and J.W.; funding acquisition, H.Y. and J.W. All authors have read and agreed to the published version of the manuscript.

Funding: This study was funded by the National Key Research and Development Program of China (2018YFD0900604), the Guangdong Forestry Science and Technology Innovation Project (2021KJCX012), Provincial Science and Technology Special Fund Project for Zhongshan City (major special project + Task list management mode) (2021), Rural Science and Technology Commissioner Service Team project of New Rural Development Institute, South China Agricultural University (2021XNYNYKJHZGJ022), Provincial Projects with Special Funds for Promoting Economic Development of Marine and Fisheries Department of Guangdong (SDYY-2018-05), the Project of Guangzhou Association for Science & Technology (K20200102008), the Guangdong Province Universities and Colleges Pearl River Scholar Funded Scheme (2018), the National Natural Science Foundation of China (42077364), the Innovation Group Project of Southern Marine Science and Engineering Guangdong Laboratory (Zhuhai) (311021006), and Key Research Projects of Universities in Guangdong Province (2019KZDXM003 and 2020KZDZX1040). We appreciate the provision of SCAU Wushan Campus Teaching & Research Base.

Data Availability Statement: Not Applicable.

Conflicts of Interest: The authors declare no competing financial interests.

References

1. Capillo, G.; Savoca, S.; Panarello, G.; Mancuso, M.; Branca, C.; Romano, V.; D'Angelo, G.; Bottari, T.; Spano, N. Quali-quantitative analysis of plastics and synthetic microfibers found in demersal species from Southern Tyrrhenian Sea (Central Mediterranean). *Mar. Pollut. Bull.* **2020**, *150*, 110596. [CrossRef] [PubMed]
2. Yuan, W.; Liu, X.; Wang, W.; Di, M.; Wang, J. Microplastic abundance, distribution and composition in water, sediments, and wild fish from Poyang Lake, China. *Ecotoxicol. Environ. Saf.* **2019**, *170*, 180–187. [CrossRef] [PubMed]
3. Toumi, H.; Abidli, S.; Bejaoui, M. Microplastics in freshwater environment: The first evaluation in sediments from seven water streams surrounding the lagoon of Bizerte (Northern Tunisia). *Environ. Sci. Pollut. Res. Int.* **2019**, *26*, 14673–14682. [CrossRef] [PubMed]
4. O'Connor, D.; Pan, S.; Shen, Z.; Song, Y.; Jin, Y.; Wu, W.M.; Hou, D. Microplastics undergo accelerated vertical migration in sand soil due to small size and wet-dry cycles. *Environ. Pollut.* **2019**, *249*, 527–534. [CrossRef]
5. Kutralam-Muniasamy, G.; Perez-Guevara, F.; Elizalde-Martinez, I.; Shruti, V.C. Branded milks—Are they immune from microplastics contamination? *Sci. Total Environ.* **2020**, *714*, 136823. [CrossRef]
6. Peixoto, D.; Pinheiro, C.; Amorim, J.; Oliva-Teles, L.; Guilhermino, L.; Vieira, M.N. Microplastic pollution in commercial salt for human consumption: A review. *Estuar. Coast. Shelf Sci.* **2019**, *219*, 161–168. [CrossRef]
7. He, P.; Chen, L.; Shao, L.; Zhang, H.; Lu, F. Municipal solid waste (MSW) landfill: A source of microplastics?—Evidence of microplastics in landfill leachate. *Water Res.* **2019**, *159*, 38–45. [CrossRef]
8. Corradini, F.; Meza, P.; Eguiluz, R.; Casado, F.; Huerta-Lwanga, E.; Geissen, V. Evidence of microplastic accumulation in agricultural soils from sewage sludge disposal. *Sci. Total Environ.* **2019**, *671*, 411–420. [CrossRef]
9. Gieré, R.; Sommer, F.; Dietze, V.; Baum, A.; Maschowski, C. Tire-wear particles as a major component of microplastics in the environment. In Proceedings of the GSA Annual Meeting, Indianapolis, IN, USA, 4–7 November 2018.
10. Cole, M.; Lindeque, P.; Halsband, C.; Galloway, T.S. Microplastics as contaminants in the marine environment: A review. *Mar. Pollut. Bull.* **2011**, *62*, 2588–2597. [CrossRef]
11. Blasing, M.; Amelung, W. Plastics in soil: Analytical methods and possible sources. *Sci. Total Environ.* **2018**, *612*, 422–435. [CrossRef]
12. Lebreton, L.C.M.; van der Zwet, J.; Damsteeg, J.W.; Slat, B.; Andrady, A.; Reisser, J. River plastic emissions to the world's oceans. *Nat. Commun.* **2017**, *8*, 15611. [CrossRef]
13. Guo, J.J.; Huang, X.P.; Xiang, L.; Wang, Y.Z.; Li, Y.W.; Li, H.; Cai, Q.Y.; Mo, C.H.; Wong, M.H. Source, migration and toxicology of microplastics in soil. *Environ. Int.* **2020**, *137*, 105263. [CrossRef] [PubMed]
14. Leslie, H.A.; Brandsma, S.H.; van Velzen, M.J.; Vethaak, A.D. Microplastics en route: Field measurements in the Dutch river delta and Amsterdam canals, wastewater treatment plants, North Sea sediments and biota. *Environ. Int.* **2017**, *101*, 133–142. [CrossRef] [PubMed]
15. Cole, M.; Coppock, R.; Lindeque, P.K.; Altin, D.; Reed, S.; Pond, D.W.; Sorensen, L.; Galloway, T.S.; Booth, A.M. Effects of Nylon Microplastic on Feeding, Lipid Accumulation, and Moulting in a Coldwater Copepod. *Environ. Sci. Technol.* **2019**, *53*, 7075–7082. [CrossRef] [PubMed]
16. Lee, W.S.; Cho, H.J.; Kim, E.; Huh, Y.H.; Kim, H.J.; Kim, B.; Kang, T.; Lee, J.S.; Jeong, J. Bioaccumulation of polystyrene nanoplastics and their effect on the toxicity of Au ions in zebrafish embryos. *Nanoscale* **2019**, *11*, 3173–3185. [CrossRef]
17. Deng, Y.; Zhang, Y.; Lemos, B.; Ren, H. Tissue accumulation of microplastics in mice and biomarker responses suggest widespread health risks of exposure. *Sci. Rep.* **2017**, *7*, 46687. [CrossRef]
18. Trevisan, R.; Voy, C.; Chen, S.; Di Giulio, R.T. Nanoplastics Decrease the Toxicity of a Complex PAH Mixture but Impair Mitochondrial Energy Production in Developing Zebrafish. *Environ. Sci. Technol.* **2019**, *53*, 8405–8415. [CrossRef]
19. Wang, X.; Zheng, H.; Zhao, J.; Luo, X.; Wang, Z.; Xing, B. Photodegradation Elevated the Toxicity of Polystyrene Microplastics to Grouper (Epinephelus moara) through Disrupting Hepatic Lipid Homeostasis. *Environ. Sci. Technol.* **2020**, *54*, 6202–6212. [CrossRef]
20. Savoca, S.; Capillo, G.; Mancuso, M.; Bottari, T.; Crupi, R.; Branca, C.; Romano, V.; Faggio, C.; D'Angelo, G.; Spano, N. Microplastics occurrence in the Tyrrhenian waters and in the gastrointestinal tract of two congener species of seabreams. *Environ. Toxicol. Pharmacol.* **2019**, *67*, 35–41. [CrossRef]
21. Fournier, S.B.; D'Errico, J.N.; Adler, D.S.; Kollontzi, S.; Goedken, M.J.; Fabris, L.; Yurkow, E.J.; Stapleton, P.A. Nanopolystyrene translocation and fetal deposition after acute lung exposure during late-stage pregnancy. *Part. Fibre Toxicol.* **2020**, *17*, 55. [CrossRef]
22. Holloczki, O.; Gehrke, S. Can Nanoplastics Alter Cell Membranes? *Chemphyschem* **2020**, *21*, 9–12. [CrossRef] [PubMed]
23. Wang, J.; Peng, J.; Tan, Z.; Gao, Y.; Zhan, Z.; Chen, Q.; Cai, L. Microplastics in the surface sediments from the Beijiang River littoral zone: Composition, abundance, surface textures and interaction with heavy metals. *Chemosphere* **2017**, *171*, 248–258. [CrossRef] [PubMed]
24. Diepens, N.J.; Koelmans, A.A. Accumulation of Plastic Debris and Associated Contaminants in Aquatic Food Webs. *Environ. Sci. Technol.* **2018**, *52*, 8510–8520. [CrossRef]
25. Wardrop, P.; Shimeta, J.; Nugegoda, D.; Morrison, P.D.; Miranda, A.; Tang, M.; Clarke, B.O. Chemical Pollutants Sorbed to Ingested Microbeads from Personal Care Products Accumulate in Fish. *Environ. Sci. Technol.* **2016**, *50*, 4037–4044. [CrossRef]

26. Ebere, E.C.; Ngozi, V.E. Microplastics, an emerging concern: A review of analytical techniques for detecting and quantifying MPs. *Anal. Methods Environ. Chem. J.* **2019**, *2*, 13–30.
27. González-Pleiter, M.; Edo, C.; Aguilera, Á.; Viúdez-Moreiras, D.; Pulido-Reyes, G.; González-Toril, E.; Osuna, S.; de Diego-Castilla, G.; Leganés, F.; Fernández-Piñas, F.; et al. Occurrence and transport of microplastics sampled within and above the planetary boundary layer. *Sci. Total. Environ.* **2021**, *761*, 143213. [CrossRef]
28. Zhang, Y.; Gao, T.; Kang, S.; Allen, S.; Luo, X.; Allen, D. Microplastics in glaciers of the Tibetan Plateau: Evidence for the long-range transport of microplastics. *Sci. Total Environ.* **2021**, *758*, 143634. [CrossRef]
29. Rubio, L.; Marcos, R.; Hernandez, A. Potential adverse health effects of ingested micro- and nanoplastics on humans. Lessons learned from in vivo and in vitro mammalian models. *J. Toxicol. Environ. Health B Crit. Rev.* **2020**, *23*, 51–68. [CrossRef]
30. Dris, R.; Gasperi, J.; Rocher, V.; Saad, M.; Renault, N.; Tassin, B. Microplastic contamination in an urban area: A case study in Greater Paris. *Environ. Chem.* **2015**, *12*, 592–599. [CrossRef]
31. Dris, R.; Gasperi, J.; Mirande, C.; Mandin, C.; Guerrouache, M.; Langlois, V.; Tassin, B. A first overview of textile fibers, including microplastics, in indoor and outdoor environments. *Environ. Pollut.* **2017**, *221*, 453–458. [CrossRef] [PubMed]
32. Dris, R.; Gasperi, J.; Saad, M.; Mirande, C.; Tassin, B. Synthetic fibers in atmospheric fallout: A source of microplastics in the environment? *Mar. Pollut. Bull.* **2016**, *104*, 290–293. [CrossRef] [PubMed]
33. Cai, L.; Wang, J.; Peng, J.; Tan, Z.; Zhan, Z.; Tan, X.; Chen, Q. Characteristic of microplastics in the atmospheric fallout from Dongguan city, China: Preliminary research and first evidence. *Environ. Sci. Pollut. Res. Int.* **2017**, *24*, 24928–24935. [CrossRef] [PubMed]
34. Zhou, Q.; Tian, C.; Luo, Y. Various forms and deposition fluxes of microplastics identified in the coastal urban atmosphere. *Chin. Sci. Bull.* **2017**, *62*, 3902–3909. [CrossRef]
35. Kaya, T.A.; Yurtsever, M.; Çiftçi Bayraktar, S. Ubiquitous exposure to microfiber pollution in the air. *Eur. Phys. J. Plus* **2018**, *133*, 488. [CrossRef]
36. Catarino, A.I.; Macchia, V.; Sanderson, W.G.; Thompson, R.C.; Henry, T.B. Low levels of microplastics (MP) in wild mussels indicate that MP ingestion by humans is minimal compared to exposure via household fibres fallout during a meal. *Environ. Pollut.* **2018**, *237*, 675–684. [CrossRef] [PubMed]
37. Stanton, T.; Johnson, M.; Nathanail, P.; MacNaughtan, W.; Gomes, R.L. Freshwater and airborne textile fibre populations are dominated by 'natural', not microplastic, fibres. *Sci. Total. Environ.* **2019**, *666*, 377–389. [CrossRef]
38. Liu, K.; Wang, X.; Fang, T.; Xu, P.; Zhu, L.; Li, D. Source and potential risk assessment of suspended atmospheric microplastics in Shanghai. *Sci. Total Environ.* **2019**, *675*, 462–471. [CrossRef] [PubMed]
39. Liu, K.; Wang, X.; Wei, N.; Song, Z.; Li, D. Accurate quantification and transport estimation of suspended atmospheric microplastics in megacities: Implications for human health. *Environ. Int.* **2019**, *132*, 105127. [CrossRef]
40. Abbasi, S.; Keshavarzi, B.; Moore, F.; Turner, A.; Kelly, F.J.; Dominguez, A.O.; Jaafarzadeh, N. Distribution and potential health impacts of microplastics and microrubbers in air and street dusts from Asaluyeh County, Iran. *Environ. Pollut.* **2019**, *244*, 153–164. [CrossRef]
41. Liu, K.; Wu, T.; Wang, X.; Song, Z.; Zong, C.; Wei, N.; Li, D. Consistent Transport of Terrestrial Microplastics to the Ocean through Atmosphere. *Environ. Sci. Technol.* **2019**, *53*, 10612–10619. [CrossRef]
42. Allen, S.; Allen, D.; Phoenix, V.R.; Le Roux, G.; Durántez Jiménez, P.; Simonneau, A.; Binet, S.; Galop, D. Atmospheric transport and deposition of microplastics in a remote mountain catchment. *Nat. Geosci.* **2019**, *12*, 339–344. [CrossRef]
43. Klein, M.; Fischer, E.K. Microplastic abundance in atmospheric deposition within the Metropolitan area of Hamburg, Germany. *Sci. Total Environ.* **2019**, *685*, 96–103. [CrossRef] [PubMed]
44. Vianello, A.; Jensen, R.L.; Liu, L.; Vollertsen, J. Simulating human exposure to indoor airborne microplastics using a Breathing Thermal Manikin. *Sci. Rep.* **2019**, *9*, 8670. [CrossRef] [PubMed]
45. Wright, S.L.; Ulke, J.; Font, A.; Chan, K.L.A.; Kelly, F.J. Atmospheric microplastic deposition in an urban environment and an evaluation of transport. *Environ. Int.* **2020**, *136*, 105411. [CrossRef]
46. Wang, X.; Li, C.; Liu, K.; Zhu, L.; Song, Z.; Li, D. Atmospheric microplastic over the South China Sea and East Indian Ocean: Abundance, distribution and source. *J. Hazard. Mater.* **2020**, *389*, 121846. [CrossRef]
47. Li, Y.; Shao, L.; Wang, W.; Zhang, M.; Feng, X.; Li, W.; Zhang, D. Airborne fiber particles: Types, size and concentration observed in Beijing. *Sci. Total Environ.* **2020**, *705*, 135967. [CrossRef] [PubMed]
48. Bergmann, M.; Mützel, S.; Primpke, S.; Tekman, M.B.; Trachsel, J.; Gerdts, G. White and wonderful?Microplastics prevail in snow from the Alps to the Arctic. *Sci. Adv.* **2019**, *5*, eaax1157. [CrossRef] [PubMed]
49. Prata, J.C. Airborne microplastics: Consequences to human health? *Environ. Pollut.* **2018**, *234*, 115–126. [CrossRef]
50. Boucher, J.; Friot, D. *Primary Microplastics in the Oceans: A Global Evaluation*; IUCN: Gland, Switzerland, 2017; p. 43.
51. Fernando, H.J.S.; Lee, S.M.; Anderson, J.; Princevac, M.; Pardyjak, E.; Grossman-Clarke, S. Urban Fluid Mechanics: Air Circulation and Contaminant Dispersion in Cities. *Environ. Fluid Mech.* **2001**, *1*, 107–164. [CrossRef]
52. Ambrosini, R.; Azzoni, R.S.; Pittino, F.; Diolaiuti, G.; Franzetti, A.; Parolini, M. First evidence of microplastic contamination in the supraglacial debris of an alpine glacier. *Environ. Pollut.* **2019**, *253*, 297–301. [CrossRef] [PubMed]
53. Gasperi, J.; Wright, S.L.; Dris, R.; Collard, F.; Mandin, C.; Guerrouache, M.; Langlois, V.; Kelly, F.J.; Tassin, B. Microplastics in air: Are we breathing it in? *Curr. Opin. Environ. Sci. Health* **2018**, *1*, 1–5. [CrossRef]

54. Napper, I.E.; Thompson, R.C. Release of synthetic microplastic plastic fibres from domestic washing machines: Effects of fabric type and washing conditions. *Mar. Pollut. Bull.* **2016**, *112*, 39–45. [CrossRef] [PubMed]
55. De Falco, F.; Cocca, M.; Avella, M.; Thompson, R.C. Microfiber Release to Water, Via Laundering, and to Air, via Everyday Use: A Comparison between Polyester Clothing with Differing Textile Parameters. *Environ. Sci. Technol.* **2020**, *54*, 3288–3296. [CrossRef] [PubMed]
56. Li, G.; Li, G.; Li, Y.; Huang, J.; Zhao, J. Analysis of the Status Quo of China's Anti-pandemic Textiles and Suggestions fo Innovative Development. *China Text. Lead.* **2021**, *02*, 83–87.
57. Kole, P.J.; Lohr, A.J.; Van Belleghem, F.; Ragas, A.M.J. Wear and Tear of Tyres: A Stealthy Source of Microplastics in the Environment. *Int. J. Environ. Res. Public Health* **2017**, *14*, 1265. [CrossRef] [PubMed]
58. Eisentraut, P.; Dümichen, E.; Ruhl, A.S.; Jekel, M.; Albrecht, M.; Gehde, M.; Braun, U. Two Birds with One Stone—Fast and Simultaneous Analysis of Microplastics: Microparticles Derived from Thermoplastics and Tire Wear. *Environ. Sci. Technol. Lett.* **2018**, *5*, 608–613. [CrossRef]
59. Wagner, S.; Hüffer, T.; Klöckner, P.; Wehrhahn, M.; Hofmann, T.; Reemtsma, T. Tire wear particles in the aquatic environment—A review on generation, analysis, occurrence, fate and effects. *Water Res.* **2018**, *139*, 83–100. [CrossRef]
60. Baensch-Baltruschat, B.; Kocher, B.; Stock, F.; Reifferscheid, G. Tyre and road wear particles (TRWP)—A review of generation, properties, emissions, human health risk, ecotoxicity, and fate in the environment. *Sci. Total. Environ.* **2020**, *733*, 137823. [CrossRef]
61. Sommer, F.; Dietze, V.; Baum, A.; Sauer, J.; Gilge, S.; Maschowski, C.; Gieré, R. Tire Abrasion as a Major Source of Microplastics in the Environment. *Aerosol Air Qual. Res.* **2018**, *18*, 2014–2028. [CrossRef]
62. Liu, C.; Li, J.; Zhang, Y.; Wang, L.; Deng, J.; Gao, Y.; Yu, L.; Zhang, J.; Sun, H. Widespread distribution of PET and PC microplastics in dust in urban China and their estimated human exposure. *Environ. Int.* **2019**, *128*, 116–124. [CrossRef]
63. Kausar, A. A review of filled and pristine polycarbonate blends and their applications. *J. Plast. Film. Sheeting* **2017**, *34*, 60–97. [CrossRef]
64. Kaur, S.; Nieuwenhuijsen, M.J.; Colvile, R.N. Fine particulate matter and carbon monoxide exposure concentrations in urban street transport microenvironments. *Atmos. Environ.* **2007**, *41*, 4781–4810. [CrossRef]
65. Zhao, S.; Zhu, L.; Wang, T.; Li, D. Suspended microplastics in the surface water of the Yangtze Estuary System, China: First observations on occurrence, distribution. *Mar. Pollut. Bull.* **2014**, *86*, 562–568. [CrossRef] [PubMed]
66. Browne Mark, A.; Galloway Tamara, S.; Thompson Richard, C. Spatial patterns of plastic debris along Estuarine shorelines. *Environ. Sci. Technol.* **2010**, *44*, 3404–3409. [CrossRef] [PubMed]
67. Alzona, J.; Cohen, B.L.; Rudolph, H.e.a. Indoor-outdoor relationships for airborne particulate matter of outdoor origin. *Atmos. Environ.* **1979**, *13*, 55–60. [CrossRef]
68. Donaldson, K.; Stone, V.; Gilmour, P.S.; Brown, D.M.; MacNee, W. Ultrafine particles: Mechanisms of lung injury. *Philos. Trans. R. Soc. A Math. Phys. Eng. Sci.* **2000**, *358*, 2741–2749. [CrossRef]
69. Enyoh, C.E.; Verla, A.W.; Verla, E.N.; Ibe, F.C.; Amaobi, C.E. Airborne microplastics: A review study on method for analysis, occurrence, movement and risks. *Environ. Monit. Assess.* **2019**, *191*, 668. [CrossRef] [PubMed]
70. Rist, S.; Carney Almroth, B.; Hartmann, N.B.; Karlsson, T.M. A critical perspective on early communications concerning human health aspects of microplastics. *Sci. Total Environ.* **2018**, *626*, 720–726. [CrossRef] [PubMed]
71. Brown, D.M.; Wilson, M.R.; MacNee, W.; Stone, V.; Donaldson, K. Size-dependent proinflammatory effects of ultrafine polystyrene particles: A role for surface area and oxidative stress in the enhanced activity of ultrafines. *Toxicol. Appl. Pharmacol.* **2001**, *175*, 191–199. [CrossRef]
72. Kelly, F.J.; Fussell, J.C. Size, source and chemical composition as determinants of toxicity attributable to ambient particulate matter. *Atmos. Environ.* **2012**, *60*, 504–526. [CrossRef]
73. Valavanidis, A.; Vlachogianni, T.; Fiotakis, K.; Loridas, S. Pulmonary oxidative stress, inflammation and cancer: Respirable particulate matter, fibrous dusts and ozone as major causes of lung carcinogenesis through reactive oxygen species mechanisms. *Int. J. Environ. Res. Public Health* **2013**, *10*, 3886–3907. [CrossRef] [PubMed]
74. Xu, H.; Hoet, P.H.M.; Nemery, B. In vitro toxicity assessment of polyvinyl chloride particles and comparison of six cellular systems. *J. Toxicol. Environ. Health Part A* **2002**, *65*, 1141–1149. [CrossRef] [PubMed]
75. Lim, S.L.; Ng, C.T.; Zou, L.; Lu, Y.; Chen, J.; Bay, B.H.; Shen, H.-M.; Ong, C.N. Targeted metabolomics reveals differential biological effects of nanoplastics and nanoZnO in human lung cells. *Nanotoxicology* **2019**, *13*, 1117–1132. [CrossRef]
76. Dong, C.-D.; Chen, C.-W.; Chen, Y.-C.; Chen, H.-H.; Lee, J.-S.; Lin, C.-H. Polystyrene microplastic particles: In vitro pulmonary toxicity assessment. *J. Hazard. Mater.* **2020**, *385*, 121575. [CrossRef]
77. Paget, V.; Dekali, S.; Kortulewski, T.; Grall, R.; Gamez, C.; Blazy, K.; Aguerre-Chariol, O.; Chevillard, S.; Braun, A.; Rat, P.; et al. Specific Uptake and Genotoxicity Induced by Polystyrene Nanobeads with Distinct Surface Chemistry on Human Lung Epithelial Cells and Macrophages. *PLoS ONE* **2015**, *10*, e0123297. [CrossRef]
78. Kim, J.; Chankeshwara, S.V.; Thielbeer, F.; Jeong, J.; Donaldson, K.; Bradley, M.; Cho, W.-S. Surface charge determines the lung inflammogenicity: A study with polystyrene nanoparticles. *Nanotoxicology* **2016**, *10*, 94–101. [CrossRef] [PubMed]
79. Agarwal, D.K.; Kaw, J.L.; Srivastava, S.P.; Seth, P.K. Some biochemical and histopathological changes induced by polyvinyl chloride dust in rat lung. *Environ. Res.* **1978**, *16*, 333–341. [CrossRef]
80. Atis, S.; Tutluoglu, B.; Levent, E.; Ozturk, C.; Tunaci, A.; Sahin, K.; Saral, A.; Oktay, I.; Kanik, A.; Nemery, B. The respiratory effects of occupational polypropylene flock exposure. *Eur. Respir. J.* **2005**, *25*, 110–117. [CrossRef]

81. Pimentel, J.C.; Avila, R.; Lourenco, A.G. Respiratory disease caused by synthetic fibres: A new occupational disease. *Thorax* **1975**, *30*, 204. [CrossRef]
82. Porter, D.W.; Castranova, V.; Robinson, V.A.; Hubbs, A.F.; Mercer, R.R.; Scabilloni, J.; Goldsmith, T.; Schwegler-Berry, D.; Battelli, L.; Washko, R.; et al. Acute inflammatory reaction in rats after intratracheal instillation of material collected from a nylon flocking plant. *J. Toxicol. Environ. Health A* **1999**, *57*, 25–45.
83. Xu, H.; Verbeken, E.; Vanhooren, H.M.; Nemery, B.; Hoet, P.H. Pulmonary toxicity of polyvinyl chloride particles after a single intratracheal instillation in rats. Time course and comparison with silica. *Toxicol. Appl. Pharmacol.* **2004**, *194*, 111–121. [CrossRef] [PubMed]
84. Cavallo, D.; Campopiano, A.; Cardinali, G.; Casciardi, S.; De Simone, P.; Kovacs, D.; Perniconi, B.; Spagnoli, G.; Ursini, C.L.; Fanizza, C. Cytotoxic and oxidative effects induced by man-made vitreous fibers (MMVFs) in a human mesothelial cell line. *Toxicology* **2004**, *201*, 219–229. [CrossRef]
85. Jones, A.E.; Watts, J.A.; Debelak, J.P.; Thornton, L.R.; Younger, J.G.; Kline, J.A. Inhibition of prostaglandin synthesis during polystyrene microsphere-induced pulmonary embolism in the rat. *AJP Lung Cell. Mol. Physiol.* **2003**, *284*, L1072–L1081. [CrossRef]
86. Canesi, L.; Ciacci, C.; Bergami, E.; Monopoli, M.P.; Dawson, K.A.; Papa, S.; Canonico, B.; Corsi, I. Evidence for immunomodulation and apoptotic processes induced by cationic polystyrene nanoparticles in the hemocytes of the marine bivalve Mytilus. *Mar. Environ. Res.* **2015**, *111*, 34–40. [CrossRef] [PubMed]
87. Hwang, J.; Choi, D.; Han, S.; Choi, J.; Hong, J. An assessment of the toxicity of polypropylene microplastics in human derived cells. *Sci. Total Environ.* **2019**, *684*, 657–669. [CrossRef] [PubMed]
88. Salim, S.Y.; Kaplan, G.G.; Madsen, K.L. Air pollution effects on the gut microbiota: A link between exposure and inflammatory disease. *Gut Microbes* **2014**, *5*, 215–219. [CrossRef]
89. Linares, V.; Belles, M.; Domingo, J.L. Human exposure to PBDE and critical evaluation of health hazards. *Arch. Toxicol* **2015**, *89*, 335–356. [CrossRef]
90. Mariana, M.; Feiteiro, J.; Verde, I.; Cairrao, E. The effects of phthalates in the cardiovascular and reproductive systems: A review. *Environ. Int.* **2016**, *94*, 758–776. [CrossRef]
91. Brun, N.R.; Koch, B.E.V.; Varela, M.; Peijnenburg, W.J.G.M.; Spaink, H.P.; Vijver, M.G. Nanoparticles induce dermal and intestinal innate immune system responses in zebrafish embryos. *Environ. Sci. Nano* **2018**, *5*, 904–916. [CrossRef]
92. Grandjean, P.; Landrigan, P.J. Neurobehavioural effects of developmental toxicity. *Lancet Neurol.* **2014**, *13*, 330–338. [CrossRef]
93. Peng, J.; Wang, J.; Cai, L. Current understanding of microplastics in the environment: Occurrence, fate, risks, and what we should do. *Integr. Environ. Assess. Manag.* **2017**, *13*, 476–482. [CrossRef]
94. Wright, S.L.; Kelly, F.J. Plastic and Human Health: A Micro Issue? *Environ. Sci. Technol.* **2017**, *51*, 6634. [CrossRef]
95. Brandts, I.; Teles, M.; Goncalves, A.P.; Barreto, A.; Franco-Martinez, L.; Tvarijonaviciute, A.; Martins, M.A.; Soares, A.; Tort, L.; Oliveira, M. Effects of nanoplastics on Mytilus galloprovincialis after individual and combined exposure with carbamazepine. *Sci. Total Environ.* **2018**, *643*, 775–784. [CrossRef]
96. Avio, C.G.; Gorbi, S.; Milan, M.; Benedetti, M.; Fattorini, D.; d'Errico, G.; Pauletto, M.; Bargelloni, L.; Regoli, F. Pollutants bioavailability and toxicological risk from microplastics to marine mussels. *Environ. Pollut.* **2015**, *198*, 211–222. [CrossRef]
97. Dong, S.; Qu, M.; Rui, Q.; Wang, D. Combinational effect of titanium dioxide nanoparticles and nanopolystyrene particles at environmentally relevant concentrations on nematode Caenorhabditis elegans. *Ecotoxicol. Environ. Saf.* **2018**, *161*, 444–450. [CrossRef] [PubMed]
98. Chen, Q.; Yin, D.; Jia, Y.; Schiwy, S.; Legradi, J.; Yang, S.; Hollert, H. Enhanced uptake of BPA in the presence of nanoplastics can lead to neurotoxic effects in adult zebrafish. *Sci. Total Environ.* **2017**, *609*, 1312–1321. [CrossRef]
99. Lang, I.A.; Galloway, T.S.; Scarlett, A.; Henley, W.E.; Depledge, M.; Wallace, R.B.; Melzer, D. Association of urinary bisphenol A concentration with medical disorders and laboratory abnormalities in adults. *JAMA* **2008**, *300*, 1303–1310. [CrossRef] [PubMed]
100. Munari, C.; Infantini, V.; Scoponi, M.; Rastelli, E.; Corinaldesi, C.; Mistri, M. Microplastics in the sediments of Terra Nova Bay (Ross Sea, Antarctica). *Mar. Pollut. Bull.* **2017**, *122*, 161–165. [CrossRef] [PubMed]
101. Lusher, A.L.; McHugh, M.; Thompson, R.C. Occurrence of microplastics in the gastrointestinal tract of pelagic and demersal fish from the English Channel. *Mar. Pollut. Bull.* **2013**, *67*, 94–99. [CrossRef]
102. Gallagher, A.; Rees, A.; Rowe, R.; Stevens, J.; Wright, P. Microplastics in the Solent estuarine complex, UK: An initial assessment. *Mar. Pollut. Bull.* **2016**, *102*, 243–249. [CrossRef]
103. Zhao, J.; Ran, W.; Teng, J.; Liu, Y.; Liu, H.; Yin, X.; Cao, R.; Wang, Q. Microplastic pollution in sediments from the Bohai Sea and the Yellow Sea, China. *Sci. Total Environ.* **2018**, *640-641*, 637–645. [CrossRef]
104. Imhof, H.K.; Laforsch, C.; Wiesheu, A.C.; Schmid, J.; Anger, P.M.; Niessner, R.; Ivleva, N.P. Pigments and plastic in limnetic ecosystems: A qualitative and quantitative study on microparticles of different size classes. *Water Res.* **2016**, *98*, 64–74. [CrossRef]
105. Lenz, R.; Enders, K.; Stedmon, C.A.; Mackenzie, D.M.A.; Nielsen, T.G. A critical assessment of visual identification of marine microplastic using Raman spectroscopy for analysis improvement. *Mar. Pollut. Bull.* **2015**, *100*, 82–91. [CrossRef]
106. Zhao, S.; Zhu, L.; Li, D. Microplastic in three urban estuaries, China. *Environ. Pollut.* **2015**, *206*, 597–604. [CrossRef]
107. Erni-Cassola, G.; Gibson, M.I.; Thompson, R.C.; Christie-Oleza, J.A. Lost, but Found with Nile Red: A Novel Method for Detecting and Quantifying Small Microplastics (1 mm to 20 mum) in Environmental Samples. *Environ. Sci. Technol.* **2017**, *51*, 13641–13648. [CrossRef]

108. Huppertsberg, S.; Knepper, T.P. Instrumental analysis of microplastics-benefits and challenges. *Anal. Bioanal. Chem.* **2018**, *410*, 6343–6352. [CrossRef]
109. Majewsky, M.; Bitter, H.; Eiche, E.; Horn, H. Determination of microplastic polyethylene (PE) and polypropylene (PP) in environmental samples using thermal analysis (TGA-DSC). *Sci. Total Environ.* **2016**, *568*, 507–511. [CrossRef]
110. Rocha-Santos, T.; Duarte, A.C. A critical overview of the analytical approaches to the occurrence, the fate and the behavior of microplastics in the environment. *TrAC Trends Anal. Chem.* **2015**, *65*, 47–53. [CrossRef]
111. Shim, W.J.; Hong, S.H.; Eo, S.E. Identification methods in microplastic analysis: A review. *Anal. Methods* **2017**, *9*, 1384–1391. [CrossRef]
112. Silva, A.B.; Bastos, A.S.; Justino, C.I.L.; da Costa, J.P.; Duarte, A.C.; Rocha-Santos, T.A.P. Microplastics in the environment: Challenges in analytical chemistry—A review. *Anal. Chim. Acta* **2018**, *1017*, 1–19. [CrossRef] [PubMed]
113. Nuelle, M.-T.; Dekiff, J.H.; Remy, D.; Fries, E. A new analytical approach for monitoring microplastics in marine sediments. *Environ. Pollut.* **2014**, *184*, 161–169. [CrossRef] [PubMed]
114. Hanvey, J.S.; Lewis, P.J.; Lavers, J.L.; Crosbie, N.D.; Pozo, K.; Clarke, B.O. A review of analytical techniques for quantifying microplastics in sediments. *Anal. Methods* **2017**, *9*, 1369–1383. [CrossRef]
115. Dekiff, J.H.; Remy, D.; Klasmeier, J.; Fries, E. Occurrence and spatial distribution of microplastics in sediments from Norderney. *Environ. Pollut.* **2014**, *186*, 248–256. [CrossRef] [PubMed]
116. Dumichen, E.; Eisentraut, P.; Bannick, C.G.; Barthel, A.K.; Senz, R.; Braun, U. Fast identification of microplastics in complex environmental samples by a thermal degradation method. *Chemosphere* **2017**, *174*, 572–584. [CrossRef]
117. Cooper, D.A.; Corcoran, P.L. Effects of mechanical and chemical processes on the degradation of plastic beach debris on the island of Kauai, Hawaii. *Mar. Pollut. Bull.* **2010**, *60*, 650–654. [CrossRef] [PubMed]
118. Ding, J.; Li, J.; Sun, C.; Jiang, F.; Ju, P.; Qu, L.; Zheng, Y.; He, C. Detection of microplastics in local marine organisms using a multi-technology system. *Anal. Methods* **2019**, *11*, 78–87. [CrossRef]
119. Kirstein, I.V.; Kirmizi, S.; Wichels, A.; Garin-Fernandez, A.; Erler, R.; Löder, M.; Gerdts, G. Dangerous hitchhikers? Evidence for potentially pathogenic Vibrio spp. on microplastic particles. *Mar. Environ. Res.* **2016**, *120*, 1–8. [CrossRef]
120. Weidner, S.M.; Sarah, T. Mass spectrometry of synthetic polymers. *Anal. Chem.* **2010**, *82*, 4811–4829. [CrossRef]
121. Foerch, R.; Beamson, G.; Briggs, D. XPS valence band analysis of plasma-treated polymers. *Surf. Interface Anal.* **1991**, *17*, 842–846. [CrossRef]
122. Magri, D.; Sanchez-Moreno, P.; Caputo, G.; Gatto, F.; Veronesi, M.; Bardi, G.; Catelani, T.; Guarnieri, D.; Athanassiou, A.; Pompa, P.P.; et al. Laser Ablation as a Versatile Tool To Mimic Polyethylene Terephthalate Nanoplastic Pollutants: Characterization and Toxicology Assessment. *ACS Nano* **2018**, *12*, 7690–7700. [CrossRef]
123. Malyuskin, O. Microplastic Detection in Soil and Water Using Resonance Microwave Spectroscopy: A Feasibility Study. *IEEE Sens. J.* **2020**, *20*, 14817–14826. [CrossRef]

MDPI

Review

Impact of Microplastics and Nanoplastics on Human Health

Maxine Swee-Li Yee [1,*], Ling-Wei Hii [2,3,4], Chin King Looi [2,3], Wei-Meng Lim [2,4], Shew-Fung Wong [5,6], Yih-Yih Kok [5,7], Boon-Keat Tan [5,6], Chiew-Yen Wong [5,7] and Chee-Onn Leong [2,4,*]

1 Centre of Nanotechnology and Advanced Materials, Faculty of Engineering, University of Nottingham Malaysia Campus, Jalan Broga, Semenyih 43500, Malaysia
2 Center for Cancer and Stem Cell Research, Institute for Research, Development and Innovation (IRDI), International Medical University, Kuala Lumpur 57000, Malaysia; lingweihii@imu.edu.my (L.-W.H.); 00000028295@student.imu.edu.my (C.K.L.); weimeng_lim@imu.edu.my (W.-M.L.)
3 School of Postgraduate Studies, International Medical University, Kuala Lumpur 57000, Malaysia
4 School of Pharmacy, International Medical University, Kuala Lumpur 57000, Malaysia
5 Center for Environmental and Population Health, Institute for Research, Development and Innovation (IRDI), International Medical University, Kuala Lumpur 57000, Malaysia; shewfung_wong@imu.edu.my (S.-F.W.); yihyih_kok@imu.edu.my (Y.-Y.K.); boonkeat_tan@imu.edu.my (B.-K.T.); wongchiewyen@imu.edu.my (C.-Y.W.)
6 School of Medicine, International Medical University, Kuala Lumpur 57000, Malaysia
7 School of Health Sciences, International Medical University, 126, Jalan Jalil Perkasa 19, Bukit Jalil, Kuala Lumpur 57000, Malaysia
* Correspondence: Maxine.Yee@nottingham.edu.my (M.S.-L.Y.); cheeonn_leong@imu.edu.my (C.-O.L.)

Abstract: Plastics have enormous impacts to every aspect of daily life including technology, medicine and treatments, and domestic appliances. Most of the used plastics are thrown away by consumers after a single use, which has become a huge environmental problem as they will end up in landfill, oceans and other waterways. These plastics are discarded in vast numbers each day, and the breaking down of the plastics from micro- to nano-sizes has led to worries about how toxic these plastics are to the environment and humans. While, there are several earlier studies reported the effects of micro- and nano-plastics have on the environment, there is scant research into their impact on the human body at subcellular or molecular levels. In particular, the potential of how nano-plastics move through the gut, lungs and skin epithelia in causing systemic exposure has not been examined thoroughly. This review explores thoroughly on how nanoplastics are created, how they behave/breakdown within the environment, levels of toxicity and pollution of these nanoplastics, and the possible health impacts on humans, as well as suggestions for additional research. This paper aims to inspire future studies into core elements of micro- and nano-plastics, the biological reactions caused by their specific and unusual qualities.

Keywords: nanoplastics; nanotoxicity; nanomaterials; toxicology; plastics; health impacts; environmental impacts; pollution

Citation: Yee, M.S.-L.; Hii, L.-W.; Looi, C.K.; Lim, W.-M.; Wong, S.-F.; Kok, Y.-Y.; Tan, B.-K.; Wong, C.-Y.; Leong, C.-O. Impact of Microplastics and Nanoplastics on Human Health. *Nanomaterials* **2021**, *11*, 496. https://doi.org/10.3390/nano11020496

Academic Editor: Vivian Hsiu-Chuan Liao

Received: 24 January 2021
Accepted: 13 February 2021
Published: 16 February 2021

1. Introduction

Worldwide, plastic use is growing year by year, with current figures showing plastic production exceeding 368 million tons in 2019 [1]. Furthermore, the waste produced is not disposed in the correct way. As plastic pervades every aspect of life and then breaks down into smaller particles, the possible impacts of micro- and nano-plastics on the human body and the environment are of global concerns [2].

Plastics are made of natural materials that have undergone several chemical processes and physical reactions. The main processes used are polymerization and polycondensation, during which the core elements are fundamentally transformed into polymer chains [3]. This process is rarely reversible; the plastics must go through more chemical processes in order to be recycled into new types of plastics [4]. The use of industrial additives, such as pigments, plasticizers and stabilizers, allows plastics to be engineered to various application requirements [5]. Due to the chemical stability of the conventional plastics, environmental accumulation is on the rise.

Once disposed of, plastic waste is exposed to biological, chemical and environmental elements, and will break down into huge amounts of microplastics (measuring < 5 mm) and nanoplastics (<0.1 μm) [6–9]. Previous studies into plastic waste have looked at the effect of microplastics on the environment, and this has been widely discussed in both the scientific community and the media, including appeals for institutional policies to implement hazardous classifications of the harmful plastics [10]. However, there is very little research into the quantities, varieties and toxicity of nanoplastics and the impacts on human health. A solitary microplastic particle will break down into billions of nanoplastic particles suggesting that nanoplastic pollution will be prevalent across the globe [11–13]. It is probable that nanoplastics are more damaging than microplastics as they are small enough to permeate through biological membranes. Despite of this, the potential human health effects of nanoplastic exposure remains under-studied.

Hence, this review aims to explore thoroughly on how nanoplastics are created, how they behave/breakdown within the environment, levels of toxicity and pollution of these nanoplastics, and the possible health impacts on humans, as well as suggestions for additional research.

2. Sources and Fate of Microplastics and Nanoplastics in the Environment

Over 80% of microplastics are produced on land, with less than 20% originating from the sea. As microplastics are uniquely light, indestructible and able to float, they can travel far across the globe [14,15]. The majority of plastics that polluting the marine environment originate from terrestrial sources, fishing and other aquaculture activities, and from coastal tourism [14,16,17]; indeed, it is estimated that over 800 million tons of plastics in the sea originated from land [18]. As micro- and nanoplastics are incredibly small, wastewater treatment processes are not able to filter them out and, therefore, such plastic particles will be introduced into the rivers and oceans, as well as the fresh water supply system [19]. Furthermore, micro- and nanoplastics are present in soil and, through natural erosion, they will also get into rivers and oceans in this way [20]. Figures from the United Nations Environment Program show that 275 million tons of plastic waste were produced in 2010 with an estimated 4.8–12.7 million tons leaching way into the water systems [21].

Micro- and nanoplastics are generated from both primary and secondary sources (Figure 1) [22]. Primary sources are those that deliberately created micro- and nanoplastics for consumer and industrial uses, such as exfoliants in cleansers, cosmetics, as drug delivery particles in medicines, and industrial air blasting [15]. Macroplastic products that disintegrate into micron-sized and smaller particles are the secondary source of micro- and nanoplastics; they occur both terrestrially and in the aquatic environment [15].

Plastics can break down into micro- and nanoplastics in various ways which can be defined as either via biodegradation or non-biodegradation process (Figure 1). Processes such as thermal degradation, physical degradation, photodegradation, thermo-oxidative degradation, and hydrolysis are all examples of non-biodegradation [23–26]. Thermal, or heat degradation is a non-natural, commercial process whilst physical degradation, through weathering, causes larger plastics to be fragmented into smaller pieces. On the other hand, hydrolysis and photodegradation are naturally occurring chemical processes which use water molecules, and UV-visible light, respectively, to break down the chemical bonds in plastics, converting them into monomeric forms. Plastic non-biodegradation processes decompose polymeric structures, altering their mechanical properties, and increasing their specific surface area, resulting in enhanced physical-chemical reactions and interactions with microorganisms [27].

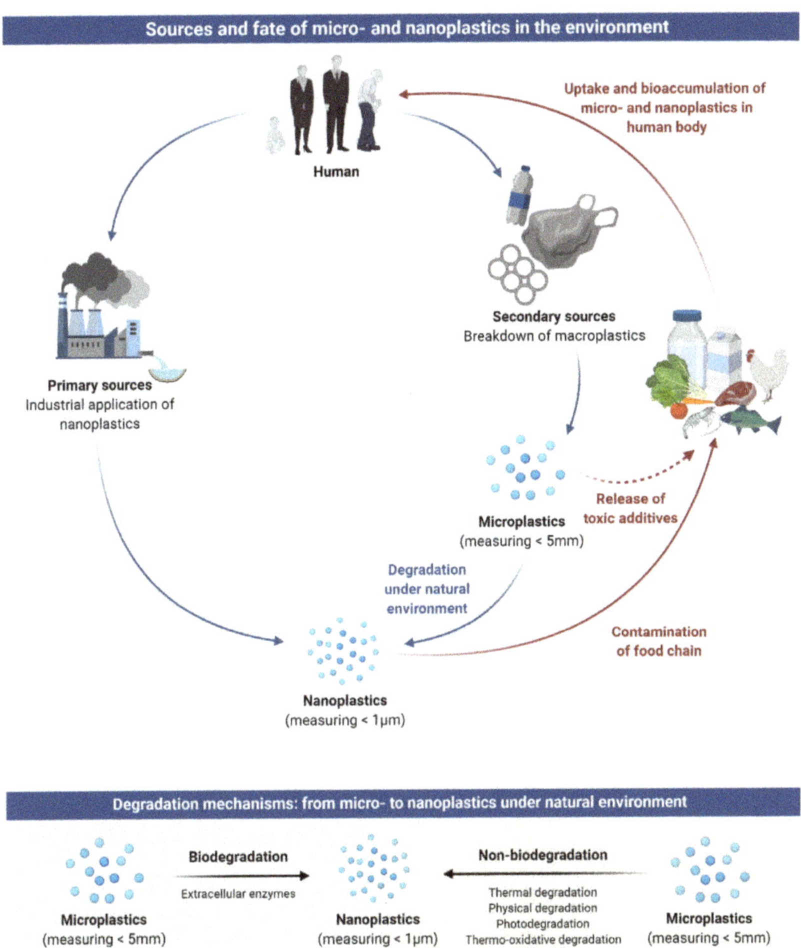

Figure 1. Sources and fate of micro- and nanoplastics in the environment. Micro- and nanoplastics are generated from primary and secondary sources through consumers and industries. Macroplastic products that disintegrate into micron-sized can break down into nanoplastics via biodegradation or non-biodegradation process. Both micro- and nanoplastics can occur in both aquatic and terrestrial environment, and eventually enter the food chain and water supplies, leading to the uptake and bioaccumulation of these plastic particles in the human body.

Environmental bacteria and other microorganisms can also mediate the biodegradation process of plastics [28]. Extracellular enzymes produced by these living organisms have the ability to break down the chemical bonds within plastics [29]. Smaller plastic particles, with altered molecular structures are created in this process, eventually resulting in nano-sized plastics; a single gram of macroplastic can yield billions of nanoplastic particles with greatly increased surface area. As there is a vast quantity of plastic entering the oceans daily, it is clear that these nanoplastics must be present in enormous quantities in the marine environment.

In addition, plastic waste fragmentation is thought to occur faster on the coast than in the oceans. One of the primary methods of degradation of plastic is oxidation triggered by solar UV irradiation. This process speeds up on the coast when plastic is more directly exposed to UV radiation and higher temperatures than when it is in the oceans [30].

Further, plastic degradation is quickened in the presence of salt at these coastal areas [31]. In comparison to terrestrial ecosystems, high saline content, along with naturally occurring microorganisms in marine areas, will cause plastics to break down at a faster rate [20].

3. Occurrence of Microplastics and Nanoplastics in the Food Chain

As plastic waste increases, the presence of micro- and nanoplastics in the food chain creates a risk to human health [22,32,33]. Due to their wide bioavailability and ubiquity in both aquatic and terrestrial areas, it is highly probable that micro- and nanoplastics are present in many food products.

Several studies have shown that micro- and nanoplastics enter into the human food chain in various ways: Animals consuming them in their natural environment [34]; contamination during the food production processes [35]; and/or through leaching from plastic packaging of the food and drinks [36]. To date, micro- and nanoplastic fragments have been detected in honey, beer, salt, sugar, fish, shrimps and bivalves [37–42]. Experimental sampling using Fourier-transform infrared spectroscopy (FTIR) performed on tap, bottled and spring waters showed that microplastics are present in all these water sources. Tap water from 159 global sources was tested and 81% were found to contain microplastic particles measuring less than 5 mm [43]. Tests were conducted on 259 individual bottles of water from 11 different brands and 27 different batches, and the results demonstrated that 93% contained microplastic particles [36]. Statistics show the following average levels of microplastic pollution in food: seafood = 1.48 particles/g, sugar = 0.44 particles/g, honey = 0.10 particles/g, salt = 0.11 particles/g, alcohol = 32.27 particles/L, bottled water = 94.37 particles/L, tap water = 4.23 particles/L, and air = 9.80 particles/m^3 [9,44]. From these figures, it is possible to extrapolate that the average human is consuming around 39,000 to 52,000 microplastic particles per year, with age and gender impacting the total amount. If inhalation of plastic particles is included in the figures, then the amounts rise to between 74,000 and 121,000 particles per year. Further, an individual who only ingest bottled water is potentially consuming an extra 90,000 particles in comparison to people who only drink tap water, who will ingest only 4000 extra particles [44]. These results indicate that the human food chain is, indeed, a major source of microplastic consumption by humans.

There are currently no data regarding the presence of nanoplastics in food as the analytical tools are not yet available [9,45]. It appears clear, though, that nanoplastics will occur in the food chain due to the degradation of microplastic waste [25,45]. Scientific tests on polystyrene drinking cup lids showed that nanoplastics were formed over time as the material broke down [28]. There is also evidence to suggest that microbial degradation will occur in oceans due to the presence of hydrocarbon-degrading microorganisms that have been shown to flourish on plastic waste, forming a "plastisphere" ecosystem [26]. The huge scale of plastic waste dispersal in the oceans indicates that microplastics will continue to degrade once they enter the sea, forming more nanoplastic particles [46]. There are also a number of products that use commercially manufactured nanoplastics, and these will also become plastic waste in the seas and on land, and eventually find their way into the food supply chain [9,45].

4. Uptake and Bioaccumulation of Microplastics and Nanoplastics in the Human Body

There are three key routes for microplastics and nanoplastics to end up in the human body: Inhalation, ingestion and skin contact (Figure 2) [47,48]. Inhaled airborne microplastics originate from urban dust, and include synthetic textiles and rubber tyres [49]. As discussed above, microplastics will be ingested as they are prevalent in the food chain and water supplies [50]. While, the skin membrane was too fine for microplastics or nanoplastics to pass through, it is possible for them to enter through wounds, sweat glands or hair follicles [51]. Although all three routes contribute to the total amount of microplastics and nanoplastics present in the human body, it is the particles in seafood and the environment that constitute the greatest risk of absolute exposure. This is due to long-term weathering of polymers, leaching of polymer chemical additives, residual monomers, exposure to pollutants and pathogenic microorganisms all being active in these environments [52–56].

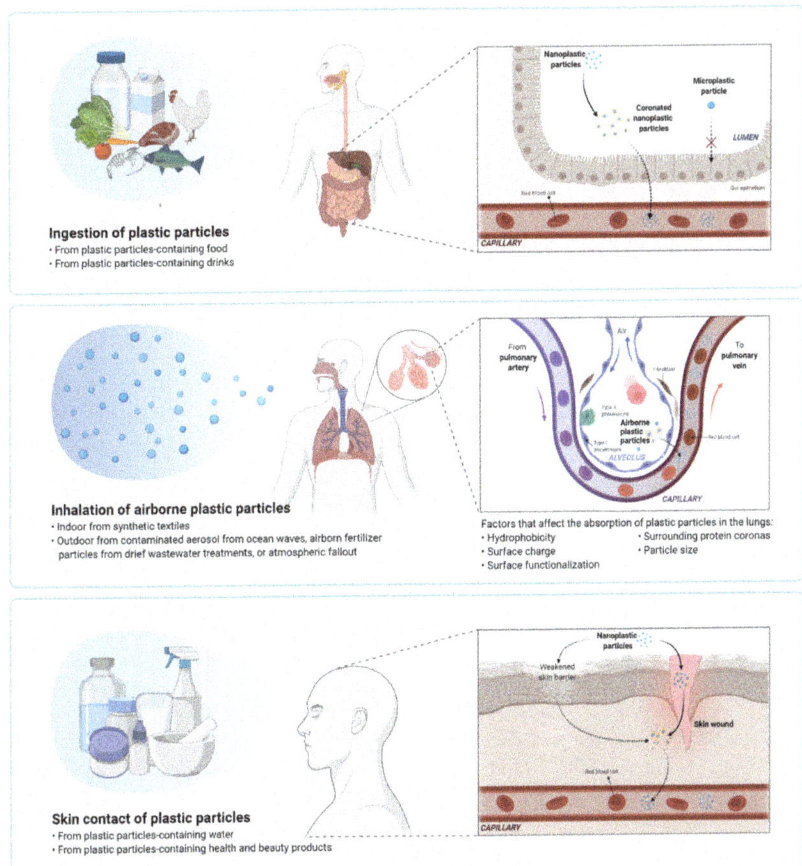

Figure 2. Routes of plastic particles entry into human body. There are three key routes for micro- and nanoplastics entry into the human body: Inhalation, ingestion and skin contact. Nanoplastics may interact with proteins, lipids, carbohydrates, nucleic acids, ions, and water in human body, leading to the formation of coronated nanoplastic particles for absorption. The plastic particles can enter human body through ingestion of contaminated food and water supplies, or inhalation of the airborne plastic particles that originate from synthetic textiles and polluted outdoor air. While, the skin membrane is too fine for these plastic particles to pass through, nanoplastics may penetrate through wound and weakened skin barrier, directly or indirectly.

4.1. *Gastric Exposure*

Recent studies into microplastics and nanoplastics exposure and toxicity have indicated that the most significant way humans consume plastic particles is via ingestion [57]. While, there are no studies looking specifically at nanoplastic toxicity in humans, there is research showing that microplastics are being ingested through food and drink [36]. The initial analysis of human stool samples showed that plastic particles were being excreted, which supports the theory that humans are ingesting these particles via food and water. These results, along with research into ingestion uptake in environmental models, clearly show that humans will be regularly consuming microplastics and nanoplastics [58]. However, no studies have yet investigated what happens to the micro- and nanoplastic particles once they enter the gastrointestinal (GI) tract. It would be pertinent to examine their route through the GI tract and whether particles remain in the gut lumen, or they translocate across the gut epithelia.

It is unlikely that microplastics are able to permeate at a paracellular level as the relevant pores at the tight junction channels have a maximum functional size of approximately 1.5 nm [59]. It is more likely that they enter through lymphatic tissue, and it is particularly possible that they enter via phagocytosis or endocytosis and infiltrate the microfold (M) cells in the Peyer's patches [45]. Following intraperitoneal injections in mice, the peritoneal macrophages were seen to phagocytose 1, 5 and 12 μm polymethacrylate and polystyrene particles [60,61]. Nevertheless, the results indicate that absorption via intestinal tracts in rodent models is low at 0.04–0.3% [61].

The potential of nanoplastics to permeate the gut epithelium, leading to systemic exposure in humans, is a significant issue. Historically, studies have used polystyrene nanoparticles for in vivo and in vitro tests on a variety of animals. The probable oral bioavailability level of 50 nm polystyrene nanoparticles is ten to one hundred times greater than the level of microplastics (2–7%) [62,63]. Similar to the results seen with microplastics, there is no straightforward correlation between the absorption, size and structure of nanoplastics [62]. Previous research has shown that the absorption rates of nanoparticles (50–500 nm) vary greatly across different in vitro intestinal models, with figures of 1.5–10% according to the size and chemical structure of the nanoparticles as well as the type of in vitro model used [62,64,65].

The lumen of the GI tract presents a challenge when researching nanoplastic absorption rates. Once consumed, nanoparticles undergo transformation, and this will impact absorption ability and rates. There are several molecules within the GI tract that nanoparticles may interact with, such as proteins, lipids, carbohydrates, nucleic acids, ions, and water [9]. This then leads to the nanoparticles being encompassed by a collection of proteins known as a 'corona' [66]. Polystyrene nanoparticles may develop into varying forms of complex coronas, according to the conditions they are in [67]. Studies have shown that protein corona changes within an in vitro model representing human digestion, and this leads to greater translocation of nanoparticles [62]. Furthermore, organic matters found in bodies of water will adhere to the surface of nanoparticles; a recent review has examined how dispersed organic materials react with metal (oxide) nanoparticles and determined that these interactions have a significant impact on agglomeration and deposition [68].

It is worth noting that most of the reported studies were based on experiments using polystyrene nanoparticle models, and excluded samples gathered from marine and terrestrial environments. Other plastics such as polypropylene (PP), polyethylene (PE), and polyethylene terephthalate (PET) are, however, the main polymeric materials present in these environments. Thus, it is critical to qualify any extrapolations made from the findings of the research discussed above, which relies solely on polystyrene. Instead, new model studies should include experiments with PP, PE and PET.

4.2. *Pulmonary Exposure*

The second most likely method of human exposure to nanoplastics is through inhalation. Indoor environments contain airborne plastic particles, primarily from synthetic textiles, leading to unintended inhalation or occupational exposure [69]. In outdoor environments, exposure could happen through breathing in contaminated aerosols from ocean waves or airborne fertilizer particles from dried wastewater treatments [57]. The alveolar surface area of the lungs is vast, being approximately 150 m^2 and has an incredibly fine tissue barrier measuring less than 1 μm. This barrier is thin enough for nanoparticles to permeate through it and into the capillary blood system, thus, meaning that nanoparticles can disperse through the entire human body [57].

There are several negative health concerns resulting from the absorption of plastic particles, particularly micro- and nanoplastics, such as particle toxicity, chemical toxicity, and the introduction of pathogens and parasite vectors [70,71]. Particles within this range of sizes can potentially be embedded deep into the lung and then stay on the alveolar surface or translocate to other parts of the body [69,72]. The absorption of plastic particles through inhalation could lead to lung damage. There are a number of factors that affect absorption and expel micro- and nanoplastics in the lungs, such as hydrophobicity, surface charge, surface functionalization, surrounding protein coronas, and particle size [73]. In addition, research into absorption rates in animals indicate a positive correlation between occupational exposures and higher rates of pulmonary inflammation and cancer [49]. Research looking at absorption rates of polystyrene particles in alveolar epithelial cells, in vitro, suggests that absorption varies according to the size of the plastic particles [74–77].

Recent studies into the human inhalation of plastic particles have indicated that atmospheric fallout in urban areas is a significant cause of the particles [78]. The major constituent atmospheric fallout of microplastics from both urban and suburban areas of Paris was found to be synthetic fibre particles, where 29% of those fibres contained petrochemicals. By considering the average atmospheric flux of total fibres, the fibre dimensions and fibre densities, an estimated 3–10 tons of microplastics are deposited annually as a result from atmospheric fallout. Urban areas recorded double the average atmospheric flux, compared with suburban areas, with rainfall having a demonstrable impact on the observed depositions [78]. Dris et al. also examined the levels of microplastic particles in indoor and outdoor air at two private apartments and one office. The results showed a concentration of between 1 and 60 fibres/m^3 in the indoor samples. These readings were considerably greater than the outdoor samples which had levels of between 0.3 and 1.5 fibres/m^3. Approximately one-third of the indoor samples were of petrochemical origins, with the majority composed of polypropylene, while the rest were cellulose [79]. To date, there is no information regarding the amount or concentration of airborne nanoplastics.

4.3. Dermal Exposure

Health and beauty products are another key source of nanoplastics, particularly in the body and facial scrubs that are used topically on the skin [11]. Nanocarriers for drug delivery via dermal application is another important exposure route. Although there are no conclusive data showing the effects of nanocarriers, small particle size and stressed skin conditions are critical factors to skin penetration [51]. There is no current research that has specifically looked at the ability of nanoplastics to penetrate the surface of the skin. Only one study reported the likelihood of engineered nanoparticles from textiles in permeating the skin barrier at very minute quantity [80].

The skin is protected by the stratum corneum, the outermost layer, which forms a barrier against injuries, chemicals and microbial agents. The stratum corneum consists of corneocytes that are surrounded by lamellae of hydrophilic lipids including ceramide, long-chain free fatty acids and cholesterol [81]. Plastic particles may be introduced on the skin through health and beauty products, or through contact with nanoplastic-contaminated water. As micro- and nanoplastics are hydrophobic, it is predicted that absorption through the stratum corneum through contaminated water is unlikely, though plastic particles could enter the body via sweat glands, skin wounds or hair follicles [51].

Alvarez-Roman et al. looked at how plastic particles enter the body and how they are then distributed throughout the skin tissue. They used fluorescent polystyrene particles between 20 and 200 nm in diameter and skin tissue from a pig to conduct their experiment [82]. A confocal laser scanning micrograph of the skin revealed that a greater number of 20 nm polystyrene nanoplastics concentrated in the hair follicles than those of 200 nm nanoplastics. However, neither particles were able to permeate the stratum corneum in order to embed themselves into the deeper skin tissue. Campbell et al. supported these findings and established that polystyrene particles with diameters of 20–200 nm can infiltrate only the top layers of the skin at a depth of 2–3 μm [83]. Vogt et al. were able to distinguish 40 nm-diameter fluorescent polystyrene nanoparticles in the perifollicular tissue of skin explants that had been treated with cyanoacrylate follicular stripping. This work ascertained that when particles were applied transcutaneously, they were then absorbed by the Langerhans cells [84].

The mechanical production method used to manufacture the microbeads of the health and beauty products, including facial and body scrubs, increases the likelihood of the breakdown of microbeads into even more harmful nanoplastics. Hernandez et al. investigated the amount of nanoplastics present in facial scrubs containing 200 μm polyethene microbeads. The results from scanning electron microscopy confirmed that nanoparticles were observed at sizes between 24 ± 6 nm and 52 ± 14 nm. Then, they examined the chemical composition of these nanoparticles by using X-ray photoelectron spectroscopy and Fourier transform infrared spectroscopy and established that they consisted of polyethylene [11].

Data from these previous studies can be used to determine the likelihood of nanoplastics penetrating the stratum corneum. Exposure to UV rays causes skin damage, which means that the skin barrier becomes weaker [85]. A study on the effects of UV irradiation on murine skin found an increase in the skin permeation by carboxylated quantum dots. The intercellular adhesion in the irradiated skin was compromised by the perturbed expression of tight-junction-related proteins: Zonula occludens-1, claudin-1, and occludin [86]. There are a number of compounds, e.g., short chain- and long chain-alcohols, cyclic amides, esters, fatty acids, glycols, pyrrolidones, sulphoxides, surfactants and terpenes, that are used to enhance the chemical permeation of drugs and formulations through the skin barrier [87]. Widely used ingredients in body lotions, such as urea, glycerol and α-hydroxyl acids also enhanced the ability of nanoparticles to permeate the skin barrier [88].

Kuo et al. highlighted how oleic acid, ethanol and oleic acid-ethanol enhancers impact the transdermal delivery of 10 nm zinc oxide nanoparticles. They determined that each chemical had the potential to improve the effectiveness of zinc oxide nanoparticles in penetrating the skin barrier, through the multilamellar lipid regions between corneocytes [89]. Through crystalline structure analysis of various compositions of lipid lamellae in stratum corneum samples taken from human and porcine sources, Bouwstra et al. presented a three-layer "sandwich model", which likely prevents large nanoparticles from permeating undamaged skin [81].

To summarize, both in vitro and in vivo studies, discussed above, have established that micro- and nanoplastics can be absorbed into the human body through the skin barrier. That being said, these studies all rely on polystyrene particle models. Further studies conducted with samples collected from the environment would be helpful to fully understand the permeation qualities of micro- and nanoplastics. Such samples would include a variety of plastic particles with wide-ranging characteristics.

5. Cellular Uptake and Intracellular Fate of Microplastic and Nanoplastic Particles

After permeating into the stratum corneum and absorbed into the human body, microplastics and nanoplastics are then able to interact with numerous target cells. The quantity of nanoparticles that are absorbed and subsequently react with cells depends on a number of factors, such as their size, surface chemistry or charge of the biological elements they encounter, including proteins, phospholipids and carbohydrates [90]. As nanoparticles absorb proteins from the human body, they create 'protein coronas' around themselves [91]. This means that nanoparticles that interact with organs or skin cells will usually already be surrounded with protein corona as opposed to an exposed nanoparticle. The protein coating will lead to modified characteristics of the nanoparticle. Previous in vitro experiments have determined that protein coronas will surround polystyrene nanoparticles and that this enables the nanoparticles to translocate at greater rates [62]. These studies also showed that the protein coronas would alter their form according to their environment [67] and that there were more occurences of cell interactions and increased toxicity [92]. Finally, it was shown that protein coronas around the polystyrene nanoparticles led to them being accumulated in the gut.

Microplastics and nanoplastics can be absorbed by cells via a number of routes [93]. The primary route is via endocytotic nanoparticle uptake where adhesive interaction of nanoparticles (or inactive permeation of the cell membrane) with channel- or transport-protein occurs. Several endocytotic pathways have been identified, such as phagocytosis and macropinocytosis, along with clathrin- and caveolae-mediated endocytosis (Figure 3) [94–96].

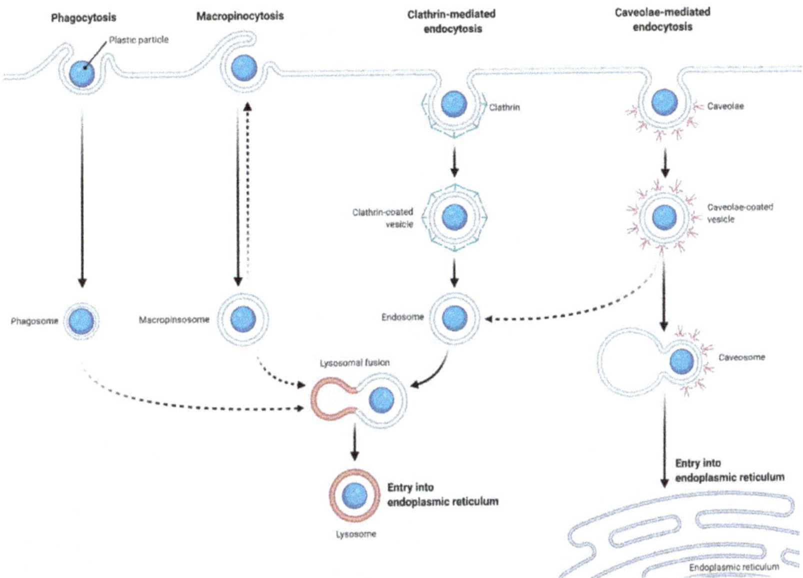

Figure 3. Routes of cellular uptake of plastic particles. Phagocytosis, macropinocytosis, clathrin- and caveolae-mediated endocytosis are the common endocytotic pathways that have been identified for cellular uptake of plastic particles. Micro- and nanoplastics can be absorbed by cells through different routes, of which endocytotic nanoparticle uptake is the primary route where adhesive interaction of nanoparticles (or inactive permeation of the cell membrane) with channel- or transport-protein occurs.

The initial barrier to nanoparticle incursion into the skin is the outer cell membrane. Coarse-grained molecular simulations of polystyrene (PS) particles in interacting with biological membranes revealed that polystyrene nanoparticles effortlessly penetrated the lipid bilayer membranes, causing changes to the structure of the cell membrane, ultimately disrupting the cell function [97]. Uptake inhibition studies on the absorption rate of 44 nm polystyrene particles to human colon fibroblasts and bovine oviductal epithelial cells indicated that polystyrene nanoparticles were primarily absorbed through a clathrin-independent uptake mechanism [98].

The cellular absorption of carboxylated polystyrene nanoparticles of 40 nm and 200 nm was studied using several tumor cell lines, including human cervical HeLa cells, human glial astrocytoma 1321N1, and adenocarcinomic human alveolar basal epithelial A549 in the presence of various transport inhibitors [99]. The results indicated that the nanoparticles were always absorbed by cells through an active and energy-dependent method, suggesting that the nanoparticle uptake for different cell types utilized different pathways. Actin depolymerization significantly influenced the nanoparticle uptake in HeLa and 1321N1 cell lines, while clathrin-mediated endocytosis of the nanoparticles in 1321N1 was significantly reduced by the inhibitor chlorpromazine. Nanoparticle uptake in A549 cell line via caveolae-mediated endocytosis was notably diminished due to disruption of microtubule formation by the inhibitor genistein [99].

The routes of nanoparticle absorption in human cells are dependent on the size and surface chemistry of the particles, but also vary according to the type of cell penetrated. This is seen when 120 nm polystyrene nanoparticles altered with amidine groups were seen to permeate rat alveolar epithelial monolayers using non-endocytic pathways, whilst MDCK-II cells use energy-dependent processes to absorb nanoparticles [76,100]. Macrophages and epithelial cells were found to use various combinations of endocytotic uptake mechanisms to absorb 40 nm carboxylated polystyrene nanoparticles.

Using several endocytotic pathway inhibitors, J774A.1 macrophages were shown to absorb nanoplastics via macropinocytosis, phagocytosis, and clathrin-mediated endocytosis pathways, whilst absorption in A549 cells relied on caveolae- and clathrin-mediated endocytosis pathways [101]. If nanoplastic particles are entering the human body via non-vesicular pathways, then they may be able to interact with intracellular molecules or discharge persistent organic pollutants (POPs) straight into the cytoplasm. This, in turn, implies that POPs may be stored in human cells which could have a negative toxicological impact [102].

Plastic particles enter cells via the intracellular endocytotic pathway engaging with early and late endosomes before combining with lysosomes. Polystyrene nanoparticles have been reported to accumulate in the lysosome [103], including the observance of intracellular localization of 40–50 nm polystyrene nanoparticles in A549 cells [77]. No lysosomal leakage or any fragmentation of the nanoparticles were reported when acidic conditions were applied [104].

A comparison of two types of nanoparticles made of polystyrene and mesoporous silica demonstrated clear differences in cellular uptake mechanisms in ovarian cancer cells. Data gathered demonstrated that ovarian cancer cells absorbed both types of nanoparticles with different endocytotic pathways [105]. Caveola-mediated endocytosis was the pathway used by mesoporous silica particles to permeate the cells and these then, according to the size of the particle, either remained in the lysosome (50 nm) or translocated into the cytoplasm (10 nm). On the other hand, polystyrene nanoparticles were absorbed through a caveola-independent pathway. Localized amine-modified 50 nm polystyrene particles showed toxicity to the lysosome after 4–8 h, whereas 30 nm carboxyl-modified polystyrene particles showed no signs of toxicity and did not enter cells via the standard acidic endocytotic route.

Previous research has also determined that cytotoxicity is greater with a positive surface charge and this also leads to increased absorption of nanoparticles via non-specific binding, where they end up on the negatively charged sugar moieties on cell surfaces. In contrast, the repellent interactions of negatively charged particles will impede endocytosis [103,106]. Studies have determined that there are several cellular absorption pathways and intracellular localization of polystyrene nanoparticles depending on their physicochemical characteristics. Despite this, there is little quantitative data concerning how nanoplastics enter cells and their eventual fates.

The size of plastic particles also affects their interaction with human cells [107]. Due to their high specific surface areas, the nanoparticle-cell interaction is vastly different compared to larger particles. Furthermore, the charge of the particle can also affect its interaction with the cell and its structure [107]. Since most of the in vitro studies described here have used polystyrene particles, it is difficult to extrapolate the results to other kinds of plastic particles. It is important that future research looks at cellular uptake and behaviours of other types of plastic particles.

6. Potential Toxic Effects of Microplastics and Nanoplastics on Human Health

Several in vitro and in vivo studies have shown that micro- and nanoplastics were able to cause serious impacts on the human body, including physical stress and damage, apoptosis, necrosis, inflammation, oxidative stress and immune responses (Table 1) [108–111].

6.1. Inflammation

An in vitro study using various sizes of polystyrene particles found that larger particles (202 nm and 535 nm) produced inflammatory effects on human A549 lung cells. There was higher IL-8 expression by the lung cells treated with the larger particles, in comparison with the same cells exposed to 64 nm particles [112]. Furthermore, unaltered or carboxylated polystyrene nanoparticles brought on a substantial up-regulation of IL-6 and IL-8 genes in human gastric adenocarcinoma, leukemia, and histiocytic lymphoma cells, which suggests that the increase in inflammatory reactions to polystyrene particles is likely to be due to the constitution of the particle, or down to simple particle occurrence rather than because of the particle charge [113,114].

A study on the impact of carboxylated and amino-modified polystyrene particles (120 nm) on the polarization of human macrophages into M1 or M2 phenotypes revealed no change in the expression of M1 markers like CD86, NOS2, TNFα, and IL-1β [115]. However, the introduction of both types of nanoparticles negatively impacted the expression of scavenger receptors CD163 and CD200R, and the release of IL-10 by M2 cells. There was a reduction of *Escherichia coli* phagocytosis by both M1 and M2 macrophages with the introduction of amino-modified particles. On the other hand, phagocytosis by M2 was unaffected by the carboxylated particles. The carboxylated particles also caused increases in the protein mass in M1 and M2, enhanced the release of TGFβ1 by M1, and heightened levels of ATP in M2 [115]. Similarly, in vitro study also shown that unmodified polyethylene particles measuring between 0.3 and 10 μm caused murine macrophages to produce significant levels of cytokines, such as IL-6, IL-1β, and TNFα [116,117].

Furthermore, a number of previous studies indicated that when polyethylene components are used as prostheses, they can fragmentize as a result of wear and tear, and form debris in the joints [118–120]. The polyethylene wear particles trigger the TNFα and IL-1 pro-inflammatory factors, as well as pro-osteoclastic factors, including the receptor activator of NF-κB ligand (RANKL), which causes periprosthetic bone resorption and could eventually result in the patient losing the prosthesis [121]. High levels of plastic particles measuring between 0.2 and 10 μm have been observed in the periprosthetic tissue of ultrahigh molecular weight polyethylene-based implants. In addition, the presence of macrophages in the tissues in the vicinity of the implant area indicates the stimulation of the inflammatory response [121–123]. A study of failed titanium alloy total hip arthroplasty cases found that a majority of wear debris in the interfacial membranes consisted of polyethylene particles with an average diameter of 530 nm [124]. To overcome the observed negative effects, surgeons are now increasingly employing metal-on-metal joint replacements.

Table 1. Summary of potential toxic effects of micro- and nanoplastics on human health.

Toxic Effects	Characteristics of Plastic Particles	Particle Size	Details	References
Inflammation	Polystyrene particles	202 nm and 535 nm	• Upregulation of IL-8 expression. • Induced inflammation in human A549 lung cells.	[112]
	Unaltered/Carboxylated polystyrene nanoparticles	20 nm, 44 nm, 500 nm, and 1000 nm	• Upregulation of IL-6 and IL-8 expression. • Enhanced inflammation in multiple human malignancies.	[113,114]
	Carboxylated and amino-modified polystyrene particles	120 nm	• Altered expression of scavenger receptors. • M2 cells increased IL-10 production. • Increased TGFβ1 (M1) and energy metabolism (M2).	[115]
	Unaltered polyethylene particles	0.3 µm, 10 µm	• Increased the secretion of IL-6, IL-1β, and TNFα in murine macrophages.	[117]
	Polyethylene particles from plastic prosthetic implants	0.2 µm and 10 µm	• Induced the expression of TNFα, IL-1, and RANKL. • Resulted in periprosthetic bone resorption. • Induced inflammatory response at the implant area.	[121] [121–123]
	Polystyrene microplastics particles	5 µm and 20 µm	• Induced inflammation in the liver. • Induced adverse effects on neurotransmission.	[125]
Oxidative stress and apoptosis	Amine-modified polystyrene nanoparticles	60 nm	• Strong interaction and aggregation with mucin. • Induced apoptosis in all intestinal epithelial cells.	[126]
	Cationic polystyrene nanoparticles	60 nm	• Induced ROS generation and ER stress • Induced autophagic cell death of mouse macrophages and lung epithelial cells.	[127,128]
	Unaltered or functionalized polystyrene	20 nm, 40 nm, 50 nm, and 100 nm	• Induced apoptosis of several human cell types.	[129–132]
	polyvinyl chloride (PVC) and poly (methyl methacrylate) (PMMA)	120 nm, 140 nm	• Reduced cell viability with a reduction of ATP and increase of ROS concentrations.	[133]
Metabolic homeostasis	Pristine and fluorescent polystyrene microplastics	5 µm	• Changes in amino acid and bile acid metabolism. • Induced gut microbiota dysbiosis and intestinal barrier dysfunction.	[134,135]
	Anionic carboxylated polystyrene nanoparticles	20 nm	• Altered ion channel function and ionic homeostasis • Activated basolateral K^+ channels. • Induced Cl^- and HCO^{3-} ion efflux.	[136]
	Polystyrene nanoparticles	30 nm	• Blocked vesicle transport and the distribution of cytokinesis-associated proteins.	[137]
	Cationic polystyrene nanoparticles	50 nm and 200 nm	• Disrupted intestinal iron transport and cellular uptake.	[138]
	Pristine polystyrene microparticles	5 µm and 20 µm	• Reduction in hepatic ATP levels. • Impairment of energy metabolism.	[125,139,140]
	Microplastics	0.5 µm and 5 µm	• Metabolic disorder associated with gut microbiota dysbiosis and gut barrier dysfunction. • Increased the risks of metabolic disorder in the offspring.	[135,141]

6.2. Oxidative Stress and Apoptosis

A number of in vitro studies have shown that different polystyrene nanoparticles can induce oxidative stress, apoptosis and autophagic cell death in cell context-dependent manner. For instance, amine-modified polystyrene nanoparticles were shown to interact and aggregate with mucin strongly, and induce apoptosis of mucin- and non-mucin-secreting intestinal epithelial cells [126]. Cationic polystyrene nanoparticles were shown to induce reactive oxygen species (ROS) production and endoplasmic reticulum (ER) stress in mouse macrophages and lung epithelial cells via aggregation of misfolded protein, leading to autophagic cell death of RAW 264.7 mouse macrophages and BEAS-2B lung epithelial cells [127,128]. While, unmodified or functionalized polystyrene was shown to induce apoptosis in several human cell types, including primary human alveolar macrophages (MAC), primary human alveolar type 2 (AT2) epithelial cells, human monocytic leukemia cell line (THP-1), human immortalized alveolar epithelial type 1 cells (TT1), human colon carcinoma cells (Caco-2), and human lung cancer cells (Calu-3) [129–132]; and polystyrene nanoparticles were shown to regulate ROS via long non-coding RNAs (e.g., *linc-61, linc-50, linc-9, and linc-2*) in *Caenorhabditis elegans* [142].

Despite the toxic effects observed in the in vitro models, no obvious severe toxicity was observed in liver, duodenum, ileum, jejunum, large intestine, testes, lungs, heart, spleen, and kidneys of mice following oral exposure of a mixture of microplastics [139]. While, other studies have demonstrated that oral exposure (either through oral gavage or drinking water) caused liver inflammation [125], neurotoxic responses [125], reduced body and liver weight [140], reduced mucin excretion in colon [134,140], altered amino acid and bile acid metabolism [134,135], and altered microbiota composition [134,140,141,143]. Interestingly, some of the effects such as altered lipid metabolism was observed even in the offspring of the mice following microplastic exposure [141].

6.3. Metabolic Homeostasis

In addition to the induction of inflammation and apoptosis, recent studies have revealed that microplastics and nanoplastics can impair cellular metabolism in both in vitro and in vivo models. Polystyrene-based nanoparticles influence signaling systems in airway epithelial cells due to nanoparticle-cytoplasmic membrane interactions. After exposure to negatively charged carboxylated polystyrene nanoparticles measuring 20 nm, basolateral K^+ ion channels were found to be activated in human lung cells [136]. The nanoplastic particles caused persistent and concentration-dependent increases in short-circuit currents by the activation of the ion channels and the stimulation of Cl^- and $HCO_3{}^-$ ion efflux [136].

Furthermore, 30 nm polystyrene nanoparticles induced large vesicle-like structures in the endocytic route in macrophages and human cancer cell lines A549, HepG-2, and HCT116. As a result, vesicle transport and the distribution of proteins involved in cytokinesis are blocked, thus stimulating the formation of binucleated cells [137]. In addition, acute oral exposure to positively charged polystyrene nanoparticles has the potential to disrupt intestinal iron transport and cellular uptake [138].

When mice were fed with pristine polystyrene microparticles (5 μm and 20 μm) for 28 days, the microplastics were found to be distributed in the liver, kidneys and gut, with larger particles dispersed regularly across all tissues, while the smaller particles found at higher concentration in the gut [125]. Inflammation and lipid droplets were also evident in the histopathological analysis. There was evidence that showed microplastic accumulation in murine tissue caused impairment of energy metabolism, lipid metabolism, oxidative stress and neurotoxic responses. There were decreases noted in hepatic levels of ATP, total cholesterol and triglycerides, as well as reduction in catalase activity, whereas increases were observed in the activity of several biomarkers (LDH, SOD, GSH-Px and AchE) [125,139,140].

Furthermore, pregnant mice exposed to microplastics via ingestion developed gut microbiota dysbiosis, intestinal barrier dysfunction and metabolic disorders. The effects of microplastics exposure at the maternal level also conferred permanent altered metabolism in the F1 and F2 generations [135,141]. The key results from these studies showed: (1) Change in the gut microbiota; (2) change in the intestinal barrier where less mucus was secreted and lower levels of ion transporter gene expression; and (3) alterations to lipid/fatty acid metabolism, as demonstrated by the differences in serum and liver triglyceride and total cholesterol levels [134,135,140,141].

7. Leaching of Toxic Chemicals from Plastics

As discussed above, plastics usually contain chemicals from the raw monomers and various types of additives to improve their properties. In addition, plastics also absorb chemicals from their surroundings [144,145]. As a result, these chemicals have the potential to leach from the polymer and into the environment around them. For example, polycyclic aromatic hydrocarbons (PAHs) have been shown to be adsorbed by microplastics and causing various toxic effects when ingested by various organisms [145]. Chemical species diffuses from the interior of a particle to its surface, leaching into the surrounding environment, and is possibly driven by a gradient function. Although these chemical species are transient and degrade rapidly in the human body, these plastic particles provide a durable 'reservoir' for chemical leaching into tissues and body fluid [146].

To date, toxic chemical additives in plastic that are known to affect human health include bisphenol A (BPA), phthalates, triclosan, bisphenone, organotins and brominated flame retardants (BFR) (Figure 4) [147]. Although limited information is available whether these additives leach into the biological tissues directly, certain additives, such as nonylphenol and BPA, are found to be ingested by marine biota [148]. In particular, exposure to leached BPA, an additive that is commonly used to make polycarbonate (PC) plastics and epoxy resin as lining layer of food and beverage cans, has been shown to cause endocrine disorders and impact human health [147,149,150].

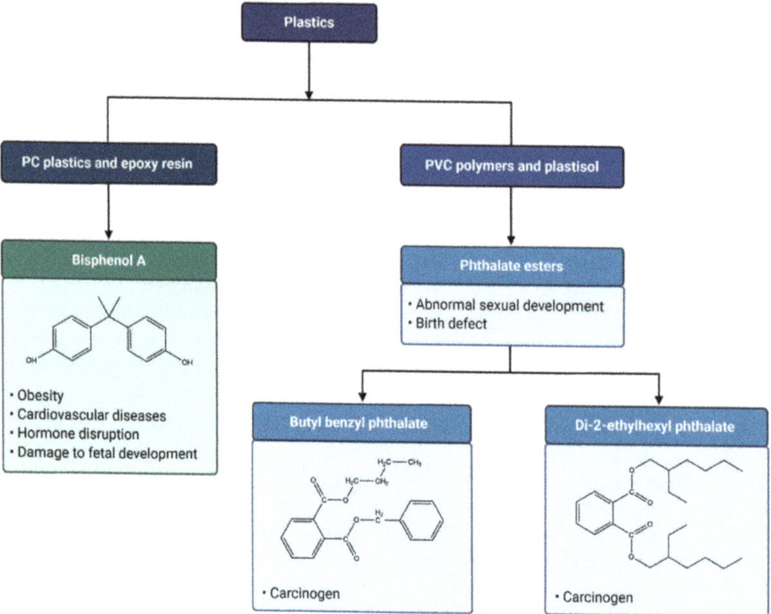

Figure 4. Overview of the toxic effects of chemicals leaching from plastics. Plastics are made up of different chemical compositions, in which some are hazardous that can leach to the surroundings upon degradation. Plastics typically contain additives that can improve their properties, such as durability and elasticity. The leaching of these additives from plastics to the surrounding environment, not only causing harmful impacts to the aquatic environment, but also human health. For instance, bisphenol A (BPA), an industrial chemical that is widely used to make polycarbonate (PC) plastics and epoxy resin

as lining layer of food and beverage containers. Studies reported that the leaching of BPA from food containers into the food and drinks can cause a series of diseases, including obesity and cardiovascular diseases. BPA also acts as a hormonal disruptor, imitating or blocking the production, action, and function of hormones in the human body. BPA also known to affect brain development in the womb, causing damage to the developing fetus. Polyvinyl chloride (PVC) polymers and plastisol generally contain phthalate esters as plasticizers, in order to increase their durability and flexibility. Human exposure to phthalate esters has been shown to associate with abnormal sexual development and changes in the levels of sex hormones. Additionally, studies have demonstrated that some phthalate esters such as butyl benzyl phthalate (BBP) and di-2-ethylhexyl phthalate (DEHP) can increase tumor incidence in human, representing potential carcinogens.

Importantly, studies have found that BPA will leach from PC into food and drinks [147,151,152], and that the toxicity of BPA causes changes in liver function and insulin resistance, damage of a developing fetus and modification of the reproductive system and neurological functions [153]. BPA acts as an agonist for estrogen receptors and inhibits thyroid hormone-mediated transcription by acting as an antagonist [154], and alters pancreatic beta cell function [155]. Increased likelihood of developing obesity and cardiovascular diseases [156–158], and several other reproductive and developmental issues have been noted when humans are exposed to BPA at concentrations of 0.2–20 ng/mL [147].

Phthalate esters are used as plasticizers in the manufacturing of PVC polymers and plastisol to achieve enhanced flexibility and durability [159]. Human exposure to phthalate esters are potentially harmful and may cause abnormal sexual development and birth defects [160]. Additionally, butyl benzyl phthalate (BBP) has been named as a probable carcinogen, and di-2-ethylhexyl phthalate (DEHP) has been cited as a possible carcinogen by U.S. EPA [15].

8. Conclusions

While, microplastics and nanoplastics are widely studied in the context of the marine environment, we have only recently recognized the potential human exposure pathways. Following exposure, uptake is plausible via ingestion and/or inhalation. The toxicity assessments of micro- and nanoplastics on human are mainly focusing on gastrointestinal and pulmonary toxicity, which involve oxidative stress, inflammatory reactions, and metabolism disorders.

Based on the findings of recent studies, further research is needed to investigate the potential mechanisms of micro- and nanoplastics toxicity in human. Moreover, it is important to understand whether microplastics and nanoplastics can be further degraded after ingestion under the acidic conditions in the gut or inside the lysosomes of the cells. Hence, the long-term fate of the ingested microplastics and nanoplastics in human body warrant further investigation.

Unfortunately, the accurate assessment of human exposure to nanoplastics remains a scientific challenge due to the lack of validated methods, certified reference materials, and standardization across the analytical procedures used [161,162]. Notably, most of the reported studies were conducted using polystyrene due to its ease in synthesis and processing into nanoparticles, while the most common commercial used of plastics are polyolefins (e.g., polyethylene and polypropylene), polyesters, and polyurethanes. Given the large variety in particle size, shape and chemical composition of plastics, the potentially hazardous effects of different types of micro- and nanoplastics to human health remain largely unknown [163]. Therefore, we recommend that future research should focus on the understanding of the potential hazards and risks of chronic exposure to diverse micro- and nanoplastics at relevant concentrations.

Author Contributions: Writing—original draft preparation, M.S.-L.Y., S.-F.W., Y.-Y.K., B.-K.T., C.-Y.W.; writing—review and editing, L.-W.H., C.K.L., W.-M.L., C.-O.L.; visualization, L.-W.H., C.K.L., W.-M.L.; supervision, C.-O.L.; project administration, C.-O.L.; funding acquisition, B.-K.T. All authors have read and agreed to the published version of the manuscript.

Funding: This research was funded by the MINISTRY OF HIGHER EDUCATION, MALAYSIA (FRGS/1/2020/SKK06/IMU/01/1 to COL, SFW, BKT, and MSLY).

Acknowledgments: All figures in this review paper were created with BioRender.com (accessed on 24 January 2021) by the authors.

Conflicts of Interest: The authors declare no conflict of interest.

References

1. Garside, M. Global Plastic Production Statistics. Retrieved from Statista. Available online: https://www.statista.com/statistics/282732/global (accessed on 1 August 2020).
2. Wagner, S.; Reemtsma, T. Things we know and don't know about nanoplastic in the environment. *Nat. Nanotechnol.* **2019**, *14*, 300–301. [CrossRef]
3. Shrivastava, A. Polymerization. In *Introduction to Plastics Engineering*; William Andrew Publishing: New York, NY, USA, 2018; pp. 17–48. [CrossRef]
4. Liu, T.; Guo, X.; Liu, W.; Hao, C.; Wang, L.; Hiscox, W.C.; Liu, C.; Jin, C.; Xin, J.; Zhang, J. Selective cleavage of ester linkages of anhydride-cured epoxy using a benign method and reuse of the decomposed polymer in new epoxy preparation. *Green Chem.* **2017**, *19*, 4364–4372. [CrossRef]
5. Ambrogi, V.; Carfagna, C.; Cerruti, P.; Marturano, V. Additives in Polymers. *Modif. Polym. Prop.* **2017**, 87–108. [CrossRef]
6. Song, Y.K.; Hong, S.H.; Jang, M.; Han, G.M.; Jung, S.W.; Shim, W.J. Combined Effects of UV Exposure Duration and Mechanical Abrasion on Microplastic Fragmentation by Polymer Type. *Environ. Sci. Technol.* **2017**, *51*, 4368–4376. [CrossRef] [PubMed]
7. Da Costa, J.P. *Micro- and Nanoplastics in the Environment: Research and Policymaking*; Elsevier: Amsterdam, The Netherlands, 2018; Volume 1, pp. 12–16.
8. Gigault, J.; Halle, A.; Baudrimont, M.; Pascal, P.-Y.; Gauffre, F.; Phi, T.-L.; El Hadri, H.; Grassl, B.; Reynaud, S. Current opinion: What is a nanoplastic? *Environ. Pollut.* **2018**, *235*, 1030–1034. [CrossRef]
9. Contam, E.P.O.C.I.T.F.C. Presence of microplastics and nanoplastics in food, with particular focus on seafood. *EFSA J.* **2016**, *14*, 14. [CrossRef]
10. Rochman, C.M.; Browne, M.A.; Halpern, B.S.; Hentschel, B.T.; Hoh, E.; Karapanagioti, H.K.; Rios-Mendoza, L.M.; Takada, H.; Teh, S.; Thompson, R.C. Classify plastic waste as hazardous. *Nat. Cell Biol.* **2013**, *494*, 169–171. [CrossRef] [PubMed]
11. Hernandez, L.M.; Yousefi, N.; Tufenkji, N. Are There Nanoplastics in Your Personal Care Products? *Environ. Sci. Technol. Lett.* **2017**, *4*, 280–285. [CrossRef]
12. Ter Halle, A.; Jeanneau, L.; Martignac, M.; Jardé, E.; Pedrono, B.; Brach, L.; Gigault, J. Nanoplastic in the North Atlantic Subtropical Gyre. *Environ. Sci. Technol.* **2017**, *51*, 13689–13697. [CrossRef]
13. Gigault, J.; Pedrono, B.; Maxit, B.; Ter Halle, A. Marine plastic litter: The unanalyzed nano-fraction. *Environ. Sci. Nano* **2016**, *3*, 346–350. [CrossRef]
14. Browne, M.A.; Crump, P.; Niven, S.J.; Teuten, E.; Tonkin, A.; Galloway, T.; Thompson, R. Accumulation of Microplastic on Shorelines Woldwide: Sources and Sinks. *Environ. Sci. Technol.* **2011**, *45*, 9175–9179. [CrossRef]
15. Karbalaei, S.; Hanachi, P.; Walker, T.R.; Cole, M. Occurrence, sources, human health impacts and mitigation of microplastic pollution. *Environ. Sci. Pollut. Res.* **2018**, *25*, 36046–36063. [CrossRef]
16. Browne, M.A.; Underwood, A.J.; Chapman, M.G.; Williams, R.; Thompson, R.C.; Van Franeker, J.A. Linking effects of anthropogenic debris to ecological impacts. *Proc. R. Soc. B Boil. Sci.* **2015**, *282*, 20142929. [CrossRef] [PubMed]
17. Thushari, G.; Senevirathna, J. Plastic pollution in the marine environment. *Heliyon* **2020**, *6*, e04709. [CrossRef]
18. Jambeck, J.R.; Geyer, R.; Wilcox, C.; Siegler, T.R.; Perryman, M.; Andrady, A.; Narayan, R.; Law, K.L. Plastic waste inputs from land into the ocean. *Science* **2015**, *347*, 768–771. [CrossRef]
19. Vance, M.E.; Kuiken, T.; Vejerano, E.P.; McGinnis, S.P.; Hochella, M.F., Jr.; Rejeski, D.; Hull, M.S. Nanotechnology in the real world: Redeveloping the nanomaterial consumer products inventory. *Beilstein J. Nanotechnol.* **2015**, *6*, 1769–1780. [CrossRef]
20. Horton, A.A.; Walton, A.; Spurgeon, D.J.; Lahive, E.; Svendsen, C. Microplastics in freshwater and terrestrial environments: Evaluating the current understanding to identify the knowledge gaps and future research priorities. *Sci. Total Environ.* **2017**, *586*, 127–141. [CrossRef]
21. Mattsson, K.; Jocic, S.; Doverbratt, I.; Hansson, L.-A. Nanoplastics in the Aquatic Environment. In *Microplastic Contamination in Aquatic Environments*; Elsevier: Amsterdam, The Netherlands, 2018; pp. 379–399. [CrossRef]
22. Cole, M.; Lindeque, P.; Halsband, C.; Galloway, T.S. Microplastics as contaminants in the marine environment: A review. *Mar. Pollut. Bull.* **2011**, *62*, 2588–2597. [CrossRef]
23. Turner, A.; Holmes, L.; Thompson, R.C.; Fisher, A.S. Metals and marine microplastics: Adsorption from the environment versus addition during manufacture, exemplified with lead. *Water Res.* **2020**, *173*, 115577. [CrossRef] [PubMed]

24. Kazour, M.; Terki, S.; Rabhi, K.; Jemaa, S.; Khalaf, G.; Amara, R. Sources of microplastics pollution in the marine environment: Importance of wastewater treatment plant and coastal landfill. *Mar. Pollut. Bull.* **2019**, *146*, 608–618. [CrossRef] [PubMed]
25. Andrady, A.L. Microplastics in the marine environment. *Mar. Pollut. Bull.* **2011**, *62*, 1596–1605. [CrossRef]
26. Zettler, E.R.; Mincer, T.J.; Amaral-Zettler, L.A. Life in the "Plastisphere": Microbial Communities on Plastic Marine Debris. *Environ. Sci. Technol.* **2013**, *47*, 7137–7146. [CrossRef]
27. Lucas, N.; Bienaime, C.; Belloy, C.; Queneudec, M.; Silvestre, F.; Nava-Saucedo, J.-E. Polymer biodegradation: Mechanisms and estimation techniques—A review. *Chemosphere* **2008**, *73*, 429–442. [CrossRef]
28. Lambert, S.; Wagner, M. Characterisation of nanoplastics during the degradation of polystyrene. *Chemosphere* **2016**, *145*, 265–268. [CrossRef]
29. Yuan, J.; Ma, J.; Sun, Y.; Zhou, T.; Zhao, Y.; Yu, F. Microbial degradation and other environmental aspects of microplastics/plastics. *Sci. Total Environ.* **2020**, *715*, 136968. [CrossRef]
30. Andrady, A.L. Persistence of Plastic Litter in the Oceans. In *Marine Anthropogenic Litter*; Springer: Cham, Switzerland, 2015; pp. 57–72.
31. Corcoran, P.L.; Biesinger, M.C.; Grifi, M. Plastics and beaches: A degrading relationship. *Mar. Pollut. Bull.* **2009**, *58*, 80–84. [CrossRef] [PubMed]
32. Kolandhasamy, P.; Su, L.; Li, J.; Qu, X.; Jabeen, K.; Shi, H. Adherence of microplastics to soft tissue of mussels: A novel way to uptake microplastics beyond ingestion. *Sci. Total Environ.* **2018**, 635–640. [CrossRef] [PubMed]
33. Wang, Y.-L.; Lee, Y.-H.; Chiu, I.-J.; Lin, Y.-F.; Chiu, H.-W. Potent Impact of Plastic Nanomaterials and Micromaterials on the Food Chain and Human Health. *Int. J. Mol. Sci.* **2020**, *21*, 1727. [CrossRef] [PubMed]
34. Santillo, D.; Miller, K.; Johnston, P. Microplastics as contaminants in commercially important seafood species. *Integr. Environ. Assess. Manag.* **2017**, *13*, 516–521. [CrossRef]
35. Karami, A.; Golieskardi, A.; Choo, C.K.; Larat, V.; Galloway, T.S.; Salamatinia, B. The presence of microplastics in commercial salts from different countries. *Sci. Rep.* **2017**, *7*, 1–11. [CrossRef]
36. Mason, S.A.; Welch, V.G.; Neratko, J. Synthetic Polymer Contamination in Bottled Water. *Front. Chem.* **2018**, *6*, 407. [CrossRef] [PubMed]
37. Li, J.; Yang, D.; Li, L.; Jabeen, K.; Shi, H. Microplastics in commercial bivalves from China. *Environ. Pollut.* **2015**, *207*, 190–195. [CrossRef] [PubMed]
38. Neves, D.; Sobral, P.; Ferreira, J.L.; Pereira, T. Ingestion of microplastics by commercial fish off the Portuguese coast. *Mar. Pollut. Bull.* **2015**, *101*, 119–126. [CrossRef]
39. Devriese, L.I.; Van Der Meulen, M.D.; Maes, T.; Bekaert, K.; Paul-Pont, I.; Frère, L.; Robbens, J.; Vethaak, A.D. Microplastic contamination in brown shrimp (Crangon crangon, Linnaeus 1758) from coastal waters of the Southern North Sea and Channel area. *Mar. Pollut. Bull.* **2015**, *98*, 179–187. [CrossRef] [PubMed]
40. Yang, D.; Shi, H.; Li, L.; Li, J.; Jabeen, K.; Kolandhasamy, P. Microplastic Pollution in Table Salts from China. *Environ. Sci. Technol.* **2015**, *49*, 13622–13627. [CrossRef]
41. Liebezeit, G.; Liebezeit, E. Non-pollen particulates in honey and sugar. *Food Addit. Contam. Part A* **2013**, *30*, 2136–2140. [CrossRef] [PubMed]
42. Liebezeit, G.; Liebezeit, E. Synthetic particles as contaminants in German beers. *Food Addit. Contam. Part A* **2014**, *31*, 1574–1578. [CrossRef]
43. Kosuth, M.; Mason, S.A.; Wattenberg, E.V. Anthropogenic contamination of tap water, beer, and sea salt. *PLoS ONE* **2018**, *13*, e0194970. [CrossRef]
44. Cox, K.D.; Covernton, G.A.; Davies, H.L.; Dower, J.F.; Juanes, F.; Dudas, S.E. Human Consumption of Microplastics. *Environ. Sci. Technol.* **2019**, *53*, 7068–7074. [CrossRef] [PubMed]
45. Bergmann, M.; Gutow, L.; Klages, M. *Marine Anthropogenic Litter*; Springer: Cham, Switzerland, 2015.
46. Cózar, A.; Echevarría, F.; González-Gordillo, J.I.; Irigoien, X.; Úbeda, B.; Hernández-León, S.; Palma, Á.T.; Navarro, S.; García-De-Lomas, J.; Ruiz, A.; et al. Plastic debris in the open ocean. *Proc. Natl. Acad. Sci. USA* **2014**, *111*, 10239–10244. [CrossRef]
47. Prata, J.C.; Da Costa, J.P.; Lopes, I.; Duarte, A.C.; Rocha-Santos, T. Environmental exposure to microplastics: An overview on possible human health effects. *Sci. Total Environ.* **2020**, *702*, 134455. [CrossRef]
48. Rahman, A.; Sarkar, A.; Yadav, O.P.; Achari, G.; Slobodnik, J. Potential human health risks due to environmental exposure to nano- and microplastics and knowledge gaps: A scoping review. *Sci. Total Environ.* **2021**, *757*, 143872. [CrossRef]
49. Prata, J.C. Airborne microplastics: Consequences to human health? *Environ. Pollut.* **2018**, *234*, 115–126. [CrossRef]
50. Carbery, M.; O'Connor, W.; Palanisami, T. Trophic transfer of microplastics and mixed contaminants in the marine food web and implications for human health. *Environ. Int.* **2018**, *115*, 400–409. [CrossRef]
51. Schneider, M.; Stracke, F.; Hansen, S.; Schaefer, U.F. Nanoparticles and their interactions with the dermal barrier. *Dermato-Endocrinology* **2009**, *1*, 197–206. [CrossRef]
52. Brennecke, D.; Duarte, B.; Paiva, F.; Caçador, I.; Canning-Clode, J. Microplastics as vector for heavy metal contamination from the marine environment. *Estuar. Coast. Shelf Sci.* **2016**, *178*, 189–195. [CrossRef]
53. Camacho, M.; Herrera, A.; Gómez, M.; Acosta-Dacal, A.; Martínez, I.; Henríquez-Hernández, L.A.; Luzardo, O.P. Organic pollutants in marine plastic debris from Canary Islands beaches. *Sci. Total Environ.* **2019**, *662*, 22–31. [CrossRef] [PubMed]

54. Li, J.; Zhang, K.; Zhang, H. Adsorption of antibiotics on microplastics. *Environ. Pollut.* **2018**, *237*, 460–467. [CrossRef] [PubMed]
55. Rochman, C.M.; Kurobe, T.; Flores, I.; Teh, S.J. Early warning signs of endocrine disruption in adult fish from the ingestion of polyethylene with and without sorbed chemical pollutants from the marine environment. *Sci. Total Environ.* **2014**, *493*, 656–661. [CrossRef] [PubMed]
56. Viršek, M.K.; Lovšin, M.N.; Koren, Š.; Kržan, A.; Peterlin, M. Microplastics as a vector for the transport of the bacterial fish pathogen species Aeromonas salmonicida. *Mar. Pollut. Bull.* **2017**, *125*, 301–309. [CrossRef] [PubMed]
57. Lehner, R.; Weder, C.; Petri-Fink, A.; Rothen-Rutishauser, B. Emergence of Nanoplastic in the Environment and Possible Impact on Human Health. *Environ. Sci. Technol.* **2019**, *53*, 1748–1765. [CrossRef]
58. Ge, H.; Yan, Y.; Wu, D.; Huang, Y.; Tian, F. Potential role of LINC00996 in colorectal cancer: A study based on data mining and bioinformatics. *OncoTargets Ther.* **2018**, *11*, 4845–4855. [CrossRef] [PubMed]
59. Alberts, B.; Johnson, A.; Lewis, J.; Raff, M.; Roberts, K.; Walter, P. Cell Junctions. In *Molecular Biology of the Cell*, 4th ed.; Garland Science: New York, NY, USA, 2002.
60. Tomazic-Jezic, V.J.; Merritt, K.; Umbreit, T.H. Significance of the type and the size of biomaterial particles on phagocytosis and tissue distribution. *J. Biomed. Mater. Res.* **2001**, *55*, 523–529. [CrossRef]
61. Carr, K.E.; Smyth, S.H.; McCullough, M.T.; Morris, J.F.; Moyes, S.M. Morphological aspects of interactions between microparticles and mammalian cells: Intestinal uptake and onward movement. *Prog. Histochem. Cytochem.* **2012**, *46*, 185–252. [CrossRef] [PubMed]
62. Walczak, A.P.; Kramer, E.; Hendriksen, P.J.M.; Tromp, P.; Helsper, J.P.F.G.; Van Der Zande, M.; Rietjens, I.M.C.M.; Bouwmeester, H. Translocation of differently sized and charged polystyrene nanoparticles in in vitro intestinal cell models of increasing complexity. *Nanotoxicology* **2014**, *9*, 453–461. [CrossRef] [PubMed]
63. Jani, P.; Halbert, G.W.; Langridge, J.; Florence, A.T. Nanoparticle Uptake by the Rat Gastrointestinal Mucosa: Quantitation and Particle Size Dependency. *J. Pharm. Pharmacol.* **1990**, *42*, 821–826. [CrossRef]
64. Des Rieux, A.; Fievez, V.; Théate, I.; Mast, J.; Préat, V.; Schneider, Y.-J. An improved in vitro model of human intestinal follicle-associated epithelium to study nanoparticle transport by M cells. *Eur. J. Pharm. Sci.* **2007**, *30*, 380–391. [CrossRef]
65. Kulkarni, S.A.; Feng, S.-S. Effects of Particle Size and Surface Modification on Cellular Uptake and Biodistribution of Polymeric Nanoparticles for Drug Delivery. *Pharm. Res.* **2013**, *30*, 2512–2522. [CrossRef] [PubMed]
66. Lundqvist, M.; Stigler, J.; Elia, G.; Lynch, I.; Cedervall, T.; Dawson, K.A. Nanoparticle size and surface properties determine the protein corona with possible implications for biological impacts. *Proc. Natl. Acad. Sci. USA* **2008**, *105*, 14265–14270. [CrossRef]
67. Tenzer, S.; Docter, D.; Kuharev, J.; Musyanovych, A.; Fetz, V.; Hecht, R.; Schlenk, F.; Fischer, D.; Kiouptsi, K.; Reinhardt, C.; et al. Rapid formation of plasma protein corona critically affects nanoparticle pathophysiology. *Nat. Nanotechnol.* **2013**, *8*, 772–781. [CrossRef]
68. Philippe, A.; Schaumann, G.E. Interactions of Dissolved Organic Matter with Natural and Engineered Inorganic Colloids: A Review. *Environ. Sci. Technol.* **2014**, *48*, 8946–8962. [CrossRef] [PubMed]
69. Stapleton, P. Toxicological considerations of nano-sized plastics. *AIMS Environ. Sci.* **2019**, *6*, 367–378. [CrossRef] [PubMed]
70. Vethaak, A.D.; Leslie, H.A. Plastic Debris Is a Human Health Issue. *Environ. Sci. Technol.* **2016**, *50*, 6825–6826. [CrossRef] [PubMed]
71. Ohlwein, S.; Kappeler, R.; Kutlar Joss, M.; Künzli, N.; Hoffmann, B. Health effects of ultrafine particles: A systematic literature review update of epidemiological evidence. *Int. J. Public Health* **2019**, *64*, 547–559. [CrossRef] [PubMed]
72. Porter, D.W.; Hubbs, A.F.; Mercer, R.R.; Wu, N.; Wolfarth, M.G.; Sriram, K.; Leonard, S.; Battelli, L.; Schwegler-Berry, D.; Friend, S. Mouse pulmonary dose- and time course-responses induced by exposure to multi-walled carbon nanotubes. *Toxicology* **2010**, *269*, 136–147. [CrossRef]
73. Rist, S.; Almroth, B.C.; Hartmann, N.B.; Karlsson, T.M. A critical perspective on early communications concerning human health aspects of microplastics. *Sci. Total Environ.* **2018**, *626*, 720–726. [CrossRef]
74. Varela, J.A.; Bexiga, M.G.; Åberg, C.; Simpson, J.C.; Dawson, K.A. Quantifying size-dependent interactions between fluorescently labeled polystyrene nanoparticles and mammalian cells. *J. Nanobiotechnol.* **2012**, *10*, 39. [CrossRef]
75. Deville, S.; Penjweini, R.; Smisdom, N.; Notelaers, K.; Nelissen, I.; Hooyberghs, J.; Ameloot, M. Intracellular dynamics and fate of polystyrene nanoparticles in A549 Lung epithelial cells monitored by image (cross-) correlation spectroscopy and single particle tracking. *Biochim. Biophys. Acta Mol. Cell Res.* **2015**, *1853*, 2411–2419. [CrossRef]
76. Yacobi, N.R.; DeMaio, L.; Xie, J.; Hamm-Alvarez, S.F.; Borok, Z.; Kim, K.-J.; Crandall, E.D. Polystyrene nanoparticle trafficking across alveolar epithelium. *Nanomed. Nanotechnol. Biol. Med.* **2008**, *4*, 139–145. [CrossRef] [PubMed]
77. Salvati, A.; Åberg, C.; Dos Santos, T.; Varela, J.; Pinto, P.; Lynch, I.; Dawson, K.A. Experimental and theoretical comparison of intracellular import of polymeric nanoparticles and small molecules: Toward models of uptake kinetics. *Nanomed. Nanotechnol. Biol. Med.* **2011**, *7*, 818–826. [CrossRef]
78. Dris, R.; Gasperi, J.; Saad, M.; Mirande, C.; Tassin, B. Synthetic fibers in atmospheric fallout: A source of microplastics in the environment? *Mar. Pollut. Bull.* **2016**, *104*, 290–293. [CrossRef]
79. Dris, R.; Gasperi, J.; Mirande, C.; Mandin, C.; Guerrouache, M.; Langlois, V.; Tassin, B. A first overview of textile fibers, including microplastics, in indoor and outdoor environments. *Environ. Pollut.* **2017**, *221*, 453–458. [CrossRef] [PubMed]
80. Som, C.; Wick, P.; Krug, H.; Nowack, B. Environmental and health effects of nanomaterials in nanotextiles and façade coatings. *Environ. Int.* **2011**, *37*, 1131–1142. [CrossRef] [PubMed]

81. Bouwstra, J.; Pilgram, G.; Gooris, G.; Koerten, H.; Ponec, M. New aspects of the skin barrier organization. *Ski. Pharmacol. Appl. Ski. Physiol.* **2001**, *14*, 52–62. [CrossRef]
82. Alvarez-Román, R.; Naik, A.; Kalia, Y.; Guy, R.; Fessi, H. Skin penetration and distribution of polymeric nanoparticles. *J. Control. Release* **2004**, *99*, 53–62. [CrossRef]
83. Campbell, C.S.J.; Contreras-Rojas, L.R.; Delgado-Charro, M.B.; Guy, R.H. Objective assessment of nanoparticle disposition in mammalian skin after topical exposure. *J. Control. Release* **2012**, *162*, 201–207. [CrossRef]
84. Vogt, A.; Combadiere, B.; Hadam, S.; Stieler, K.M.; Lademann, J.; Schaefer, H.E.; Autran, B.; Sterry, W.; Blume-Peytavi, U. 40 nm, but not 750 or 1500 nm, Nanoparticles Enter Epidermal CD1a+ Cells after Transcutaneous Application on Human Skin. *J. Investig. Dermatol.* **2006**, *126*, 1316–1322. [CrossRef]
85. Biniek, K.; Levi, K.; Dauskardt, R.H. Solar UV radiation reduces the barrier function of human skin. *Proc. Natl. Acad. Sci. USA* **2012**, *109*, 17111–17116. [CrossRef] [PubMed]
86. Mortensen, L.J.; Oberdörster, G.; Pentland, A.P.; DeLouise, L.A. In Vivo Skin Penetration of Quantum Dot Nanoparticles in the Murine Model: The Effect of UVR. *Nano Lett.* **2008**, *8*, 2779–2787. [CrossRef]
87. Lane, M.E. Skin penetration enhancers. *Int. J. Pharm.* **2013**, *447*, 12–21. [CrossRef]
88. Jatana, S.; Callahan, L.M.; Pentland, A.P.; DeLouise, L.A. Impact of Cosmetic Lotions on Nanoparticle Penetration through ex vivo C57BL/6 Hairless Mouse and Human Skin: A Comparison Study. *Cosmetics* **2016**, *3*, 6. [CrossRef] [PubMed]
89. Kuo, T.-R.; Wu, C.-L.; Hsu, C.-T.; Lo, W.; Chiang, S.-J.; Lin, S.-J.; Dong, C.-Y.; Chen, C.-C. Chemical enhancer induced changes in the mechanisms of transdermal delivery of zinc oxide nanoparticles. *Biomaterials* **2009**, *30*, 3002–3008. [CrossRef]
90. Mahmoudi, M.; Lynch, I.; Ejtehadi, M.R.; Monopoli, M.P.; Bombelli, F.B.; Laurent, S. Protein−Nanoparticle Interactions: Opportunities and Challenges. *Chem. Rev.* **2011**, *111*, 5610–5637. [CrossRef]
91. Treuel, L.; Brandholt, S.; Maffre, P.; Wiegele, S.; Shang, L.; Nienhaus, G.U. Impact of Protein Modification on the Protein Corona on Nanoparticles and Nanoparticle–Cell Interactions. *ACS Nano* **2014**, *8*, 503–513. [CrossRef] [PubMed]
92. Nasser, F.; Lynch, I. Secreted protein eco-corona mediates uptake and impacts of polystyrene nanoparticles on Daphnia magna. *J. Proteom.* **2016**, *137*, 45–51. [CrossRef] [PubMed]
93. Krug, H.F.; Wick, P. Nanotoxicology: An Interdisciplinary Challenge. *Angew. Chem. Int. Ed.* **2011**, *50*, 1260–1278. [CrossRef] [PubMed]
94. Doherty, G.J.; McMahon, H.T. Mechanisms of Endocytosis. *Annu. Rev. Biochem.* **2009**, *78*, 857–902. [CrossRef]
95. Kaksonen, M.; Roux, A. Mechanisms of clathrin-mediated endocytosis. *Nat. Rev. Mol. Cell Biol.* **2018**, *19*, 313–326. [CrossRef]
96. Mani, I.; Pandey, K.N. Emerging concepts of receptor endocytosis and concurrent intracellular signaling: Mechanisms of guanylyl cyclase/natriuretic peptide receptor-A activation and trafficking. *Cell. Signal.* **2019**, *60*, 17–30. [CrossRef]
97. Rossi, G.; Barnoud, J.; Monticelli, L. Polystyrene Nanoparticles Perturb Lipid Membranes. *J. Phys. Chem. Lett.* **2014**, *5*, 241–246. [CrossRef]
98. Fiorentino, I.; Gualtieri, R.; Barbato, V.; Mollo, V.; Braun, S.; Angrisani, A.; Turano, M.; Furia, M.; Netti, P.A.; Guarnieri, D.; et al. Energy independent uptake and release of polystyrene nanoparticles in primary mammalian cell cultures. *Exp. Cell Res.* **2015**, *330*, 240–247. [CrossRef]
99. Dos Santos, T.; Varela, J.; Lynch, I.; Salvati, A.; Dawson, K.A. Effects of Transport Inhibitors on the Cellular Uptake of Carboxylated Polystyrene Nanoparticles in Different Cell Lines. *PLoS ONE* **2011**, *6*, e24438. [CrossRef]
100. Fazlollahi, F.; Angelow, S.; Yacobi, N.R.; Marchelletta, R.; Yu, A.S.; Hamm-Alvarez, S.F.; Borok, Z.; Kim, K.-J.; Crandall, E.D. Polystyrene nanoparticle trafficking across MDCK-II. *Nanomed. Nanotechnol. Biol. Med.* **2011**, *7*, 588–594. [CrossRef]
101. Kuhn, D.A.; Vanhecke, D.; Michen, B.; Blank, F.; Gehr, P.; Petri-Fink, A.; Rothen-Rutishauser, B. Different endocytotic uptake mechanisms for nanoparticles in epithelial cells and macrophages. *Beilstein J. Nanotechnol.* **2014**, *5*, 1625–1636. [CrossRef] [PubMed]
102. Koelmans, A.A.; Besseling, E.; Wegner, A.; Foekema, E.M. Plastic as a Carrier of POPs to Aquatic Organisms: A Model Analysis. *Environ. Sci. Technol.* **2013**, *47*, 7812–7820. [CrossRef] [PubMed]
103. Fröhlich, E. The role of surface charge in cellular uptake and cytotoxicity of medical nanoparticles. *Int. J. Nanomed.* **2012**, *7*, 5577–5591. [CrossRef] [PubMed]
104. Fröhlich, E.; Meindl, C.; Roblegg, E.; Ebner, B.; Absenger, M.; Pieber, T.R. Action of polystyrene nanoparticles of different sizes on lysosomal function and integrity. *Part. Fibre Toxicol.* **2012**, *9*, 26. [CrossRef]
105. Isidoro, C.; Maneerat, E.; Giovia, A.; Carlo, F.; Caputo, G. Biocompatibility, endocytosis, and intracellular trafficking of mesoporous silica and polystyrene nanoparticles in ovarian cancer cells: Effects of size and surface charge groups. *Int. J. Nanomed.* **2012**, *7*, 4147–4158. [CrossRef]
106. Malek, A.; Merkel, O.; Fink, L.; Czubayko, F.; Kissel, T.; Aigner, A. In vivo pharmacokinetics, tissue distribution and underlying mechanisms of various PEI(–PEG)/siRNA complexes. *Toxicol. Appl. Pharmacol.* **2009**, *236*, 97–108. [CrossRef] [PubMed]
107. Najahi-Missaoui, W.; Arnold, R.D.; Cummings, B.S. Safe Nanoparticles: Are We There Yet? *Int. J. Mol. Sci.* **2020**, *22*, 385. [CrossRef] [PubMed]
108. Koelmans, B.; Pahl, S.; Backhaus, T.; Bessa, F.; van Calster, G.; Contzen, N.; Cronin, R.; Galloway, T.; Hart, A.; Henderson, L. *A Scientific Perspective on Microplastics in Nature and Society*; SAPEA: Berlin, Germany, 2019.
109. Wright, S.L.; Kelly, F.J. Plastic and Human Health: A Micro Issue? *Environ. Sci. Technol.* **2017**, *51*, 6634–6647. [CrossRef]

110. Oliveira, M.; Almeida, M.; Miguel, I. *A Micro(nano)plastic Boomerang Tale: A Never Ending Story?* Elsevier: Amsterdam, The Netherlands, 2019; Volume 112, pp. 196–200.
111. Qiao, R.; Sheng, C.; Lu, Y.; Zhang, Y.; Ren, H.; Lemos, B. Microplastics induce intestinal inflammation, oxidative stress, and disorders of metabolome and microbiome in zebrafish. *Sci. Total Environ.* **2019**, *662*, 246–253. [CrossRef]
112. Brown, D.M.; Wilson, M.R.; MacNee, W.; Stone, V.; Donaldson, K. Size-Dependent Proinflammatory Effects of Ultrafine Polystyrene Particles: A Role for Surface Area and Oxidative Stress in the Enhanced Activity of Ultrafines. *Toxicol. Appl. Pharmacol.* **2001**, *175*, 191–199. [CrossRef] [PubMed]
113. Forte, M.; Iachetta, G.; Tussellino, M.; Carotenuto, R.; Prisco, M.; De Falco, M.; Laforgia, V.; Valiante, S. Polystyrene nanoparticles internalization in human gastric adenocarcinoma cells. *Toxicol. In Vitro* **2016**, *31*, 126–136. [CrossRef]
114. Prietl, B.; Meindl, C.; Roblegg, E.; Pieber, T.R.; Lanzer, G.; Fröhlich, E. Nano-sized and micro-sized polystyrene particles affect phagocyte function. *Cell Biol. Toxicol.* **2014**, *30*, 1–16. [CrossRef]
115. Fuchs, A.-K.; Syrovets, T.; Haas, K.A.; Loos, C.; Musyanovych, A.; Mailänder, V.; Landfester, K.; Simmet, T. Carboxyl- and amino-functionalized polystyrene nanoparticles differentially affect the polarization profile of M1 and M2 macrophage subsets. *Biomaterials* **2016**, *85*, 78–87. [CrossRef] [PubMed]
116. Ingram, J.H.; Stone, M.; Fisher, J.; Ingham, E. The influence of molecular weight, crosslinking and counterface roughness on TNF-alpha production by macrophages in response to ultra high molecular weight polyethylene particles. *Biomaterials* **2004**, *25*, 3511–3522. [CrossRef]
117. Green, T.R.; Fisher, J.; Stone, M.; Wroblewski, B.M.; Ingham, E. Polyethylene particles of a 'critical size' are necessary for the induction of cytokines by macrophages in vitro. *Biomaterials* **1998**, *19*, 2297–2302. [CrossRef]
118. Veruva, S.Y.; Lanman, T.H.; Isaza, J.E.; Freeman, T.A.; Kurtz, S.M.; Steinbeck, M.J. Periprosthetic UHMWPE Wear Debris Induces Inflammation, Vascularization, and Innervation After Total Disc Replacement in the Lumbar Spine. *Clin. Orthop. Relat. Res.* **2017**, *475*, 1369–1381. [CrossRef]
119. Suñer, S.; Gowland, N.; Craven, R.; Joffe, R.; Emami, N.; Tipper, J.L. Ultrahigh molecular weight polyethylene/graphene oxide nanocomposites: Wear characterization and biological response to wear particles. *J. Biomed. Mater. Res. Part B Appl. Biomater.* **2016**, *106*, 183–190. [CrossRef] [PubMed]
120. Massin, P.; Achour, S. Wear products of total hip arthroplasty: The case of polyethylene. *Morphologie* **2017**, *101*, 1–8. [CrossRef] [PubMed]
121. Nich, C.; Goodman, S.B. Role of Macrophages in the Biological Reaction to Wear Debris from Joint Replacements. *J. Autom. Inf. Sci.* **2014**, *24*, 259–265. [CrossRef] [PubMed]
122. Devane, P.A.; Bourne, R.B.; Rorabeck, C.H.; Hardie, R.M.; Horne, J.G. Measurement of polyethylene wear in metal-backed acetabular cups. I. Three-dimensional technique. *Clin. Orthop. Relat. Res.* **1995**, *319*, 303–316.
123. Devane, P.A.; Bourne, R.B.; Rorabeck, C.H.; Macdonald, S.; Robinson, E.J. Measurement of polyethylene wear in metal-backed acetabular cups. II. Clinical application. *Clin. Orthop. Relat. Res.* **1995**, *319*, 317–326.
124. Shanbhag, A.; Jacobs, J.; Glant, T.; Gilbert, J.; Black, J.; Galante, J. Composition and morphology of wear debris in failed uncemented total hip replacement. *J. Bone Jt. Surgery. Br. Vol.* **1994**, *76*, 60–67. [CrossRef]
125. Deng, Y.; Zhang, Y.; Lemos, B.; Ren, H. Tissue accumulation of microplastics in mice and biomarker responses suggest widespread health risks of exposure. *Sci. Rep.* **2017**, *7*, 46687. [CrossRef]
126. Inkielewicz-Stepniak, I.; Tajber, L.; Behan, G.; Zhang, H.; Radomski, M.W.; Medina, C.; Santos-Martinez, M.J. The Role of Mucin in the Toxicological Impact of Polystyrene Nanoparticles. *Materials* **2018**, *11*, 724. [CrossRef]
127. Chiu, H.-W.; Xia, T.; Lee, Y.-H.; Chen, C.-W.; Tsai, J.-C.; Wang, Y.-J. Cationic polystyrene nanospheres induce autophagic cell death through the induction of endoplasmic reticulum stress. *Nanoscale* **2015**, *7*, 736–746. [CrossRef] [PubMed]
128. Xia, T.; Kovochich, M.; Liong, M.; Zink, J.I.; Nel, A.E. Cationic Polystyrene Nanosphere Toxicity Depends on Cell-Specific Endocytic and Mitochondrial Injury Pathways. *ACS Nano* **2008**, *2*, 85–96. [CrossRef]
129. Liu, X.; Tian, X.; Xu, X.; Lu, J. Design of a phosphinate-based bioluminescent probe for superoxide radical anion imaging in living cells. *Luminescence* **2018**, *33*, 1101–1106. [CrossRef]
130. Paget, V.; Dekali, S.; Kortulewski, T.; Grall, R.; Gamez, C.; Blazy, K.; Aguerre-Chariol, O.; Chevillard, S.; Braun, A.; Rat, P.; et al. Specific Uptake and Genotoxicity Induced by Polystyrene Nanobeads with Distinct Surface Chemistry on Human Lung Epithelial Cells and Macrophages. *PLoS ONE* **2015**, *10*, e0123297. [CrossRef] [PubMed]
131. Ruenraroengsak, P.; Tetley, T.D. Differential bioreactivity of neutral, cationic and anionic polystyrene nanoparticles with cells from the human alveolar compartment: Robust response of alveolar type 1 epithelial cells. *Part. Fibre Toxicol.* **2015**, *12*, 1–20. [CrossRef] [PubMed]
132. Thubagere, A.; Reinhard, B.M. Nanoparticle-Induced Apoptosis Propagates through Hydrogen-Peroxide-Mediated Bystander Killing: Insights from a Human Intestinal EpitheliumIn VitroModel. *ACS Nano* **2010**, *4*, 3611–3622. [CrossRef] [PubMed]
133. Mahadevan, G.; Valiyaveettil, S. Understanding the interactions of poly(methyl methacrylate) and poly(vinyl chloride) nanoparticles with BHK-21 cell line. *Sci. Rep.* **2021**, *11*, 1–15. [CrossRef] [PubMed]
134. Jin, Y.; Lu, L.; Tu, W.; Luo, T.; Fu, Z. Impacts of polystyrene microplastic on the gut barrier, microbiota and metabolism of mice. *Sci. Total Environ.* **2019**, *649*, 308–317. [CrossRef]

135. Luo, T.; Zhang, Y.; Wang, C.; Wang, X.; Zhou, J.; Shen, M.; Zhao, Y.; Fu, Z.; Jin, Y. Maternal exposure to different sizes of polystyrene microplastics during gestation causes metabolic disorders in their offspring. *Environ. Pollut.* **2019**, *255*, 113122. [CrossRef] [PubMed]
136. McCarthy, J.; Gong, X.; Nahirney, D.; Duszyk, M.; Radomski, M. Polystyrene nanoparticles activate ion transport in human airway epithelial cells. *Int. J. Nanomed.* **2011**, *6*, 1343–1356. [CrossRef] [PubMed]
137. Xia, L.; Gu, W.; Zhang, M.; Chang, Y.-N.; Chen, K.; Bai, X.; Yu, L.; Li, J.; Li, S.; Xing, G. Endocytosed nanoparticles hold endosomes and stimulate binucleated cells formation. *Part. Fibre Toxicol.* **2016**, *13*, 63. [CrossRef]
138. Mahler, G.J.; Esch, M.B.; Tako, E.; Southard, T.L.; Archer, S.D.; Glahn, R.P.; Shuler, M.L. Oral exposure to polystyrene nanoparticles affects iron absorption. *Nat. Nanotechnol.* **2012**, *7*, 264–271. [CrossRef] [PubMed]
139. Stock, V.; Böhmert, L.; Lisicki, E.; Block, R.; Cara-Carmona, J.; Pack, L.K.; Selb, R.; Lichtenstein, D.; Voss, L.; Henderson, C.J.; et al. Uptake and effects of orally ingested polystyrene microplastic particles in vitro and in vivo. *Arch. Toxicol.* **2019**, *93*, 1817–1833. [CrossRef]
140. Lu, L.; Wan, Z.; Luo, T.; Fu, Z.; Jin, Y. Polystyrene microplastics induce gut microbiota dysbiosis and hepatic lipid metabolism disorder in mice. *Sci. Total Environ.* **2018**, *631*, 449–458. [CrossRef]
141. Luo, T.; Wang, C.; Pan, Z.; Jin, C.; Fu, Z.; Jin, Y. Maternal Polystyrene Microplastic Exposure during Gestation and Lactation Altered Metabolic Homeostasis in the Dams and Their F1 and F2 Offspring. *Environ. Sci. Technol.* **2019**, *53*, 10978–10992. [CrossRef]
142. Zhao, Y.; Xu, R.; Chen, X.; Wang, J.; Rui, Q.; Wang, D. Induction of protective response to polystyrene nanoparticles associated with dysregulation of intestinal long non-coding RNAs in Caenorhabditis elegans. *Ecotoxicol. Environ. Saf.* **2021**, *212*, 111976. [CrossRef] [PubMed]
143. Hirt, N.; Body-Malapel, M. Immunotoxicity and intestinal effects of nano- and microplastics: A review of the literature. *Part. Fibre Toxicol.* **2020**, *17*, 1–22. [CrossRef] [PubMed]
144. Campanale, C.; Massarelli, C.; Savino, I.; Locaputo, V.; Uricchio, V.F. A Detailed Review Study on Potential Effects of Microplastics and Additives of Concern on Human Health. *Int. J. Environ. Res. Public Health* **2020**, *17*, 1212. [CrossRef]
145. Sun, K.; Song, Y.; He, F.; Jing, M.; Tang, J.; Liu, R. A review of human and animals exposure to polycyclic aromatic hydrocarbons: Health risk and adverse effects, photo-induced toxicity and regulating effect of microplastics. *Sci. Total Environ.* **2021**, *773*, 145403. [CrossRef] [PubMed]
146. Engler, R.E. The Complex Interaction between Marine Debris and Toxic Chemicals in the Ocean. *Environ. Sci. Technol.* **2012**, *46*, 12302–12315. [CrossRef]
147. Galloway, T.S. Micro- and Nano-plastics and Human Health. In *Marine Anthropogenic Litter*; Springer: Cham, Switzerland, 2015; pp. 343–366.
148. Koelmans, A.A.; Besseling, E.; Foekema, E.M. Leaching of plastic additives to marine organisms. *Environ. Pollut.* **2014**, *187*, 49–54. [CrossRef]
149. Crain, D.A.; Eriksen, M.; Iguchi, T.; Jobling, S.; Laufer, H.; Leblanc, G.A.; Guillette, L.J. An ecological assessment of bisphenol-A: Evidence from comparative biology. *Reprod. Toxicol.* **2007**, *24*, 225–239. [CrossRef]
150. Rani, M.; Shim, W.J.; Han, G.M.; Jang, M.; Al-Odaini, N.A.; Song, Y.K.; Hong, S.H. Qualitative Analysis of Additives in Plastic Marine Debris and Its New Products. *Arch. Environ. Contam. Toxicol.* **2015**, *69*, 352–366. [CrossRef] [PubMed]
151. Calafat, A.M.; Ye, X.; Wong, L.-Y.; Reidy, J.A.; Needham, L.L. Exposure of the U.S. Population to Bisphenol A and 4- tertiary -Octylphenol: 2003–2004. *Environ. Health Perspect.* **2008**, *116*, 39–44. [CrossRef]
152. Guart, A.; Wagner, M.; Mezquida, A.; Lacorte, S.; Oehlmann, J.; Borrell, A. Migration of plasticisers from Tritan™ and polycarbonate bottles and toxicological evaluation. *Food Chem.* **2013**, *141*, 373–380. [CrossRef]
153. Srivastava, R.K.; Godara, S. Use of polycarbonate plastic products and human health. *Int. J. Basic Clin. Pharmacol.* **2013**, *2*, 12–17. [CrossRef]
154. Moriyama, K.; Tagami, T.; Akamizu, T.; Usui, T.; Saijo, M.; Kanamoto, N.; Hataya, Y.; Shimatsu, A.; Kuzuya, H.; Nakao, K. Thyroid Hormone Action Is Disrupted by Bisphenol A as an Antagonist. *J. Clin. Endocrinol. Metab.* **2002**, *87*, 5185–5190. [CrossRef] [PubMed]
155. Ropero, A.B.; Alonso-Magdalena, P.; García-García, E.; Ripoll, C.; Fuentes, E.; Nadal, A. Bisphenol-A disruption of the endocrine pancreas and blood glucose homeostasis. *Int. J. Androl.* **2008**, *31*, 194–200. [CrossRef] [PubMed]
156. Cipelli, R.; Harries, L.; Okuda, K.; Yoshihara, S.; Melzer, D.; Galloway, T. Bisphenol A modulates the metabolic regulator oestrogen-related receptor-α in T-cells. *Reproduction* **2014**, *147*, 419–426. [CrossRef] [PubMed]
157. Lang, I.A.; Galloway, T.S.; Scarlett, A.; Henley, W.E.; Depledge, M.; Wallace, R.B.; Melzer, D. Association of Urinary Bisphenol A Concentration with Medical Disorders and Laboratory Abnormalities in Adults. *JAMA* **2008**, *300*, 1303–1310. [CrossRef] [PubMed]
158. Melzer, D.; Osborne, N.J.; Henley, W.E.; Cipelli, R.; Young, A.; Money, C.; McCormack, P.; Luben, R.; Khaw, K.-T.; Wareham, N.J.; et al. Urinary Bisphenol A Concentration and Risk of Future Coronary Artery Disease in Apparently Healthy Men and Women. *Circulation* **2012**, *125*, 1482–1490. [CrossRef] [PubMed]
159. Gómez, C.; Gallart-Ayala, H. Metabolomics: A tool to characterize the effect of phthalates and bisphenol A. *Environ. Rev.* **2018**, *26*, 351–357. [CrossRef]

160. Cheng, Z.; Nie, X.-P.; Wang, H.-S.; Wong, M.-H. Risk assessments of human exposure to bioaccessible phthalate esters through market fish consumption. *Environ. Int.* **2013**, *57–58*, 75–80. [CrossRef]
161. Brachner, A.; Fragouli, D.; Duarte, I.F.; Farias, P.M.A.; Dembski, S.; Ghosh, M.; Barisic, I.; Zdzieblo, D.; Vanoirbeek, J.; Schwabl, P.; et al. Assessment of Human Health Risks Posed by Nano-and Microplastics Is Currently Not Feasible. *Int. J. Environ. Res. Public Health* **2020**, *17*, 8832. [CrossRef] [PubMed]
162. Paul, M.B.; Stock, V.; Cara-Carmona, J.; Lisicki, E.; Shopova, S.; Fessard, V.; Braeuning, A.; Sieg, H.; Böhmert, L. Micro- and nanoplastics—Current state of knowledge with the focus on oral uptake and toxicity. *Nanoscale Adv.* **2020**, *2*, 4350–4367. [CrossRef]
163. Leslie, H.; Depledge, M. Where is the evidence that human exposure to microplastics is safe? *Environ. Int.* **2020**, *142*, 105807. [CrossRef] [PubMed]

MDPI
St. Alban-Anlage 66
4052 Basel
Switzerland
Tel. +41 61 683 77 34
Fax +41 61 302 89 18
www.mdpi.com

Nanomaterials Editorial Office
E-mail: nanomaterials@mdpi.com
www.mdpi.com/journal/nanomaterials

www.ingramcontent.com/pod-product-compliance
Lightning Source LLC
LaVergne TN
LVHW070907170726
843515LV00005B/724

* 9 7 8 3 0 3 6 5 7 2 3 0 7 *